The Fermion

Edited by Paul F. Kisak

Contents

Chapter 1

Fermion

In particle physics, a **fermion** (a name coined by Paul Dirac[1] from the surname of Enrico Fermi) is any particle characterized by Fermi–Dirac statistics. These particles obey the Pauli exclusion principle. Fermions include all quarks and leptons, as well as any composite particle made of an odd number of these, such as all baryons and many atoms and nuclei. Fermions differ from bosons, which obey Bose–Einstein statistics.

A fermion can be an elementary particle, such as the electron, or it can be a composite particle, such as the proton. According to the spin-statistics theorem in any reasonable relativistic quantum field theory, particles with integer spin are bosons, while particles with half-integer spin are fermions.

Besides this spin characteristic, fermions have another specific property: they possess conserved baryon or lepton quantum numbers. Therefore what is usually referred as the spin statistics relation is in fact a spin statistics-quantum number relation.[2]

As a consequence of the Pauli exclusion principle, only one fermion can occupy a particular quantum state at any given time. If multiple fermions have the same spatial probability distribution, then at least one property of each fermion, such as its spin, must be different. Fermions are usually associated with matter, whereas bosons are generally force carrier particles, although in the current state of particle physics the distinction between the two concepts is unclear. Weakly interacting fermions can also display bosonic behavior under extreme conditions. At low temperature fermions show superfluidity for uncharged particles and superconductivity for charged particles.

Composite fermions, such as protons and neutrons, are the key building blocks of everyday matter.

1.1 Elementary fermions

The Standard Model recognizes two types of elementary fermions, quarks and leptons. In all, the model distinguishes 24 different fermions. There are six quarks (up, down, strange, charm, bottom and top quarks), and six leptons (electron, electron neutrino, muon, muon neutrino, tau particle and tau neutrino), along with the corresponding antiparticle of each of these.

Mathematically, fermions come in three types - Weyl fermions (massless), Dirac fermions (massive), and Majorana fermions (each its own antiparticle). Most Standard Model fermions are believed to be Dirac fermions, although it is unknown at this time whether the neutrinos are Dirac or Majorana fermions. Dirac fermions can be treated as a combination of two Weyl fermions.[3]:106 In July 2015, Weyl fermions have been experimentally realized in Weyl semimetals.

1.2 Composite fermions

See also: List of particles § Composite particles

1

Enrico Fermi

Composite particles (such as hadrons, nuclei, and atoms) can be bosons or fermions depending on their constituents. More precisely, because of the relation between spin and statistics, a particle containing an odd number of fermions is itself a fermion. It will have half-integer spin.

Examples include the following:

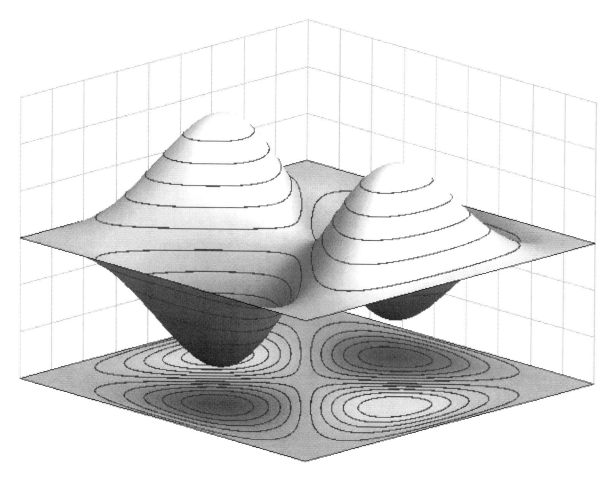

Antisymmetric wavefunction for a (fermionic) 2-particle state in an infinite square well potential.

- A baryon, such as the proton or neutron, contains three fermionic quarks and thus it is a fermion.

- The nucleus of a carbon-13 atom contains six protons and seven neutrons and is therefore a fermion.

- The atom helium-3 (^{3}He) is made of two protons, one neutron, and two electrons, and therefore it is a fermion.

The number of bosons within a composite particle made up of simple particles bound with a potential has no effect on whether it is a boson or a fermion.

Fermionic or bosonic behavior of a composite particle (or system) is only seen at large (compared to size of the system) distances. At proximity, where spatial structure begins to be important, a composite particle (or system) behaves according to its constituent makeup.

Fermions can exhibit bosonic behavior when they become loosely bound in pairs. This is the origin of superconductivity and the superfluidity of helium-3: in superconducting materials, electrons interact through the exchange of phonons, forming Cooper pairs, while in helium-3, Cooper pairs are formed via spin fluctuations.

The quasiparticles of the fractional quantum Hall effect are also known as composite fermions, which are electrons with an even number of quantized vortices attached to them.

1.2.1 Skyrmions

Main article: Skyrmion

In a quantum field theory, there can be field configurations of bosons which are topologically twisted. These are coherent states (or solitons) which behave like a particle, and they can be fermionic even if all the constituent particles are bosons. This was discovered by Tony Skyrme in the early 1960s, so fermions made of bosons are named skyrmions after him.

Skyrme's original example involved fields which take values on a three-dimensional sphere, the original nonlinear sigma model which describes the large distance behavior of pions. In Skyrme's model, reproduced in the large N or string approximation to quantum chromodynamics (QCD), the proton and neutron are fermionic topological solitons of the pion field.

Whereas Skyrme's example involved pion physics, there is a much more familiar example in quantum electrodynamics with a magnetic monopole. A bosonic monopole with the smallest possible magnetic charge and a bosonic version of the electron will form a fermionic dyon.

The analogy between the Skyrme field and the Higgs field of the electroweak sector has been used[4] to postulate that all fermions are skyrmions. This could explain why all known fermions have baryon or lepton quantum numbers and provide a physical mechanism for the Pauli exclusion principle.

1.3   See also

1.4   Notes

[1] Notes on Dirac's lecture *Developments in Atomic Theory* at Le Palais de la Découverte, 6 December 1945, UKNATARCHI Dirac Papers BW83/2/257889. See note 64 on page 331 in "The Strangest Man: The Hidden Life of Paul Dirac, Mystic of the Atom" by Graham Farmelo

[2] Physical Review D volume 87, page 0550003, year 2013, author Weiner, Richard M., title "Spin-statistics-quantum number connection and supersymmetry" arxiv:1302.0969

[3] T. Morii; C. S. Lim; S. N. Mukherjee (1 January 2004). *The Physics of the Standard Model and Beyond*. World Scientific. ISBN 978-981-279-560-1.

[4] Weiner, Richard M. (2010). "The Mysteries of Fermions". *International Journal of Theoretical Physics* **49** (5): 1174–1180. arXiv:0901.3816. Bibcode:2010IJTP...49.1174W. doi:10.1007/s10773-010-0292-7.

Chapter 2

Particle physics

For other uses of the word "particle" in physics and elsewhere, see particle (disambiguation).

Particle physics is the branch of physics that studies the nature of the particles that constitute *matter* (particles with mass) and *radiation* (massless particles). Although the word "particle" can refer to various types of very small objects (e.g. protons, gas particles, or even household dust), "particle physics" usually investigates the irreducibly smallest detectable particles and the irreducibly fundamental force fields necessary to explain them. By our current understanding, these elementary particles are excitations of the quantum fields that also govern their interactions. The currently dominant theory explaining these fundamental particles and fields, along with their dynamics, is called the Standard Model. Thus, modern particle physics generally investigates the Standard Model and its various possible extensions, e.g. to the newest "known" particle, the Higgs boson, or even to the oldest known force field, gravity.[1][2]

2.1 Subatomic particles

Modern particle physics research is focused on subatomic particles, including atomic constituents such as electrons, protons, and neutrons (protons and neutrons are composite particles called baryons, made of quarks), produced by radioactive and scattering processes, such as photons, neutrinos, and muons, as well as a wide range of exotic particles. Dynamics of particles is also governed by quantum mechanics; they exhibit wave–particle duality, displaying particle-like behaviour under certain experimental conditions and wave-like behaviour in others. In more technical terms, they are described by quantum state vectors in a Hilbert space, which is also treated in quantum field theory. Following the convention of particle physicists, the term *elementary particles* is applied to those particles that are, according to current understanding, presumed to be indivisible and not composed of other particles.[3]

All particles, and their interactions observed to date, can be described almost entirely by a quantum field theory called the Standard Model.[4] The Standard Model, as currently formulated, has 61 elementary particles.[3] Those elementary particles can combine to form composite particles, accounting for the hundreds of other species of particles that have been discovered since the 1960s. The Standard Model has been found to agree with almost all the experimental tests conducted to date. However, most particle physicists believe that it is an incomplete description of nature, and that a more fundamental theory awaits discovery (See Theory of Everything). In recent years, measurements of neutrino mass have provided the first experimental deviations from the Standard Model.

2.2 History

Main article: History of subatomic physics

The idea that all matter is composed of elementary particles dates to at least the 6th century BC.[5] In the 19th century,

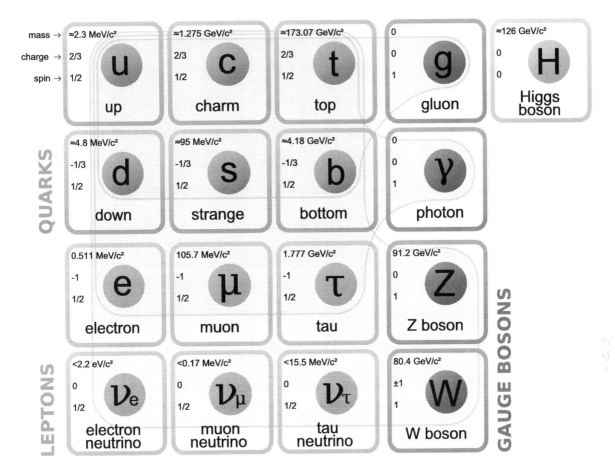

The particle content of the Standard Model of Physics

John Dalton, through his work on stoichiometry, concluded that each element of nature was composed of a single, unique type of particle.[6] The word *atom*, after the Greek word *atomos* meaning "indivisible", denotes the smallest particle of a chemical element since then, but physicists soon discovered that *atoms* are not, in fact, the fundamental particles of nature, but conglomerates of even smaller particles, such as the electron. The early 20th-century explorations of nuclear physics and quantum physics culminated in proofs of nuclear fission in 1939 by Lise Meitner (based on experiments by Otto Hahn), and nuclear fusion by Hans Bethe in that same year; both discoveries also led to the development of nuclear weapons. Throughout the 1950s and 1960s, a bewildering variety of particles were found in scattering experiments. It was referred to as the "particle zoo". That term was deprecated after the formulation of the Standard Model during the 1970s in which the large number of particles was explained as combinations of a (relatively) small number of fundamental particles.

2.3 Standard Model

Main article: Standard Model

The current state of the classification of all elementary particles is explained by the Standard Model. It describes the strong, weak, and electromagnetic fundamental interactions, using mediating gauge bosons. The species of gauge bosons are the gluons, W−, W+ and Z bosons, and the photons.[4] The Standard Model also contains 24 fundamental particles, (12 particles and their associated anti-particles), which are the constituents of all matter.[7] Finally, the Standard Model also predicted the existence of a type of boson known as the Higgs boson. Early in the morning on 4 July 2012, physicists with the Large Hadron Collider at CERN announced they had found a new particle that behaves similarly to what is expected from the Higgs boson.[8]

2.4 Experimental laboratories

In particle physics, the major international laboratories are located at the:

- Brookhaven National Laboratory (Long Island, United States). Its main facility is the Relativistic Heavy Ion Collider (RHIC), which collides heavy ions such as gold ions and polarized protons. It is the world's first heavy ion collider, and the world's only polarized proton collider.[9]

- Budker Institute of Nuclear Physics (Novosibirsk, Russia). Its main projects are now the electron-positron colliders VEPP-2000,[10] operated since 2006, and VEPP-4,[11] started experiments in 1994. Earlier facilities include the first electron-electron beam-beam collider VEP-1, which conducted experiments from 1964 to 1968; the electron-positron colliders VEPP-2, operated from 1965 to 1974; and, its successor VEPP-2M,[12] performed experiments from 1974 to 2000.[13]

- CERN, (Conseil Européen pour la Recherche Nucléaire) (Franco-Swiss border, near Geneva). Its main project is now the Large Hadron Collider (LHC), which had its first beam circulation on 10 September 2008, and is now the world's most energetic collider of protons. It also became the most energetic collider of heavy ions after it began colliding lead ions. Earlier facilities include the Large Electron–Positron Collider (LEP), which was stopped on 2 November 2000 and then dismantled to give way for LHC; and the Super Proton Synchrotron, which is being reused as a pre-accelerator for the LHC.[14]

- DESY (Deutsches Elektronen-Synchrotron) (Hamburg, Germany). Its main facility is the Hadron Elektron Ring Anlage (HERA), which collides electrons and positrons with protons.[15]

- Fermi National Accelerator Laboratory (Fermilab), (Batavia, United States). Its main facility until 2011 was the Tevatron, which collided protons and antiprotons and was the highest-energy particle collider on earth until the Large Hadron Collider surpassed it on 29 November 2009.[16]

- Institute of High Energy Physics (IHEP), (Beijing, China). IHEP manages a number of China's major particle physics facilities, including the Beijing Electron Positron Collider (BEPC), the Beijing Spectrometer (BES), the Beijing Synchrotron Radiation Facility (BSRF), the International Cosmic-Ray Observatory at Yangbajing in Tibet, the Daya Bay Reactor Neutrino Experiment, the China Spallation Neutron Source, the Hard X-ray Modulation Telescope (HXMT), and the Accelerator-driven Sub-critical System (ADS) as well as the Jiangmen Underground Neutrino Observatory (JUNO). [17]

- KEK, (Tsukuba, Japan). It is the home of a number of experiments such as the K2K experiment, a neutrino oscillation experiment and Belle, an experiment measuring the CP violation of B mesons.[18]

- SLAC National Accelerator Laboratory, (Menlo Park, United States). Its 2-mile-long linear particle accelerator began operating in 1962 and was the basis for numerous electron and positron collision experiments until 2008. Since then the linear accelerator is being used for the Linac Coherent Light Source X-ray laser as well as advanced accelerator design research. SLAC staff continue to participate in developing and building many particle physics experiments around the world.[19]

Many other particle accelerators do exist.

The techniques required to do modern, experimental, particle physics are quite varied and complex, constituting a sub-specialty nearly completely distinct from the theoretical side of the field.

2.5 Theory

Theoretical particle physics attempts to develop the models, theoretical framework, and mathematical tools to understand current experiments and make predictions for future experiments. See also theoretical physics. There are several major interrelated efforts being made in theoretical particle physics today. One important branch attempts to better understand the Standard Model and its tests. By extracting the parameters of the Standard Model, from experiments with less uncertainty, this work probes the limits of the Standard Model and therefore expands our understanding of nature's building blocks. Those efforts are made challenging by the difficulty of calculating quantities in quantum chromodynamics. Some theorists working in this area refer to themselves as **phenomenologists** and they may use the tools of quantum field theory and effective field theory. Others make use of lattice field theory and call themselves *lattice theorists*.

Another major effort is in model building where model builders develop ideas for what physics may lie beyond the Standard Model (at higher energies or smaller distances). This work is often motivated by the hierarchy problem and is constrained by existing experimental data. It may involve work on supersymmetry, alternatives to the Higgs mechanism, extra spatial dimensions (such as the Randall-Sundrum models), Preon theory, combinations of these, or other ideas.

A third major effort in theoretical particle physics is string theory. *String theorists* attempt to construct a unified description of quantum mechanics and general relativity by building a theory based on small strings, and branes rather than particles. If the theory is successful, it may be considered a "Theory of Everything", or "TOE".

There are also other areas of work in theoretical particle physics ranging from particle cosmology to loop quantum gravity.

This division of efforts in particle physics is reflected in the names of categories on the arXiv, a preprint archive:[20] hep-th (theory), hep-ph (phenomenology), hep-ex (experiments), hep-lat (lattice gauge theory).

2.6   Practical applications

In principle, all physics (and practical applications developed therefrom) can be derived from the study of fundamental particles[Debatable, belongs in philosophy]. In practice, even if "particle physics" is taken to mean only "high-energy atom smashers", many technologies have been developed during these pioneering investigations that later find wide uses in society. Cyclotrons are used to produce medical isotopes for research and treatment (for example, isotopes used in PET imaging), or used directly for certain cancer treatments. The development of Superconductors has been pushed forward by their use in particle physics. The World Wide Web and touchscreen technology were initially developed at CERN.

Additional applications are found in medicine, national security, industry, computing, science, and workforce development, illustrating a long and growing list of beneficial practical applications with contributions from particle physics.[21]

2.7   Future

The primary goal, which is pursued in several distinct ways, is to find and understand what physics may lie beyond the standard model. There are several powerful experimental reasons to expect new physics, including dark matter and neutrino mass. There are also theoretical hints that this new physics should be found at accessible energy scales.

Much of the effort to find this new physics are focused on new collider experiments. The Large Hadron Collider (LHC) was completed in 2008 to help continue the search for the Higgs boson, supersymmetric particles, and other new physics. An intermediate goal is the construction of the International Linear Collider (ILC), which will complement the LHC by allowing more precise measurements of the properties of newly found particles. In August 2004, a decision for the technology of the ILC was taken but the site has still to be agreed upon.

In addition, there are important non-collider experiments that also attempt to find and understand physics beyond the Standard Model. One important non-collider effort is the determination of the neutrino masses, since these masses may arise from neutrinos mixing with very heavy particles. In addition, cosmological observations provide many useful constraints on the dark matter, although it may be impossible to determine the exact nature of the dark matter without the colliders. Finally, lower bounds on the very long lifetime of the proton put constraints on Grand Unified Theories at energy scales much higher than collider experiments will be able to probe any time soon.

In May 2014, the Particle Physics Project Prioritization Panel released its report on particle physics funding priorities for the United States over the next decade. This report emphasized continued U.S. participation in the LHC and ILC, and expansion of the Long Baseline Neutrino Experiment, among other recommendations.

2.8 See also

- Atomic physics

- High pressure

- International Conference on High Energy Physics

- Introduction to quantum mechanics

- List of accelerators in particle physics

- List of particles

- Magnetic monopole

- Micro black hole

- Number theory

- Resonance (particle physics)

- Self-consistency principle in high energy Physics

- Non-extensive self-consistent thermodynamical theory

- Standard Model (mathematical formulation)

- Stanford Physics Information Retrieval System

- Timeline of particle physics

- Unparticle physics

- Tetraquark

2.9 References

[1] http://home.web.cern.ch/topics/higgs-boson

[2] http://www.nobelprize.org/nobel_prizes/physics/laureates/2013/advanced-physicsprize2013.pdf

[3] Braibant, S.; Giacomelli, G.; Spurio, M. (2009). *Particles and Fundamental Interactions: An Introduction to Particle Physics.* Springer. pp. 313–314. ISBN 978-94-007-2463-1.

[4] "Particle Physics and Astrophysics Research". The Henryk Niewodniczanski Institute of Nuclear Physics. Retrieved 31 May 2012.

[5] "Fundamentals of Physics and Nuclear Physics" (PDF). Retrieved 21 July 2012.

[6] "Scientific Explorer: Quasiparticles". Sciexplorer.blogspot.com. 22 May 2012. Retrieved 21 July 2012.

[7] Nakamura, K (1 July 2010). "Review of Particle Physics". *Journal of Physics G: Nuclear and Particle Physics* **37** (7A): 075021. Bibcode:2010JPhG...37g5021N. doi:10.1088/0954-3899/37/7A/075021.

[8] Mann, Adam (28 March 2013). "Newly Discovered Particle Appears to Be Long-Awaited Higgs Boson - Wired Science". Wired.com. Retrieved 6 February 2014.

[9] "Brookhaven National Laboratory – A Passion for Discovery". Bnl.gov. Retrieved 23 June 2012.

[10] "index". Vepp2k.inp.nsk.su. Retrieved 21 July 2012.

[11] "The VEPP-4 accelerating-storage complex". V4.inp.nsk.su. Retrieved 21 July 2012.

[12] "VEPP-2M collider complex" (in Russian). Inp.nsk.su. Retrieved 21 July 2012.

[13] "The Budker Institute Of Nuclear Physics". English Russia. 21 January 2012. Retrieved 23 June 2012.

[14] "Welcome to". Info.cern.ch. Retrieved 23 June 2012.

[15] "Germany's largest accelerator centre – Deutsches Elektronen-Synchrotron DESY". Desy.de. Retrieved 23 June 2012.

[16] "Fermilab | Home". Fnal.gov. Retrieved 23 June 2012.

[17] "IHEP | Home". ihep.ac.cn. Retrieved 29 November 2015.

[18] "Kek | High Energy Accelerator Research Organization". Legacy.kek.jp. Retrieved 23 June 2012.

[19] "SLAC National Accelerator Laboratory Home Page". Retrieved 19 February 2015.

[20] arxiv.org

[21] "Fermilab | Science at Fermilab | Benefits to Society". Fnal.gov. Retrieved 23 June 2012.

2.10   Further reading

Introductory reading

- Close, Frank (2004). *Particle Physics: A Very Short Introduction*. Oxford University Press. ISBN 0-19-280434-0.

- Close, Frank; Marten, Michael; Sutton, Christine (2004). *The Particle Odyssey: A Journey to the Heart of the Matter*. Oxford University Press. ISBN 9780198609438.

- Ford, Kenneth W. (2005). *The Quantum World*. Harvard University Press.

- Oerter, Robert (2006). *The Theory of Almost Everything: The Standard Model, the Unsung Triumph of Modern Physics*. Plume.

- Schumm, Bruce A. (2004). *Deep Down Things: The Breathtaking Beauty of Particle Physics*. Johns Hopkins University Press. ISBN 0-8018-7971-X.

- Close, Frank (2006). *The New Cosmic Onion*. Taylor & Francis. ISBN 1-58488-798-2.

Advanced reading

- Robinson, Matthew B.; Bland, Karen R.; Cleaver, Gerald. B.; Dittmann, Jay R. (2008). "A Simple Introduction to Particle Physics". arXiv:0810.3328 [hep-th].

- Robinson, Matthew B.; Ali, Tibra; Cleaver, Gerald B. (2009). "A Simple Introduction to Particle Physics Part II". arXiv:0908.1395 [hep-th].

- Griffiths, David J. (1987). *Introduction to Elementary Particles*. Wiley, John & Sons, Inc. ISBN 0-471-60386-4.

- Kane, Gordon L. (1987). *Modern Elementary Particle Physics*. Perseus Books. ISBN 0-201-11749-5.

- Perkins, Donald H. (1999). *Introduction to High Energy Physics*. Cambridge University Press. ISBN 0-521-62196-8.

- Povh, Bogdan (1995). *Particles and Nuclei: An Introduction to the Physical Concepts*. Springer-Verlag. ISBN 0-387-59439-6.

- Boyarkin, Oleg (2011). *Advanced Particle Physics Two-Volume Set*. CRC Press. ISBN 978-1-4398-0412-4.

2.11 External links

- *Symmetry* magazine

- Fermilab

- Particle physics – it matters – the Institute of Physics

- Nobes, Matthew (2002) "Introduction to the Standard Model of Particle Physics" on Kuro5hin: Part 1, Part 2, Part 3a, Part 3b.

- CERN – European Organization for Nuclear Research

- The Particle Adventure – educational project sponsored by the Particle Data Group of the Lawrence Berkeley National Laboratory (LBNL)

Chapter 3

Standard Model

This article is about the Standard Model of particle physics. For other uses, see Standard model (disambiguation).
This article is a non-mathematical general overview of the Standard Model. For a mathematical description, see the article Standard Model (mathematical formulation).
For the Standard Model of Big Bang cosmology, Lambda-CDM model.

 The **Standard Model** of particle physics is a theory concerning the electromagnetic, weak, and strong nuclear inter-

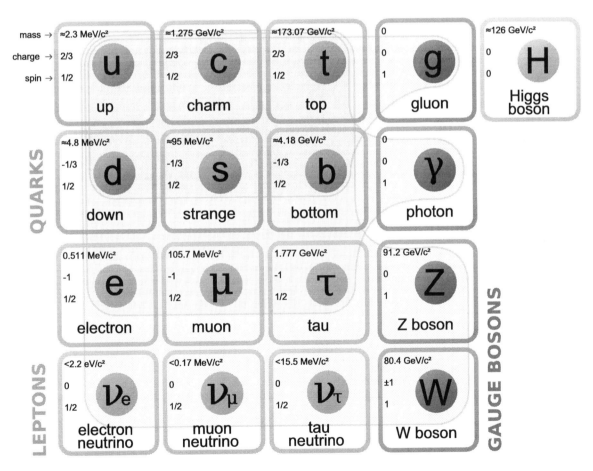

The Standard Model of elementary particles (more schematic depiction), with the three generations of matter, gauge bosons in the fourth column, and the Higgs boson in the fifth.

actions, as well as classifying all the subatomic particles known. It was developed throughout the latter half of the 20th century, as a collaborative effort of scientists around the world.[1] The current formulation was finalized in the mid-1970s upon experimental confirmation of the existence of quarks. Since then, discoveries of the top quark (1995), the tau neutrino (2000), and more recently the Higgs boson (2012), have given further credence to the Standard Model. Because of its success in explaining a wide variety of experimental results, the Standard Model is sometimes regarded as a "theory of almost everything".

Although the Standard Model is believed to be theoretically self-consistent[2] and has demonstrated huge and continued successes in providing experimental predictions, it does leave some phenomena unexplained and it falls short of being a complete theory of fundamental interactions. It does not incorporate the full theory of gravitation[3] as described by general relativity, or account for the accelerating expansion of the universe (as possibly described by dark energy). The model does not contain any viable dark matter particle that possesses all of the required properties deduced from observational cosmology. It also does not incorporate neutrino oscillations (and their non-zero masses).

The development of the Standard Model was driven by theoretical and experimental particle physicists alike. For theorists, the Standard Model is a paradigm of a quantum field theory, which exhibits a wide range of physics including spontaneous symmetry breaking, anomalies, non-perturbative behavior, etc. It is used as a basis for building more exotic models that incorporate hypothetical particles, extra dimensions, and elaborate symmetries (such as supersymmetry) in an attempt to explain experimental results at variance with the Standard Model, such as the existence of dark matter and neutrino oscillations.

3.1 Historical background

The first step towards the Standard Model was Sheldon Glashow's discovery in 1961 of a way to combine the electromagnetic and weak interactions.[4] In 1967 Steven Weinberg[5] and Abdus Salam[6] incorporated the Higgs mechanism[7][8][9] into Glashow's electroweak theory, giving it its modern form.

The Higgs mechanism is believed to give rise to the masses of all the elementary particles in the Standard Model. This includes the masses of the W and Z bosons, and the masses of the fermions, i.e. the quarks and leptons.

After the neutral weak currents caused by Z boson exchange were discovered at CERN in 1973,[10][11][12][13] the electroweak theory became widely accepted and Glashow, Salam, and Weinberg shared the 1979 Nobel Prize in Physics for discovering it. The W and Z bosons were discovered experimentally in 1981, and their masses were found to be as the Standard Model predicted.

The theory of the strong interaction, to which many contributed, acquired its modern form around 1973–74, when experiments confirmed that the hadrons were composed of fractionally charged quarks.

3.2 Overview

At present, matter and energy are best understood in terms of the kinematics and interactions of elementary particles. To date, physics has reduced the laws governing the behavior and interaction of all known forms of matter and energy to a small set of fundamental laws and theories. A major goal of physics is to find the "common ground" that would unite all of these theories into one integrated theory of everything, of which all the other known laws would be special cases, and from which the behavior of all matter and energy could be derived (at least in principle).[14]

3.3 Particle content

The Standard Model includes members of several classes of elementary particles (fermions, gauge bosons, and the Higgs boson), which in turn can be distinguished by other characteristics, such as color charge.

3.3.1 Fermions

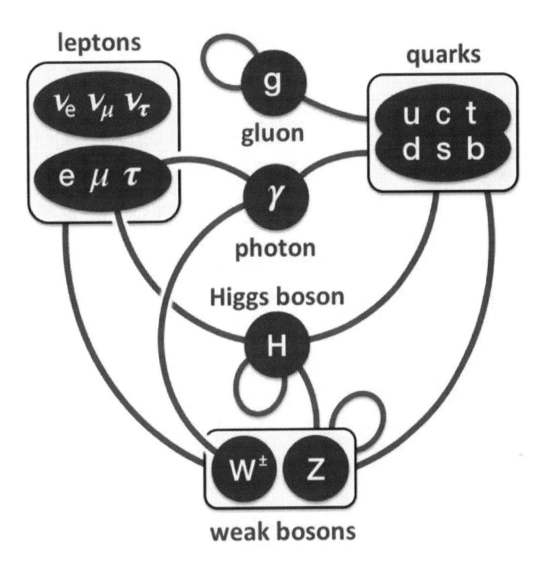

Summary of interactions between particles described by the Standard Model.

The Standard Model includes 12 elementary particles of spin-½ known as fermions. According to the spin-statistics theorem, fermions respect the Pauli exclusion principle. Each fermion has a corresponding antiparticle.

The fermions of the Standard Model are classified according to how they interact (or equivalently, by what charges they carry). There are six quarks (up, down, charm, strange, top, bottom), and six leptons (electron, electron neutrino, muon, muon neutrino, tau, tau neutrino). Pairs from each classification are grouped together to form a generation, with corresponding particles exhibiting similar physical behavior (see table).

The defining property of the quarks is that they carry color charge, and hence, interact via the strong interaction. A phenomenon called color confinement results in quarks being very strongly bound to one another, forming color-neutral composite particles (hadrons) containing either a quark and an antiquark (mesons) or three quarks (baryons). The familiar proton and the neutron are the two baryons having the smallest mass. Quarks also carry electric charge and weak isospin. Hence they interact with other fermions both electromagnetically and via the weak interaction.

The remaining six fermions do not carry colour charge and are called leptons. The three neutrinos do not carry electric charge either, so their motion is directly influenced only by the weak nuclear force, which makes them notoriously difficult to detect. However, by virtue of carrying an electric charge, the electron, muon, and tau all interact electromagnetically.

Each member of a generation has greater mass than the corresponding particles of lower generations. The first generation charged particles do not decay; hence all ordinary (baryonic) matter is made of such particles. Specifically, all atoms consist of electrons orbiting around atomic nuclei, ultimately constituted of up and down quarks. Second and third generation charged particles, on the other hand, decay with very short half lives, and are observed only in very high-energy environments. Neutrinos of all generations also do not decay, and pervade the universe, but rarely interact with baryonic matter.

3.3.2 Gauge bosons

In the Standard Model, gauge bosons are defined as force carriers that mediate the strong, weak, and electromagnetic fundamental interactions.

Interactions in physics are the ways that particles influence other particles. At a macroscopic level, electromagnetism allows particles to interact with one another via electric and magnetic fields, and gravitation allows particles with mass to attract one another in accordance with Einstein's theory of general relativity. The Standard Model explains such forces as resulting from matter particles exchanging other particles, generally referred to as *force mediating particles*. When a force-mediating particle is exchanged, at a macroscopic level the effect is equivalent to a force influencing both of them, and the particle is therefore said to have *mediated* (i.e., been the agent of) that force. The Feynman diagram calculations, which are a graphical representation of the perturbation theory approximation, invoke "force mediating particles", and when applied to analyze high-energy scattering experiments are in reasonable agreement with the data. However, perturbation theory (and with it the concept of a "force-mediating particle") fails in other situations. These include low-energy quantum chromodynamics, bound states, and solitons.

The gauge bosons of the Standard Model all have spin (as do matter particles). The value of the spin is 1, making them bosons. As a result, they do not follow the Pauli exclusion principle that constrains fermions: thus bosons (e.g. photons) do not have a theoretical limit on their spatial density (number per volume). The different types of gauge bosons are described below.

- Photons mediate the electromagnetic force between electrically charged particles. The photon is massless and is well-described by the theory of quantum electrodynamics.

- The W+, W−, and Z gauge bosons mediate the weak interactions between particles of different flavors (all quarks and leptons). They are massive, with the Z being more massive than the W±. The weak interactions involving the W± exclusively act on *left-handed* particles and *right-handed* antiparticles. Furthermore, the W± carries an electric charge of +1 and −1 and couples to the electromagnetic interaction. The electrically neutral Z boson interacts with both left-handed particles and antiparticles. These three gauge bosons along with the photons are grouped together, as collectively mediating the electroweak interaction.

- The eight gluons mediate the strong interactions between color charged particles (the quarks). Gluons are massless. The eightfold multiplicity of gluons is labeled by a combination of color and anticolor charge (e.g. red–antigreen).[nb 1] Because the gluons have an effective color charge, they can also interact among themselves. The gluons and their interactions are described by the theory of quantum chromodynamics.

The interactions between all the particles described by the Standard Model are summarized by the diagrams on the right of this section.

3.3.3 Higgs boson

Main article: Higgs boson

The Higgs particle is a massive scalar elementary particle theorized by Robert Brout, François Englert, Peter Higgs, Gerald Guralnik, C. R. Hagen, and Tom Kibble in 1964 (see 1964 PRL symmetry breaking papers) and is a key building

Standard Model Interactions
(Forces Mediated by Gauge Bosons)

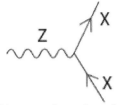

X is any fermion in
the Standard Model.

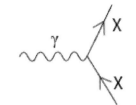

X is electrically charged.

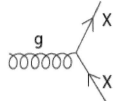

X is any quark.

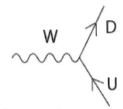

U is a up-type quark;
D is a down-type quark.

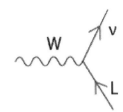

L is a lepton and ν is the
corresponding neutrino.

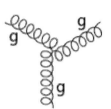

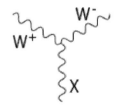

X is a photon or Z-boson.

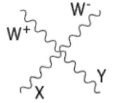

X and Y are any two
electroweak bosons such
that charge is conserved.

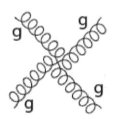

The above interactions form the basis of the standard model. Feynman diagrams in the standard model are built from these vertices. Modifications involving Higgs boson interactions and neutrino oscillations are omitted. The charge of the W bosons is dictated by the fermions they interact with; the conjugate of each listed vertex (i.e. reversing the direction of arrows) is also allowed.

block in the Standard Model.[7][8][9][15] It has no intrinsic spin, and for that reason is classified as a boson (like the gauge bosons, which have integer spin).

The Higgs boson plays a unique role in the Standard Model, by explaining why the other elementary particles, except the photon and gluon, are massive. In particular, the Higgs boson explains why the photon has no mass, while the W and Z bosons are very heavy. Elementary particle masses, and the differences between electromagnetism (mediated by the photon) and the weak force (mediated by the W and Z bosons), are critical to many aspects of the structure of microscopic (and hence macroscopic) matter. In electroweak theory, the Higgs boson generates the masses of the leptons (electron, muon, and tau) and quarks. As the Higgs boson is massive, it must interact with itself.

Because the Higgs boson is a very massive particle and also decays almost immediately when created, only a very high-energy particle accelerator can observe and record it. Experiments to confirm and determine the nature of the Higgs boson using the Large Hadron Collider (LHC) at CERN began in early 2010, and were performed at Fermilab's Tevatron until its closure in late 2011. Mathematical consistency of the Standard Model requires that any mechanism capable of generating the masses of elementary particles become visible at energies above 1.4 TeV;[16] therefore, the LHC (designed to collide two 7 to 8 TeV proton beams) was built to answer the question of whether the Higgs boson actually exists.[17]

On 4 July 2012, the two main experiments at the LHC (ATLAS and CMS) both reported independently that they found a new particle with a mass of about 125 GeV/c^2 (about 133 proton masses, on the order of 10^{-25} kg), which is "consistent with the Higgs boson." Although it has several properties similar to the predicted "simplest" Higgs,[18] they acknowledged that further work would be needed to conclude that it is indeed the Higgs boson, and exactly which version of the Standard Model Higgs is best supported if confirmed.[19][20][21][22][23]

On 14 March 2013 the Higgs Boson was tentatively confirmed to exist.[24]

3.3.4   Total particle count

Counting particles by a rule that distinguishes between particles and their corresponding antiparticles, and among the many color states of quarks and gluons, gives a total of 61 elementary particles.[25]

3.4   Theoretical aspects

Main article: Standard Model (mathematical formulation)

3.4.1   Construction of the Standard Model Lagrangian

Technically, quantum field theory provides the mathematical framework for the Standard Model, in which a Lagrangian controls the dynamics and kinematics of the theory. Each kind of particle is described in terms of a dynamical field that pervades space-time. The construction of the Standard Model proceeds following the modern method of constructing most field theories: by first postulating a set of symmetries of the system, and then by writing down the most general renormalizable Lagrangian from its particle (field) content that observes these symmetries.

The global Poincaré symmetry is postulated for all relativistic quantum field theories. It consists of the familiar translational symmetry, rotational symmetry and the inertial reference frame invariance central to the theory of special relativity. The local SU(3)×SU(2)×U(1) gauge symmetry is an internal symmetry that essentially defines the Standard Model. Roughly, the three factors of the gauge symmetry give rise to the three fundamental interactions. The fields fall into different representations of the various symmetry groups of the Standard Model (see table). Upon writing the most general Lagrangian, one finds that the dynamics depend on 19 parameters, whose numerical values are established by experiment. The parameters are summarized in the table above (note: with the Higgs mass is at 125 GeV, the Higgs self-coupling strength $\lambda \sim 1/8$).

Quantum chromodynamics sector

Main article: Quantum chromodynamics

The quantum chromodynamics (QCD) sector defines the interactions between quarks and gluons, with SU(3) symmetry, generated by T^a. Since leptons do not interact with gluons, they are not affected by this sector. The Dirac Lagrangian of the quarks coupled to the gluon fields is given by

$$\mathcal{L}_{QCD} = i\overline{U}(\partial_\mu - ig_s G_\mu^a T^a)\gamma^\mu U + i\overline{D}(\partial_\mu - ig_s G_\mu^a T^a)\gamma^\mu D.$$

G_μ^a is the SU(3) gauge field containing the gluons, γ^μ are the Dirac matrices, D and U are the Dirac spinors associated with up- and down-type quarks, and g_s is the strong coupling constant.

Electroweak sector

Main article: Electroweak interaction

The electroweak sector is a Yang–Mills gauge theory with the simple symmetry group U(1)×SU(2)L,

$$\mathcal{L}_{\text{EW}} = \sum_\psi \bar{\psi}\gamma^\mu \left(i\partial_\mu - g'\frac{1}{2}Y_{\text{W}}B_\mu - g\frac{1}{2}\vec{\tau}_{\text{L}}\vec{W}_\mu \right)\psi$$

where $B\mu$ is the U(1) gauge field; Y_W is the weak hypercharge—the generator of the U(1) group; $\vec{W}_\mu$ is the three-component SU(2) gauge field; $\vec{\tau}_L$ are the Pauli matrices—infinitesimal generators of the SU(2) group. The subscript L indicates that they only act on left fermions; g' and g are coupling constants.

Higgs sector

Main article: Higgs mechanism

In the Standard Model, the Higgs field is a complex scalar of the group SU(2)L:

$$\varphi = \frac{1}{\sqrt{2}}\begin{pmatrix} \varphi^+ \\ \varphi^0 \end{pmatrix},$$

where the indices + and 0 indicate the electric charge (Q) of the components. The weak isospin (Y_W) of both components is 1.

Before symmetry breaking, the Higgs Lagrangian is:

$$\mathcal{L}_{\text{H}} = \varphi^\dagger \left(\partial^\mu - \frac{i}{2}\left(g'Y_{\text{W}}B^\mu + g\vec{\tau}\vec{W}^\mu \right) \right)\left(\partial_\mu + \frac{i}{2}\left(g'Y_{\text{W}}B_\mu + g\vec{\tau}\vec{W}_\mu \right) \right)\varphi - \frac{\lambda^2}{4}\left(\varphi^\dagger\varphi - v^2 \right)^2,$$

which can also be written as:

$$\mathcal{L}_{\text{H}} = \left| \left(\partial_\mu + \frac{i}{2}\left(g'Y_{\text{W}}B_\mu + g\vec{\tau}\vec{W}_\mu \right) \right)\varphi \right|^2 - \frac{\lambda^2}{4}\left(\varphi^\dagger\varphi - v^2 \right)^2.$$

3.5   Fundamental forces

Main article: Fundamental interaction

The Standard Model classified all four fundamental forces in nature. In the Standard Model, a force is described as an exchange of bosons between the objects affected, such as a photon for the electromagnetic force and a gluon for the strong interaction. Those particles are called force carriers.[26]

3.6 Tests and predictions

The Standard Model (SM) predicted the existence of the W and Z bosons, gluon, and the top and charm quarks before these particles were observed. Their predicted properties were experimentally confirmed with good precision. To give an idea of the success of the SM, the following table compares the measured masses of the W and Z bosons with the masses predicted by the SM:

The SM also makes several predictions about the decay of Z bosons, which have been experimentally confirmed by the Large Electron-Positron Collider at CERN.

In May 2012 BaBar Collaboration reported that their recently analyzed data may suggest possible flaws in the Standard Model of particle physics.[28][29] These data show that a particular type of particle decay called "B to D-star-tau-nu" happens more often than the Standard Model says it should. In this type of decay, a particle called the B-bar meson decays into a D meson, an antineutrino and a tau-lepton. While the level of certainty of the excess (3.4 sigma) is not enough to claim a break from the Standard Model, the results are a potential sign of something amiss and are likely to impact existing theories, including those attempting to deduce the properties of Higgs bosons.[30]

On December 13, 2012, physicists reported the constancy, over space and time, of a basic physical constant of nature that supports the *standard model of physics*. The scientists, studying methanol molecules in a distant galaxy, found the change ($\Delta\mu/\mu$) in the proton-to-electron mass ratio μ to be equal to "$(0.0 \pm 1.0) \times 10^{-7}$ at redshift z = 0.89" and consistent with "a null result".[31][32]

3.7 Challenges

See also: Physics beyond the Standard Model

Self-consistency of the Standard Model (currently formulated as a non-abelian gauge theory quantized through path-integrals) has not been mathematically proven. While regularized versions useful for approximate computations (for example lattice gauge theory) exist, it is not known whether they converge (in the sense of S-matrix elements) in the limit that the regulator is removed. A key question related to the consistency is the Yang–Mills existence and mass gap problem.

Experiments indicate that neutrinos have mass, which the classic Standard Model did not allow.[33] To accommodate this finding, the classic Standard Model can be modified to include neutrino mass.

If one insists on using only Standard Model particles, this can be achieved by adding a non-renormalizable interaction of leptons with the Higgs boson.[34] On a fundamental level, such an interaction emerges in the seesaw mechanism where heavy right-handed neutrinos are added to the theory. This is natural in the left-right symmetric extension of the Standard Model[35][36] and in certain grand unified theories.[37] As long as new physics appears below or around 10^{14} GeV, the neutrino masses can be of the right order of magnitude.

Theoretical and experimental research has attempted to extend the Standard Model into a Unified field theory or a Theory of everything, a complete theory explaining all physical phenomena including constants. Inadequacies of the Standard Model that motivate such research include:

- It does not attempt to explain gravitation, although a theoretical particle known as a graviton would help explain it, and unlike for the strong and electroweak interactions of the Standard Model, there is no known way of describing general relativity, the canonical theory of gravitation, consistently in terms of quantum field theory. The reason for this is, among other things, that quantum field theories of gravity generally break down before reaching the Planck scale. As a consequence, we have no reliable theory for the very early universe;

- Some consider it to be *ad hoc* and inelegant, requiring 19 numerical constants whose values are unrelated and arbitrary. Although the Standard Model, as it now stands, can explain why neutrinos have masses, the specifics of neutrino mass are still unclear. It is believed that explaining neutrino mass will require an additional 7 or 8 constants, which are also arbitrary parameters;

- The Higgs mechanism gives rise to the hierarchy problem if some new physics (coupled to the Higgs) is present at high energy scales. In these cases in order for the weak scale to be much smaller than the Planck scale, severe fine tuning of the parameters is required; there are, however, other scenarios that include quantum gravity in which such fine tuning can be avoided.[38] There are also issues of Quantum triviality, which suggests that it may not be possible to create a consistent quantum field theory involving elementary scalar particles.

- It should be modified so as to be consistent with the emerging "Standard Model of cosmology." In particular, the Standard Model cannot explain the observed amount of cold dark matter (CDM) and gives contributions to dark energy which are many orders of magnitude too large. It is also difficult to accommodate the observed predominance of matter over antimatter (matter/antimatter asymmetry). The isotropy and homogeneity of the visible universe over large distances seems to require a mechanism like cosmic inflation, which would also constitute an extension of the Standard Model.

- The existence of ultra-high-energy cosmic rays are difficult to explain under the Standard Model.

Currently, no proposed Theory of Everything has been widely accepted or verified.

3.8 See also

- Fundamental interaction:

    - Quantum electrodynamics
    - Strong interaction: Color charge, Quantum chromodynamics, Quark model
    - Weak interaction: Electroweak theory, Fermi theory of beta decay, Weak hypercharge, Weak isospin

- Gauge theory: Nontechnical introduction to gauge theory

- Generation

- Higgs mechanism: Higgs boson, Higgsless model

- J. C. Ward

- J. J. Sakurai Prize for Theoretical Particle Physics

- Lagrangian

- Open questions: BTeV experiment, CP violation, Neutrino masses, Quark matter, Quantum triviality

- Penguin diagram

- Quantum field theory

- Standard Model: Mathematical formulation of, Physics beyond the Standard Model

3.9 Notes and references

[1] Technically, there are nine such color–anticolor combinations. However, there is one color-symmetric combination that can be constructed out of a linear superposition of the nine combinations, reducing the count to eight.

3.10 References

[1] R. Oerter (2006). *The Theory of Almost Everything: The Standard Model, the Unsung Triumph of Modern Physics* (Kindle ed.). Penguin Group. p. 2. ISBN 0-13-236678-9.

[2] In fact, there are mathematical issues regarding quantum field theories still under debate (see e.g. Landau pole), but the predictions extracted from the Standard Model by current methods applicable to current experiments are all self-consistent. For a further discussion see e.g. Chapter 25 of R. Mann (2010). *An Introduction to Particle Physics and the Standard Model.* CRC Press. ISBN 978-1-4200-8298-2.

[3] Sean Carroll, Ph.D., Cal Tech, 2007, The Teaching Company, *Dark Matter, Dark Energy: The Dark Side of the Universe*, Guidebook Part 2 page 59, Accessed Oct. 7, 2013, "...Standard Model of Particle Physics: The modern theory of elementary particles and their interactions ... It does not, strictly speaking, include gravity, although it's often convenient to include gravitons among the known particles of nature..."

[4] S.L. Glashow (1961). "Partial-symmetries of weak interactions". *Nuclear Physics* **22**(4): 579–588. Bibcode:1961NucPh..22..57 doi:10.1016/0029-5582(61)90469-2.

[5] S. Weinberg (1967). "A Model of Leptons". *Physical Review Letters* **19** (21): 1264–1266. Bibcode:1967PhRvL..19.1264W. doi:10.1103/PhysRevLett.19.1264.

[6] A. Salam (1968). N. Svartholm, ed. *Elementary Particle Physics: Relativistic Groups and Analyticity.* Eighth Nobel Symposium. Stockholm: Almquvist and Wiksell. p. 367.

[7] F. Englert, R. Brout (1964). "Broken Symmetry and the Mass of Gauge Vector Mesons". *Physical Review Letters* **13** (9): 321–323. Bibcode:1964PhRvL..13..321E. doi:10.1103/PhysRevLett.13.321.

[8] P.W. Higgs (1964). "Broken Symmetries and the Masses of Gauge Bosons". *Physical Review Letters* **13** (16): 508–509. Bibcode:1964PhRvL..13..508H. doi:10.1103/PhysRevLett.13.508.

[9] G.S. Guralnik, C.R. Hagen, T.W.B. Kibble (1964). "Global Conservation Laws and Massless Particles". *Physical Review Letters* **13** (20): 585–587. Bibcode:1964PhRvL..13..585G. doi:10.1103/PhysRevLett.13.585.

[10] F.J. Hasert; et al. (1973). "Search for elastic muon-neutrino electron scattering". *Physics Letters B* **46**(1): 121. Bibcode:1973PhL doi:10.1016/0370-2693(73)90494-2.

[11] F.J. Hasert; et al. (1973). "Observation of neutrino-like interactions without muon or electron in the Gargamelle neutrino experiment". *Physics Letters B* **46** (1): 138. Bibcode:1973PhLB...46..138H. doi:10.1016/0370-2693(73)90499-1.

[12] F.J. Hasert; et al. (1974). "Observation of neutrino-like interactions without muon or electron in the Gargamelle neutrino experiment". *Nuclear Physics B* **73** (1): 1. Bibcode:1974NuPhB..73....1H. doi:10.1016/0550-3213(74)90038-8.

[13] D. Haidt (4 October 2004). "The discovery of the weak neutral currents". *CERN Courier.* Retrieved 8 May 2008.

[14] "Details can be worked out if the situation is simple enough for us to make an approximation, which is almost never, but often we can understand more or less what is happening." from *The Feynman Lectures on Physics*, Vol 1. pp. 2–7

[15] G.S. Guralnik (2009). "The History of the Guralnik, Hagen and Kibble development of the Theory of Spontaneous Symmetry Breaking and Gauge Particles". *International Journal of Modern Physics A* **24** (14): 2601–2627. arXiv:0907.3466. Bibcode:2009IJMPA..24.2601G. doi:10.1142/S0217751X09045431.

[16] B.W. Lee, C. Quigg, H.B. Thacker (1977). "Weak interactions at very high energies: The role of the Higgs-boson mass". *Physical Review D* **16** (5): 1519–1531. Bibcode:1977PhRvD..16.1519L. doi:10.1103/PhysRevD.16.1519.

[17] "Huge $10 billion collider resumes hunt for 'God particle'". CNN. 11 November 2009. Retrieved 2010-05-04.

[18] M. Strassler (10 July 2012). "Higgs Discovery: Is it a Higgs?". Retrieved 2013-08-06.

[19] "CERN experiments observe particle consistent with long-sought Higgs boson". CERN. 4 July 2012. Retrieved 2012-07-04.

[20] "Observation of a New Particle with a Mass of 125 GeV". CERN. 4 July 2012. Retrieved 2012-07-05.

[21] "ATLAS Experiment". ATLAS. 1 January 2006. Retrieved 2012-07-05.

[22] "Confirmed: CERN discovers new particle likely to be the Higgs boson". *YouTube*. Russia Today. 4 July 2012. Retrieved 2013-08-06.

[23] D. Overbye (4 July 2012). "A New Particle Could Be Physics' Holy Grail". *New York Times*. Retrieved 2012-07-04.

[24] "New results indicate that new particle is a Higgs boson". CERN. 14 March 2013. Retrieved 2013-08-06.

[25] S. Braibant, G. Giacomelli, M. Spurio (2009). *Particles and Fundamental Interactions: An Introduction to Particle Physics*. Springer. pp. 313–314. ISBN 978-94-007-2463-1.

[26] http://home.web.cern.ch/about/physics/standard-model Official CERN website

[27] http://www.pha.jhu.edu/~{}dfehling/particle.gif

[28] "BABAR Data in Tension with the Standard Model". SLAC. 31 May 2012. Retrieved 2013-08-06.

[29] BaBar Collaboration (2012). "Evidence for an excess of $B \to D^{(*)} \tau^- \nu\tau$ decays". *Physical Review Letters* **109** (10): 101802. arXiv:1205.5442. Bibcode:2012PhRvL.109j1802L. doi:10.1103/PhysRevLett.109.101802.

[30] "BaBar data hint at cracks in the Standard Model". *e! Science News*. 18 June 2012. Retrieved 2013-08-06.

[31] J. Bagdonaite; et al. (2012). "A Stringent Limit on a Drifting Proton-to-Electron Mass Ratio from Alcohol in the Early Universe". *Science* **339** (6115): 46. Bibcode:2013Sci...339...46B. doi:10.1126/science.1224898.

[32] C. Moskowitz (13 December 2012). "Phew! Universe's Constant Has Stayed Constant". Space.com. Retrieved 2012-12-14.

[33] "Particle chameleon caught in the act of changing". CERN. 31 May 2010. Retrieved 2012-07-05.

[34] S. Weinberg (1979). "Baryon and Lepton Nonconserving Processes".*Physical Review Letters***43**(21): 1566.Bibcode:1979PhR doi:10.1103/PhysRevLett.43.1566.

[35] P.Minkowski(1977). "$\mu \to e\gamma$ at a Rate of One Out of 10_9Muon Decays?".*Physics Letters B***67**(4):421.Bibcode:1977PhLB... doi:10.1016/0370-2693(77)90435-X.

[36] R. N. Mohapatra, G. Senjanovic (1980). "Neutrino Mass and Spontaneous Parity Nonconservation". *Physical Review Letters* **44** (14): 912–915. Bibcode:1980PhRvL..44..912M. doi:10.1103/PhysRevLett.44.912.

[37] M. Gell-Mann, P. Ramond and R. Slansky (1979). F. van Nieuwenhuizen and D. Z. Freedman, ed. *Supergravity*. North Holland. pp. 315–321. ISBN 0-444-85438-X.

[38] Salvio, Strumia (2014-03-17)."Agravity".*JHEP 1406 (2014) 080*.arXiv:1403.4226.Bibcode:2014JHEP...06..080S.doi:10

3.11   Further reading

- R. Oerter (2006). *The Theory of Almost Everything: The Standard Model, the Unsung Triumph of Modern Physics*. Plume.

- B.A. Schumm (2004). *Deep Down Things: The Breathtaking Beauty of Particle Physics*. Johns Hopkins University Press. ISBN 0-8018-7971-X.

- "The Standard Model of Particle Physics Interactive Graphic".

Introductory textbooks

- I. Aitchison, A. Hey (2003). *Gauge Theories in Particle Physics: A Practical Introduction*. Institute of Physics. ISBN 978-0-585-44550-2.

- W. Greiner, B. Müller (2000). *Gauge Theory of Weak Interactions*. Springer. ISBN 3-540-67672-4.

- G.D. Coughlan, J.E. Dodd, B.M. Gripaios (2006). *The Ideas of Particle Physics: An Introduction for Scientists*. Cambridge University Press.

- D.J. Griffiths (1987). *Introduction to Elementary Particles.* John Wiley & Sons. ISBN 0-471-60386-4.

- G.L. Kane (1987). *Modern Elementary Particle Physics.* Perseus Books. ISBN 0-201-11749-5.

Advanced textbooks

- T.P. Cheng, L.F. Li (2006). *Gauge theory of elementary particle physics.* Oxford University Press. ISBN 0-19-851961-3. Highlights the gauge theory aspects of the Standard Model.

- J.F. Donoghue, E. Golowich, B.R. Holstein (1994). *Dynamics of the Standard Model.* Cambridge University Press. ISBN 978-0-521-47652-2. Highlights dynamical and phenomenological aspects of the Standard Model.

- L. O'Raifeartaigh (1988). *Group structure of gauge theories.* Cambridge University Press. ISBN 0-521-34785-8.

- Nagashima Y. Elementary Particle Physics: Foundations of the Standard Model, Volume 2. (Wiley 2013) 920 рапуы

- Schwartz, M.D. Quantum Field Theory and the Standard Model (Cambridge University Press 2013) 952 pages

- Langacker P. The standard model and beyond. (CRC Press, 2010) 670 pages Highlights group-theoretical aspects of the Standard Model.

Journal articles

- E.S. Abers, B.W. Lee (1973). "Gauge theories".*Physics Reports***9**: 1–141.Bibcode:1973PhR.....9....1A.doi:10.10 1573(73)90027-6.

- M. Baak; et al. (2012). "The Electroweak Fit of the Standard Model after the Discovery of a New Boson at the LHC". *The European Physical Journal C* **72** (11). arXiv:1209.2716. Bibcode:2012EPJC...72.2205B. doi:10.1140/epjc/s10052-012-2205-9.

- Y. Hayato; et al. (1999). "Search for Proton Decay through $p \to \nu K^+$ in a Large Water Cherenkov Detector". *Physical Review Letters***83**(8): 1529.arXiv:hep-ex/9904020.Bibcode:1999PhRvL..83.1529H.doi:10.1103/Phy

- S.F. Novaes (2000). "Standard Model: An Introduction". arXiv:hep-ph/0001283 [hep-ph].

- D.P. Roy (1999). "Basic Constituents of Matter and their Interactions — A Progress Report".arXiv:hep-ph/99125 [hep-ph].

- F. Wilczek (2004). "The Universe Is A Strange Place". *Nuclear Physics B - Proceedings Supplements* **134**: 3. arXiv:astro-ph/0401347. Bibcode:2004NuPhS.134....3W. doi:10.1016/j.nuclphysbps.2004.08.001.

3.12 External links

- "The Standard Model explained in Detail by CERN's John Ellis" omega tau podcast.

- "The Standard Model" The Standard Model on the CERN web site explains how the basic building blocks of matter interact, governed by four fundamental forces.

- "Standard Model" on YouTube

Chapter 4

Fermi–Dirac statistics

In quantum statistics, a branch of physics, **Fermi–Dirac statistics** describes a distribution of particles over energy states in systems consisting of many identical particles that obey the Pauli exclusion principle. It is named after Enrico Fermi and Paul Dirac, each of whom discovered it independently (although Fermi defined the statistics earlier than Dirac).[1][2]

Fermi–Dirac (F–D) statistics applies to identical particles with half-integer spin in a system in thermodynamic equilibrium. Additionally, the particles in this system are assumed to have negligible mutual interaction. This allows the many-particle system to be described in terms of single-particle energy states. The result is the F–D distribution of particles over these states and includes the condition that no two particles can occupy the same state, which has a considerable effect on the properties of the system. Since F–D statistics applies to particles with half-integer spin, these particles have come to be called fermions. It is most commonly applied to electrons, which are fermions with spin 1/2. Fermi–Dirac statistics is a part of the more general field of statistical mechanics and uses the principles of quantum mechanics.

4.1 History

Before the introduction of Fermi–Dirac statistics in 1926, understanding some aspects of electron behavior was difficult due to seemingly contradictory phenomena. For example, the electronic heat capacity of a metal at room temperature seemed to come from 100 times fewer electrons than were in the electric current.[3] It was also difficult to understand why the emission currents, generated by applying high electric fields to metals at room temperature, were almost independent of temperature.

The difficulty encountered by the electronic theory of metals at that time was due to considering that electrons were (according to classical statistics theory) all equivalent. In other words it was believed that each electron contributed to the specific heat an amount on the order of the Boltzmann constant k. This statistical problem remained unsolved until the discovery of F–D statistics.

F–D statistics was first published in 1926 by Enrico Fermi[1] and Paul Dirac.[2] According to an account, Pascual Jordan developed in 1925 the same statistics which he called *Pauli statistics*, but it was not published in a timely manner.[4] According to Dirac, it was first studied by Fermi, and Dirac called it Fermi statistics and the corresponding particles fermions.[5]

F–D statistics was applied in 1926 by Fowler to describe the collapse of a star to a white dwarf.[6] In 1927 Sommerfeld applied it to electrons in metals[7] and in 1928 Fowler and Nordheim applied it to field electron emission from metals.[8] Fermi–Dirac statistics continues to be an important part of physics.

4.2 Fermi–Dirac distribution

For a system of identical fermions, the average number of fermions in a single-particle state i is given by a logistic function, the Fermi–Dirac (F–D) distribution,[9]

$$\bar{n}_i = \frac{1}{e^{(\epsilon_i - \mu)/kT} + 1}$$

where k is Boltzmann's constant, T is the absolute temperature, ϵ_i is the energy of the single-particle state i, and μ is the total chemical potential.

At zero temperature, μ is equal to the Fermi energy plus the potential energy per electron. For the case of electrons in a semiconductor, μ, the point of symmetry, is typically called the Fermi level or electrochemical potential.[10][11]

The F–D distribution is only valid if the number of fermions in the system is large enough so that adding one more fermion to the system has negligible effect on μ.[12] Since the F–D distribution was derived using the Pauli exclusion principle, which allows at most one electron to occupy each possible state, a result is that $0 < \bar{n}_i < 1$.[13]

- Fermi–Dirac distribution

- **Energy dependence.** More gradual at higher T. = 0.5 when = . Not shown is that decreases for higher T.[1]

  - **Temperature dependence** for .

1. ^ (Kittel 1971, p. 245, Figs. 4 and 5)

(Click on a figure to enlarge.)

4.2.1 Distribution of particles over energy

The above Fermi–Dirac distribution gives the distribution of identical fermions over single-particle energy states, where no more than one fermion can occupy a state. Using the F–D distribution, one can find the distribution of identical fermions over energy, where more than one fermion can have the same energy.[14]

The average number of fermions with energy ϵ_i can be found by multiplying the F–D distribution $\bar{n}_i$ by the degeneracy g_i (i.e. the number of states with energy ϵ_i),[15]

$$\bar{n}(\epsilon_i) = g_i\, \bar{n}_i$$
$$= \frac{g_i}{e^{(\epsilon_i - \mu)/kT} + 1}$$

When $g_i \geq 2$, it is possible that $\bar{n}(\epsilon_i) > 1$ since there is more than one state that can be occupied by fermions with the same energy ϵ_i .

When a quasi-continuum of energies ϵ has an associated density of states $g(\epsilon)$ (i.e. the number of states per unit energy range per unit volume[16]) the average number of fermions per unit energy range per unit volume is,

$$\bar{\mathcal{N}}(\epsilon) = g(\epsilon)\, F(\epsilon)$$

where $F(\epsilon)$ is called the Fermi function and is the same function that is used for the F–D distribution $\bar{n}_i$,[17]

$$F(\epsilon) = \frac{1}{e^{(\epsilon - \mu)/kT} + 1}$$

so that,

$$\bar{\mathcal{N}}(\epsilon) = \frac{g(\epsilon)}{e^{(\epsilon-\mu)/kT} + 1}$$

4.3   Quantum and classical regimes

The classical regime, where Maxwell–Boltzmann statistics can be used as an approximation to Fermi–Dirac statistics, is found by considering the situation that is far from the limit imposed by the Heisenberg uncertainty principle for a particle's position and momentum. Using this approach, it can be shown that the classical situation occurs if the concentration of particles corresponds to an average interparticle separation $\bar{R}$ that is much greater than the average de Broglie wavelength $\bar{\lambda}$ of the particles,[18]

$$\bar{R} \gg \bar{\lambda} \approx \frac{h}{\sqrt{3mkT}}$$

where h is Planck's constant, and m is the mass of a particle.

For the case of conduction electrons in a typical metal at $T = 300K$ (i.e. approximately room temperature), the system is far from the classical regime because $\bar{R} \approx \bar{\lambda}/25$. This is due to the small mass of the electron and the high concentration (i.e. small $\bar{R}$) of conduction electrons in the metal. Thus Fermi–Dirac statistics is needed for conduction electrons in a typical metal.[18]

Another example of a system that is not in the classical regime is the system that consists of the electrons of a star that has collapsed to a white dwarf. Although the white dwarf's temperature is high (typically $T = 10,000K$ on its surface[19]), its high electron concentration and the small mass of each electron precludes using a classical approximation, and again Fermi–Dirac statistics is required.[6]

4.4   Derivations of the Fermi–Dirac distribution

4.4.1   Grand canonical ensemble

The Fermi–Dirac distribution, which applies only to a quantum system of non-interacting fermions, is easily derived from the grand canonical ensemble.[20] In this ensemble, the system is able to exchange energy and exchange particles with a reservoir (temperature T and chemical potential μ fixed by the reservoir).

Due to the non-interacting quality, each available single-particle level (with energy level ϵ) forms a separate thermodynamic system in contact with the reservoir. In other words, each single-particle level is a separate, tiny grand canonical ensemble. By the Pauli exclusion principle there are only two possible microstates for the single-particle level: no particle (energy $E=0$), or one particle (energy $E=\epsilon$). The resulting partition function for that single-particle level therefore has just two terms:

$$\begin{aligned} \mathcal{Z} &= \exp(0(\mu - \epsilon)/k_BT) + \exp(1(\mu - \epsilon)/k_BT) \\ &= 1 + \exp((\mu - \epsilon)/k_BT) \end{aligned}$$

and the average particle number for that single-particle substate is given by

$$\langle N \rangle = k_BT \frac{1}{\mathcal{Z}} \left(\frac{\partial \mathcal{Z}}{\partial \mu} \right)_{V,T} = \frac{1}{\exp((\epsilon - \mu)/k_BT) + 1}$$

This result applies for each single-particle level, and thus gives the Fermi–Dirac distribution for the entire state of the system.[20]

The variance in particle number (due to thermal fluctuations) may also be derived (the particle number has a simple Bernoulli distribution):

$$\langle (\Delta N)^2 \rangle = k_B T \left(\frac{d\langle N \rangle}{d\mu} \right)_{V,T} = \langle N \rangle (1 - \langle N \rangle)$$

This quantity is important in transport phenomena such as the Mott relations for electrical conductivity and thermoelectric coefficient for an electron gas,[21] where the ability of an energy level to contribute to transport phenomena is proportional to $\langle (\Delta N)^2 \rangle$.

4.4.2 Canonical ensemble

It is also possible to derive Fermi–Dirac statistics in the canonical ensemble. Consider a many-particle system composed of N identical fermions that have negligible mutual interaction and are in thermal equilibrium.[12] Since there is negligible interaction between the fermions, the energy E_R of a state R of the many-particle system can be expressed as a sum of single-particle energies,

$$E_R = \sum_r n_r \epsilon_r$$

where n_r is called the occupancy number and is the number of particles in the single-particle state r with energy ϵ_r. The summation is over all possible single-particle states r.

The probability that the many-particle system is in the state R, is given by the normalized canonical distribution,[22]

$$P_R = \frac{e^{-\beta E_R}}{\displaystyle\sum_{R'} e^{-\beta E_{R'}}}$$

where $\beta = 1/kT$, k is Boltzmann's constant, T is the absolute temperature, $e^{-\beta E_R}$ is called the Boltzmann factor, and the summation is over all possible states R' of the many-particle system. The average value for an occupancy number n_i is[22]

$$\bar{n}_i = \sum_R n_i P_R$$

Note that the state R of the many-particle system can be specified by the particle occupancy of the single-particle states, i.e. by specifying $n_1, n_2, \ldots$, so that

$$P_R = P_{n_1, n_2, \ldots} = \frac{e^{-\beta(n_1 \epsilon_1 + n_2 \epsilon_2 + \ldots)}}{\displaystyle\sum_{n_1', n_2', \ldots} e^{-\beta(n_1' \epsilon_1 + n_2' \epsilon_2 + \ldots)}}$$

and the equation for $\bar{n}_i$ becomes

$$\bar{n}_i = \sum_{n_1, n_2, \ldots} n_i \, P_{n_1, n_2, \ldots}$$

$$= \frac{\displaystyle\sum_{n_1, n_2, \ldots} n_i \, e^{-\beta(n_1 \epsilon_1 + n_2 \epsilon_2 + \cdots + n_i \epsilon_i + \cdots)}}{\displaystyle\sum_{n_1, n_2, \ldots} e^{-\beta(n_1 \epsilon_1 + n_2 \epsilon_2 + \cdots + n_i \epsilon_i + \cdots)}}$$

where the summation is over all combinations of values of $n_1, n_2, \ldots$ which obey the Pauli exclusion principle, and $n_r = 0$ or 1 for each r. Furthermore, each combination of values of $n_1, n_2, \ldots$ satisfies the constraint that the total number of particles is N,

$$\sum_r n_r = N$$

Rearranging the summations,

$$\bar{n}_i = \frac{\displaystyle\sum_{n_i=0}^{1} n_i \, e^{-\beta(n_i \epsilon_i)} \sum_{n_1, n_2, \ldots}^{(i)} e^{-\beta(n_1 \epsilon_1 + n_2 \epsilon_2 + \cdots)}}{\displaystyle\sum_{n_i=0}^{1} e^{-\beta(n_i \epsilon_i)} \sum_{n_1, n_2, \ldots}^{(i)} e^{-\beta(n_1 \epsilon_1 + n_2 \epsilon_2 + \cdots)}}$$

where the $^{(i)}$ on the summation sign indicates that the sum is not over n_i and is subject to the constraint that the total number of particles associated with the summation is $N_i = N - n_i$. Note that $\Sigma^{(i)}$ still depends on n_i through the N_i constraint, since in one case $n_i = 0$ and $\Sigma^{(i)}$ is evaluated with $N_i = N$, while in the other case $n_i = 1$ and $\Sigma^{(i)}$ is evaluated with $N_i = N - 1$. To simplify the notation and to clearly indicate that $\Sigma^{(i)}$ still depends on n_i through $N - n_i$, define

$$Z_i(N - n_i) \equiv \sum_{n_1, n_2, \ldots}^{(i)} e^{-\beta(n_1 \epsilon_1 + n_2 \epsilon_2 + \cdots)}$$

so that the previous expression for $\bar{n}_i$ can be rewritten and evaluated in terms of the Z_i,

$$\bar{n}_i = \frac{\displaystyle\sum_{n_i=0}^{1} n_i \, e^{-\beta(n_i \epsilon_i)} \, Z_i(N - n_i)}{\displaystyle\sum_{n_i=0}^{1} e^{-\beta(n_i \epsilon_i)} \qquad Z_i(N - n_i)}$$

$$= \frac{0 \quad + e^{-\beta \epsilon_i} \, Z_i(N - 1)}{Z_i(N) + e^{-\beta \epsilon_i} \, Z_i(N - 1)}$$

$$= \frac{1}{[Z_i(N)/Z_i(N - 1)] \, e^{\beta \epsilon_i} + 1} \quad .$$

The following approximation[23] will be used to find an expression to substitute for $Z_i(N)/Z_i(N - 1)$.

$$\ln Z_i(N-1) \simeq \ln Z_i(N) - \frac{\partial \ln Z_i(N)}{\partial N}$$
$$= \ln Z_i(N) - \alpha_i$$

where $\alpha_i \equiv \frac{\partial \ln Z_i(N)}{\partial N}$.

If the number of particles N is large enough so that the change in the chemical potential μ is very small when a particle is added to the system, then $\alpha_i \simeq -\mu/kT$. [24] Taking the base e antilog[25] of both sides, substituting for α_i , and rearranging,

$$Z_i(N)/Z_i(N-1) = e^{-\mu/kT}$$

Substituting the above into the equation for $\bar{n}_i$, and using a previous definition of β to substitute $1/kT$ for β , results in the Fermi–Dirac distribution.

$$\bar{n}_i = \frac{1}{e^{(\epsilon_i - \mu)/kT}+1}$$

4.4.3 Microcanonical ensemble

A result can be achieved by directly analyzing the multiplicities of the system and using Lagrange multipliers.[26]

Suppose we have a number of energy levels, labeled by index i, each level having energy ϵi and containing a total of *ni* particles. Suppose each level contains *gi* distinct sublevels, all of which have the same energy, and which are distinguishable. For example, two particles may have different momenta (i.e. their momenta may be along different directions), in which case they are distinguishable from each other, yet they can still have the same energy. The value of *gi* associated with level *i* is called the "degeneracy" of that energy level. The Pauli exclusion principle states that only one fermion can occupy any such sublevel.

The number of ways of distributing *ni* indistinguishable particles among the *gi* sublevels of an energy level, with a maximum of one particle per sublevel, is given by the binomial coefficient, using its combinatorial interpretation

$$w(n_i, g_i) = \frac{g_i!}{n_i!(g_i - n_i)!} \ .$$

For example, distributing two particles in three sublevels will give population numbers of 110, 101, or 011 for a total of three ways which equals 3!/(2!1!). The number of ways that a set of occupation numbers *ni* can be realized is the product of the ways that each individual energy level can be populated:

$$W = \prod_i w(n_i, g_i) = \prod_i \frac{g_i!}{n_i!(g_i - n_i)!}.$$

Following the same procedure used in deriving the Maxwell–Boltzmann statistics, we wish to find the set of *ni* for which W is maximized, subject to the constraint that there be a fixed number of particles, and a fixed energy. We constrain our solution using Lagrange multipliers forming the function:

$$f(n_i) = \ln(W) + \alpha(N - \sum n_i) + \beta(E - \sum n_i \epsilon_i).$$

Using Stirling's approximation for the factorials, taking the derivative with respect to *ni*, setting the result to zero, and solving for *ni* yields the Fermi–Dirac population numbers:

$$n_i = \frac{g_i}{e^{\alpha + \beta \epsilon_i} + 1}.$$

By a process similar to that outlined in the Maxwell–Boltzmann statistics article, it can be shown thermodynamically that $\beta = \frac{1}{kT}$ and $\alpha = -\frac{\mu}{kT}$ where μ is the chemical potential, k is Boltzmann's constant and T is the temperature, so that finally, the probability that a state will be occupied is:

$$\bar{n}_i = \frac{n_i}{g_i} = \frac{1}{e^{(\epsilon_i - \mu)/kT} + 1}.$$

4.5 See also

- Grand canonical ensemble

- Fermi level

- Maxwell–Boltzmann statistics

- Bose–Einstein statistics

- Parastatistics

- Logistic function

4.6 References

1. Reif, F. (1965). *Fundamentals of Statistical and Thermal Physics*. McGraw–Hill. ISBN 978-0-07-051800-1.

2. Blakemore, J. S. (2002). *Semiconductor Statistics*. Dover. ISBN 978-0-486-49502-6.

3. Kittel, Charles (1971). *Introduction to Solid State Physics* (4th ed.). New York: John Wiley & Sons. ISBN 0-471-14286-7. OCLC 300039591.

4.7 Footnotes

[1] Fermi, Enrico (1926). "Sulla quantizzazione del gas perfetto monoatomico". *Rendiconti Lincei* (in Italian) **3**: 145–9., translated as Zannoni, Alberto (transl.) (1999-12-14). "On the Quantization of the Monoatomic Ideal Gas". arXiv:cond-mat/9912229 [cond-mat.stat-mech].

[2] Dirac, Paul A. M. (1926). "On the Theory of Quantum Mechanics". *Proceedings of the Royal Society, Series A* **112** (762): 661–77. Bibcode:1926RSPSA.112..661D. doi:10.1098/rspa.1926.0133. JSTOR 94692.

[3] (Kittel 1971, pp. 249–50)

[4] "History of Science: The Puzzle of the Bohr–Heisenberg Copenhagen Meeting". *Science-Week* (Chicago) **4** (20). 2000-05-19. OCLC 43626035. Retrieved 2009-01-20.

[5] Dirac, Paul A. M. (1967). *Principles of Quantum Mechanics* (revised 4th ed.). London: Oxford University Press. pp. 210–1. ISBN 978-0-19-852011-5.

[6] Fowler, Ralph H. (December 1926). "On dense matter". *Monthly Notices of the Royal Astronomical Society* **87** (2): 114–22. Bibcode:1926MNRAS..87..114F. doi:10.1093/mnras/87.2.114.

[7] Sommerfeld, Arnold (1927-10-14). "Zur Elektronentheorie der Metalle" [On Electron Theory of Metals]. *Naturwissenschaften* (in German) **15** (41): 824–32. Bibcode:1927NW.....15..825S. doi:10.1007/BF01505083.

[8] Fowler, Ralph H.; Nordheim, Lothar W. (1928-05-01). "Electron Emission in Intense Electric Fields" (PDF). *Proceedings of the Royal Society A* **119** (781): 173–81. Bibcode:1928RSPSA.119..173F. doi:10.1098/rspa.1928.0091. JSTOR 95023.

[9] (Reif 1965, p. 341)

[10] (Blakemore 2002, p. 11)

[11] Kittel, Charles; Kroemer, Herbert (1980). *Thermal Physics* (2nd ed.). San Francisco: W. H. Freeman. p. 357. ISBN 978-0-7167-1088-2.

[12] (Reif 1965, pp. 340–2)

[13] Note that $\bar{n}_i$ is also the probability that the state i is occupied, since no more than one fermion can occupy the same state at the same time and $0 < \bar{n}_i < 1$.

[14] These distributions over energies, rather than states, are sometimes called the Fermi–Dirac distribution too, but that terminology will not be used in this article.

[15] Leighton, Robert B. (1959). *Principles of Modern Physics*. McGraw-Hill. p. 340. ISBN 978-0-07-037130-9.
Note that in Eq. (1), $n(\epsilon)$ and n_s correspond respectively to $\bar{n}_i$ and $\bar{n}(\epsilon_i)$ in this article. See also Eq. (32) on p. 339.

[16] (Blakemore 2002, p. 8)

[17] (Reif 1965, p. 389)

[18] (Reif 1965, pp. 246–8)

[19] Mukai, Koji; Jim Lochner (1997). "Ask an Astrophysicist". *NASA's Imagine the Universe*. NASA Goddard Space Flight Center. Archived from the original on 2009-01-20.

[20] Srivastava, R. K.; Ashok, J. (2005). "Chapter 6". *Statistical Mechanics*. New Delhi: PHI Learning Pvt. Ltd. ISBN 9788120327825.

[21] Cutler, M.; Mott, N. (1969). "Observation of Anderson Localization in an Electron Gas". *Physical Review* **181** (3): 1336. Bibcode:1969PhRv..181.1336C. doi:10.1103/PhysRev.181.1336.

[22] (Reif 1965, pp. 203–6)

[23] See for example, Derivative - Definition via difference quotients, which gives the approximation $f(a+h) \approx f(a) + f'(a)h$.

[24] (Reif 1965, pp. 341–2) See Eq. 9.3.17 and *Remark concerning the validity of the approximation*.

[25] By definition, the base e antilog of A is e^A.

[26] (Blakemore 2002, pp. 343–5)

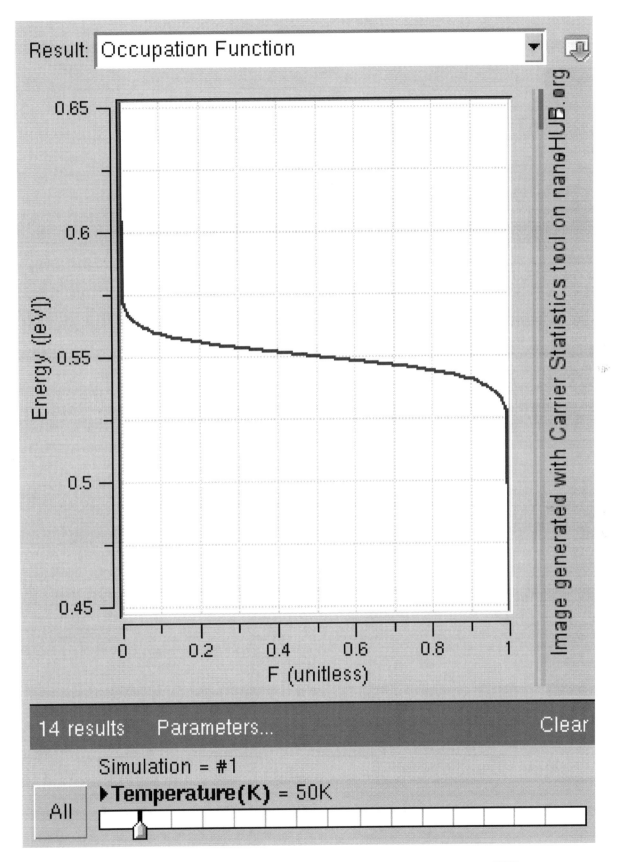

Fermi function F(ε) vs. energy ε , with μ = 0.55 eV and for various temperatures in the range 50K ≤ T ≤ 375K.

Chapter 5

Pauli exclusion principle

The **Pauli exclusion principle** is the quantum mechanical principle that states that two identical fermions (particles with half-integer spin) cannot occupy the same quantum state simultaneously. In the case of electrons, it can be stated as follows: it is impossible for two electrons of a poly-electron atom to have the same values of the four quantum numbers: n, the principal quantum number, ℓ, the angular momentum quantum number, $m\ell$, the magnetic quantum number, and ms, the spin quantum number. For two electrons residing in the same orbital, n, ℓ, and $m\ell$ are the same, so ms, the spin, must be different, and thus the electrons have opposite half-integer spins, 1/2 and $-1/2$. This principle was formulated by Austrian physicist Wolfgang Pauli in 1925.

A more rigorous statement is that the total wave function for two identical fermions is antisymmetric with respect to exchange of the particles. This means that the wave function changes its sign if the space *and* spin co-ordinates of any two particles are interchanged.

Particles with an integer spin, or bosons, are not subject to the Pauli exclusion principle: any number of identical bosons can occupy the same quantum state, as with, for instance, photons produced by a laser and Bose–Einstein condensate.

5.1 Overview

The Pauli exclusion principle governs the behavior of all fermions (particles with "half-integer spin"), while bosons (particles with "integer spin") are not subject to it. Fermions include elementary particles such as quarks (the constituent particles of protons and neutrons), electrons and neutrinos. In addition, protons and neutrons (subatomic particles composed from three quarks) and some atoms are fermions, and are therefore subject to the Pauli exclusion principle as well. Atoms can have different overall "spin", which determines whether they are fermions or bosons — for example helium-3 has spin 1/2 and is therefore a fermion, in contrast to helium-4 which has spin 0 and is a boson.[1]:123–125 As such, the Pauli exclusion principle underpins many properties of everyday matter, from its large-scale stability, to the chemical behavior of atoms.

"Half-integer spin" means that the intrinsic angular momentum value of fermions is $\hbar = h/2\pi$ (reduced Planck's constant) times a half-integer (1/2, 3/2, 5/2, etc.). In the theory of quantum mechanics fermions are described by antisymmetric states. In contrast, particles with integer spin (called bosons) have symmetric wave functions; unlike fermions they may share the same quantum states. Bosons include the photon, the Cooper pairs which are responsible for superconductivity, and the W and Z bosons. (Fermions take their name from the Fermi–Dirac statistical distribution that they obey, and bosons from their Bose–Einstein distribution).

5.2 History

In the early 20th century it became evident that atoms and molecules with even numbers of electrons are more chemically stable than those with odd numbers of electrons. In the 1916 article "The Atom and the Molecule" by Gilbert N. Lewis,

Wolfgang Pauli

for example, the third of his six postulates of chemical behavior states that the atom tends to hold an even number of electrons in the shell and especially to hold eight electrons which are normally arranged symmetrically at the eight corners of a cube (see: cubical atom).[2] In 1919 chemist Irving Langmuir suggested that the periodic table could be explained if the electrons in an atom were connected or clustered in some manner. Groups of electrons were thought to occupy a set of electron shells around the nucleus.[3] In 1922, Niels Bohr updated his model of the atom by assuming that certain numbers of electrons (for example 2, 8 and 18) corresponded to stable "closed shells".[4]:203

Pauli looked for an explanation for these numbers, which were at first only empirical. At the same time he was trying to explain experimental results of the Zeeman effect in atomic spectroscopy and in ferromagnetism. He found an essential clue in a 1924 paper by Edmund C. Stoner, which pointed out that for a given value of the principal quantum number (*n*), the number of energy levels of a single electron in the alkali metal spectra in an external magnetic field, where all degenerate energy levels are separated, is equal to the number of electrons in the closed shell of the noble gases for the same value of *n*. This led Pauli to realize that the complicated numbers of electrons in closed shells can be reduced to the simple rule of *one* electron per state, if the electron states are defined using four quantum numbers. For this purpose he introduced a new two-valued quantum number, identified by Samuel Goudsmit and George Uhlenbeck as electron spin.[5]

5.3 Connection to quantum state symmetry

The Pauli exclusion principle with a single-valued many-particle wavefunction is equivalent to requiring the wavefunction to be antisymmetric. An antisymmetric two-particle state is represented as a sum of states in which one particle is in state $|x\rangle$ and the other in state $|y\rangle$:

$$|\psi\rangle = \sum_{x,y} A(x,y)|x,y\rangle,$$

and antisymmetry under exchange means that $A(x,y) = -A(y,x)$. This implies $A(x,y) = 0$ when $x=y$, which is Pauli exclusion. It is true in any basis, since unitary changes of basis keep antisymmetric matrices antisymmetric, although strictly speaking, the quantity $A(x,y)$ is not a matrix but an antisymmetric rank-two tensor.

Conversely, if the diagonal quantities $A(x,x)$ are zero *in every basis*, then the wavefunction component

$$A(x,y) = \langle\psi|x,y\rangle = \langle\psi|(|x\rangle \otimes |y\rangle)$$

is necessarily antisymmetric. To prove it, consider the matrix element

$$\langle\psi|\Big((|x\rangle + |y\rangle) \otimes (|x\rangle + |y\rangle)\Big).$$

This is zero, because the two particles have zero probability to both be in the superposition state $|x\rangle + |y\rangle$. But this is equal to

$$\langle\psi|x,x\rangle + \langle\psi|x,y\rangle + \langle\psi|y,x\rangle + \langle\psi|y,y\rangle.$$

The first and last terms on the right side are diagonal elements and are zero, and the whole sum is equal to zero. So the wavefunction matrix elements obey:

$$\langle\psi|x,y\rangle + \langle\psi|y,x\rangle = 0,$$

or

$$A(x,y) = -A(y,x).$$

5.3.1 Pauli principle in advanced quantum theory

According to the spin-statistics theorem, particles with integer spin occupy symmetric quantum states, and particles with half-integer spin occupy antisymmetric states; furthermore, only integer or half-integer values of spin are allowed by the principles of quantum mechanics. In relativistic quantum field theory, the Pauli principle follows from applying a rotation operator in imaginary time to particles of half-integer spin.

In one dimension, bosons, as well as fermions, can obey the exclusion principle. A one-dimensional Bose gas with delta-function repulsive interactions of infinite strength is equivalent to a gas of free fermions. The reason for this is that, in one dimension, exchange of particles requires that they pass through each other; for infinitely strong repulsion this cannot happen. This model is described by a quantum nonlinear Schrödinger equation. In momentum space the exclusion principle is valid also for finite repulsion in a Bose gas with delta-function interactions,[6] as well as for interacting spins and Hubbard model in one dimension, and for other models solvable by Bethe ansatz. The ground state in models solvable by Bethe ansatz is a Fermi sphere.

5.4   Consequences

5.4.1   Atoms and the Pauli principle

The Pauli exclusion principle helps explain a wide variety of physical phenomena. One particularly important consequence of the principle is the elaborate electron shell structure of atoms and the way atoms share electrons, explaining the variety of chemical elements and their chemical combinations. An electrically neutral atom contains bound electrons equal in number to the protons in the nucleus. Electrons, being fermions, cannot occupy the same quantum state as other electrons, so electrons have to "stack" within an atom, i.e. have different spins while at the same electron orbital as described below.

An example is the neutral [helium] atom, which has two bound electrons, both of which can occupy the lowest-energy (1s) states by acquiring opposite spin; as spin is part of the quantum state of the electron, the two electrons are in different quantum states and do not violate the Pauli principle. However, the spin can take only two different values (eigenvalues). In a lithium atom, with three bound electrons, the third electron cannot reside in a 1s state, and must occupy one of the higher-energy 2s states instead. Similarly, successively larger elements must have shells of successively higher energy. The chemical properties of an element largely depend on the number of electrons in the outermost shell; atoms with different numbers of occupied electron shells but the same number of electrons in the outermost shell have similar properties, which gives rise to the periodic table of the elements.[7]:214–218

5.4.2   Solid state properties and the Pauli principle

In conductors and semiconductors, there are very large numbers of molecular orbitals which effectively form a continuous band structure of energy levels. In strong conductors (metals) electrons are so degenerate that they cannot even contribute much to the thermal capacity of a metal.[8]:133–147 Many mechanical, electrical, magnetic, optical and chemical properties of solids are the direct consequence of Pauli exclusion.

5.4.3   Stability of matter

The stability of the electrons in an atom itself is unrelated to the exclusion principle, but is described by the quantum theory of the atom. The underlying idea is that close approach of an electron to the nucleus of the atom necessarily increases its kinetic energy, an application of the uncertainty principle of Heisenberg.[9] However, stability of large systems with many electrons and many nucleons is a different matter, and requires the Pauli exclusion principle.[10]

It has been shown that the Pauli exclusion principle is responsible for the fact that ordinary bulk matter is stable and occupies volume. This suggestion was first made in 1931 by Paul Ehrenfest, who pointed out that the electrons of each atom cannot all fall into the lowest-energy orbital and must occupy successively larger shells. Atoms therefore occupy a volume and cannot be squeezed too closely together.[11]

A more rigorous proof was provided in 1967 by Freeman Dyson and Andrew Lenard, who considered the balance of attractive (electron–nuclear) and repulsive (electron–electron and nuclear–nuclear) forces and showed that ordinary matter would collapse and occupy a much smaller volume without the Pauli principle.[12][13]

The consequence of the Pauli principle here is that electrons of the same spin are kept apart by a repulsive exchange interaction, which is a short-range effect, acting simultaneously with the long-range electrostatic or Coulombic force. This effect is partly responsible for the everyday observation in the macroscopic world that two solid objects cannot be in the same place at the same time.

5.4.4 Astrophysics and the Pauli principle

Dyson and Lenard did not consider the extreme magnetic or gravitational forces which occur in some astronomical objects. In 1995 Elliott Lieb and coworkers showed that the Pauli principle still leads to stability in intense magnetic fields such as in neutron stars, although at a much higher density than in ordinary matter.[14] It is a consequence of general relativity that, in sufficiently intense gravitational fields, matter collapses to form a black hole.

Astronomy provides a spectacular demonstration of the effect of the Pauli principle, in the form of white dwarf and neutron stars. In both bodies, atomic structure is disrupted by large gravitational forces, but the stars are held in hydrostatic equilibrium by *degeneracy pressure*, also known as Fermi pressure. This exotic form of matter is known as degenerate matter. The immense gravitational force of a star's mass is normally held in equilibrium by thermal pressure caused by heat produced in thermonuclear fusion in the star's core. In white dwarfs, which do not undergo nuclear fusion, an opposing force to gravity is provided by electron degeneracy pressure. In neutron stars, subject to even stronger gravitational forces, electrons have merged with protons to form neutrons. Neutrons are capable of producing an even higher degeneracy pressure, neutron degeneracy pressure, albeit over a shorter range. This can stabilize neutron stars from further collapse, but at a smaller size and higher density than a white dwarf. Neutrons are the most "rigid" objects known; their Young modulus (or more accurately, bulk modulus) is 20 orders of magnitude larger than that of diamond. However, even this enormous rigidity can be overcome by the gravitational field of a massive star or by the pressure of a supernova, leading to the formation of a black hole.[15]:286–287

5.5 See also

- Exchange force

- Exchange interaction

- Exchange symmetry

- Hund's rule

- Fermi hole

- Pauli effect

- Fermi-Dirac distribution

5.6 References

[1] Kenneth S. Krane (5 November 1987). *Introductory Nuclear Physics*. Wiley. ISBN 978-0-471-80553-3.

[2]

[3] Langmuir, Irving (1919). "The Arrangement of Electrons in Atoms and Molecules" (PDF). *Journal of the American Chemical Society* **41** (6): 868–934. doi:10.1021/ja02227a002. Retrieved 2008-09-01.

[4] Shaviv, Glora. *The Life of Stars: The Controversial Inception and Emergence of the Theory of Stellar Structure* (2010 ed.). Springer. ISBN 978-3642020872.

[5] Straumann, Norbert (2004). "The Role of the Exclusion Principle for Atoms to Stars: A Historical Account". *Invited talk at the 12th Workshop on Nuclear Astrophysics*.

[6] A. Izergin and V. Korepin, Letter in Mathematical Physics vol 6, page 283, 1982

[7] Griffiths, David J. (2004), *Introduction to Quantum Mechanics (2nd ed.)*, Prentice Hall, ISBN 0-13-111892-7

[8] Kittel, Charles (2005), *Introduction to Solid State Physics* (8th ed.), USA: John Wiley & Sons, Inc., ISBN 978-0-471-41526-8

[9] Elliot J. Lieb *The Stability of Matter and Quantum Electrodynamics*

[10] This realization is attributed by Lieb and by GL Sewell (2002). *Quantum Mechanics and Its Emergent Macrophysics*. Princeton University Press. ISBN 0-691-05832-6. to FJ Dyson and A Lenard: *Stability of Matter, Parts I and II (J. Math. Phys.*, **8**, 423–434 (1967); *J. Math. Phys.*, **9**, 698–711 (1968)).

[11] As described by FJ Dyson (J.Math.Phys. **8**, 1538–1545 (1967)), Ehrenfest made this suggestion in his address on the occasion of the award of the Lorentz Medal to Pauli.

[12] FJ Dyson and A Lenard: *Stability of Matter, Parts I and II (J. Math. Phys.*, **8**, 423–434 (1967); *J. Math. Phys.*, **9**, 698–711 (1968))

[13] Dyson, Freeman (1967). "Ground-State Energy of a Finite System of Charged Particles". *J. Math. Phys.* **8** (8): 1538–1545. Bibcode:1967JMP.....8.1538D. doi:10.1063/1.1705389.

[14] Lieb, E. H.; Loss, M.; Solovej, J. P. (1995). "Stability of Matter in Magnetic Fields". *Phys. Rev. Letters* **75** (6): 985–9. arXiv:cond-mat/9506047. Bibcode:1995PhRvL..75..985L. doi:10.1103/PhysRevLett.75.985.

[15] Martin Bojowald (5 November 2012). *The Universe: A View from Classical and Quantum Gravity*. John Wiley & Sons. ISBN 978-3-527-66769-7.

- Dill, Dan (2006). "Chapter 3.5, Many-electron atoms: Fermi holes and Fermi heaps". *Notes on General Chemistry (2nd ed.)*. W. H. Freeman. ISBN 1-4292-0068-5.

- Liboff, Richard L. (2002). *Introductory Quantum Mechanics*. Addison-Wesley. ISBN 0-8053-8714-5.

- Massimi, Michela (2005). *Pauli's Exclusion Principle*. Cambridge University Press. ISBN 0-521-83911-4.

- Tipler, Paul; Llewellyn, Ralph (2002). *Modern Physics (4th ed.)*. W. H. Freeman. ISBN 0-7167-4345-0.

5.7   External links

- Nobel Lecture: Exclusion Principle and Quantum Mechanics Pauli's own account of the development of the Exclusion Principle.

Chapter 6

List of particles

This is a list of the different types of particles found or believed to exist in the whole of the universe. For individual lists of the different particles, see the list below...

6.1 Elementary particles

Main article: Elementary particle

Elementary particles are particles with no measurable internal structure; that is, they are not composed of other particles. They are the fundamental objects of quantum field theory. Many families and sub-families of elementary particles exist. Elementary particles are classified according to their spin. Fermions have half-integer spin while bosons have integer spin. All the particles of the Standard Model have been experimentally observed, recently including the Higgs boson.[1][2]

6.1.1 Fermions

Main article: Fermion

Fermions are one of the two fundamental classes of particles, the other being bosons. Fermion particles are described by Fermi–Dirac statistics and have quantum numbers described by the Pauli exclusion principle. They include the quarks and leptons, as well as any composite particles consisting of an odd number of these, such as all baryons and many atoms and nuclei.

Fermions have half-integer spin; for all known elementary fermions this is $\frac{1}{2}$. All known fermions, except neutrinos, are also Dirac fermions; that is, each known fermion has its own distinct antiparticle. It is not known whether the neutrino is a Dirac fermion or a Majorana fermion.[3] Fermions are the basic building blocks of all matter. They are classified according to whether they interact via the color force or not. In the Standard Model, there are 12 types of elementary fermions: six quarks and six leptons.

Quarks

Main article: Quark

Quarks are the fundamental constituents of hadrons and interact via the strong interaction. Quarks are the only known carriers of fractional charge, but because they combine in groups of three (baryons) or in groups of two with antiquarks (mesons), only integer charge is observed in nature. Their respective antiparticles are the antiquarks, which are identical except for the fact that they carry the opposite electric charge (for example the up quark carries charge $+2/3$, while the up antiquark carries charge $-2/3$), color charge, and baryon number. There are six flavors of quarks; the three positively charged quarks are called "up-type quarks" and the three negatively charged quarks are called "down-type quarks".

Leptons

Main article: Leptons

Leptons do not interact via the strong interaction. Their respective antiparticles are the antileptons which are identical, except for the fact that they carry the opposite electric charge and lepton number. The antiparticle of an electron is an antielectron, which is nearly always called a "positron" for historical reasons. There are six leptons in total; the three charged leptons are called "electron-like leptons", while the neutral leptons are called "neutrinos". Neutrinos are known to oscillate, so that neutrinos of definite flavor do not have definite mass, rather they exist in a superposition of mass eigenstates. The hypothetical heavy right-handed neutrino, called a "sterile neutrino", has been left off the list.

6.1.2   Bosons

Main article: Boson

Bosons are one of the two fundamental classes of particles, the other being fermions. Bosons are characterized by Bose–Einstein statistics and all have integer spins. Bosons may be either elementary, like photons and gluons, or composite, like mesons.

The fundamental forces of nature are mediated by gauge bosons, and mass is believed to be created by the Higgs field. According to the Standard Model the elementary bosons are:

The graviton is added to the list although it is not predicted by the Standard Model, but by other theories in the framework of quantum field theory. Furthermore, gravity is non-renormalizable. There are a total of eight independent gluons. The Higgs boson is postulated by the electroweak theory primarily to explain the origin of particle masses. In a process known as the "Higgs mechanism", the Higgs boson and the other gauge bosons in the Standard Model acquire mass via spontaneous symmetry breaking of the SU(2) gauge symmetry. The Minimal Supersymmetric Standard Model (MSSM) predicts several Higgs bosons. A new particle expected to be the Higgs boson was observed at the CERN/LHC on March 14, 2013, around the energy of 126.5GeV with an accuracy of close to five sigma (99.9999%, which is accepted as definitive). The Higgs mechanism giving mass to other particles has not been observed yet.

6.1.3   Hypothetical particles

Supersymmetric theories predict the existence of more particles, none of which have been confirmed experimentally as of 2014:

Note: just as the photon, Z boson and $W^\pm$ bosons are superpositions of the B^0, W^0, W^1, and W^2 fields – the photino, zino, and wino$^\pm$ are superpositions of the bino^0, wino^0, wino^1, and wino^2 by definition.

No matter if one uses the original gauginos or this superpositions as a basis, the only predicted physical particles are neutralinos and charginos as a superposition of them together with the Higgsinos.

Other theories predict the existence of additional bosons:

Mirror particles are predicted by theories that restore parity symmetry.

"Magnetic monopole" is a generic name for particles with non-zero magnetic charge. They are predicted by some GUTs.

"Tachyon" is a generic name for hypothetical particles that travel faster than the speed of light and have an imaginary rest mass.

Preons were suggested as subparticles of quarks and leptons, but modern collider experiments have all but ruled out their existence.

Kaluza–Klein towers of particles are predicted by some models of extra dimensions. The extra-dimensional momentum is manifested as extra mass in four-dimensional spacetime.

6.2 Composite particles

6.2.1 Hadrons

Main article: Hadron

Hadrons are defined as strongly interacting composite particles. Hadrons are either:

- Composite fermion especially 3 quarks, in which case they are called baryons.
- Composite fermion especially 2 quarks, in which case they are called mesons.

Quark models, first proposed in 1964 independently by Murray Gell-Mann and George Zweig (who called quarks "aces"), describe the known hadrons as composed of valence quarks and/or antiquarks, tightly bound by the color force, which is mediated by gluons. A "sea" of virtual quark-antiquark pairs is also present in each hadron.

Baryons

See also: List of baryons

Ordinary baryons (composite fermions) contain three valence quarks or three valence antiquarks each.

- Nucleons are the fermionic constituents of normal atomic nuclei:
  - Protons, composed of two up and one down quark (uud)
  - Neutrons, composed of two down and one up quark (ddu)
- Hyperons, such as the Λ, Σ, Ξ, and Ω particles, which contain one or more strange quarks, are short-lived and heavier than nucleons. Although not normally present in atomic nuclei, they can appear in short-lived hypernuclei.
- A number of charmed and bottom baryons have also been observed.

Some hints at the existence of exotic baryons have been found recently; however, negative results have also been reported. Their existence is uncertain.

- Pentaquarks consist of four valence quarks and one valence antiquark.

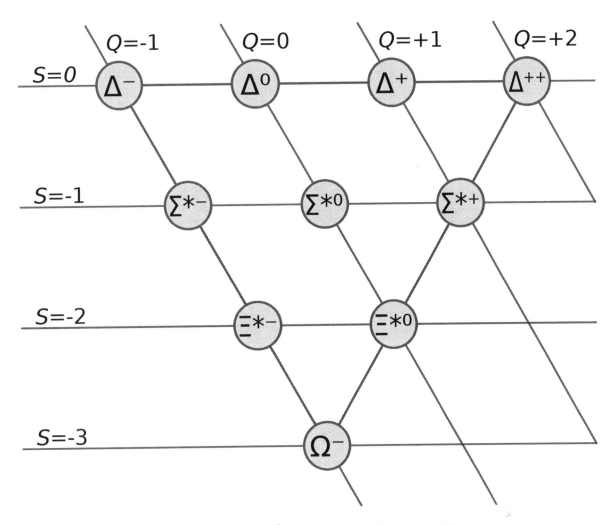

A combination of three u, d or s-quarks with a total spin of $^3/_2$ form the so-called "baryon decuplet".

Mesons

See also: List of mesons

Ordinary mesons are made up of a valence quark and a valence antiquark. Because mesons have spin of 0 or 1 and are not themselves elementary particles, they are "composite" bosons. Examples of mesons include the pion, kaon, and the J/ψ. In quantum hydrodynamic models, mesons mediate the residual strong force between nucleons.

At one time or another, positive signatures have been reported for all of the following exotic mesons but their existences have yet to be confirmed.

- A tetraquark consists of two valence quarks and two valence antiquarks;

- A glueball is a bound state of gluons with no valence quarks;

- Hybrid mesons consist of one or more valence quark-antiquark pairs and one or more real gluons.

6.2.2   Atomic nuclei

Atomic nuclei consist of protons and neutrons. Each type of nucleus contains a specific number of protons and a specific

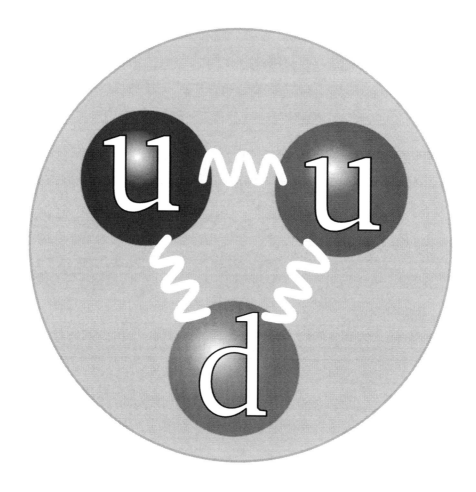

Proton quark structure: 2 up quarks and 1 down quark. The gluon tubes or flux tubes are now known to be Y shaped.

number of neutrons, and is called a "nuclide" or "isotope". Nuclear reactions can change one nuclide into another. See table of nuclides for a complete list of isotopes.

6.2.3 Atoms

Atoms are the smallest neutral particles into which matter can be divided by chemical reactions. An atom consists of a small, heavy nucleus surrounded by a relatively large, light cloud of electrons. Each type of atom corresponds to a specific chemical element. To date, 118 elements have been discovered, while only the elements 1–112,114, and 116 have received official names.

The atomic nucleus consists of protons and neutrons. Protons and neutrons are, in turn, made of quarks.

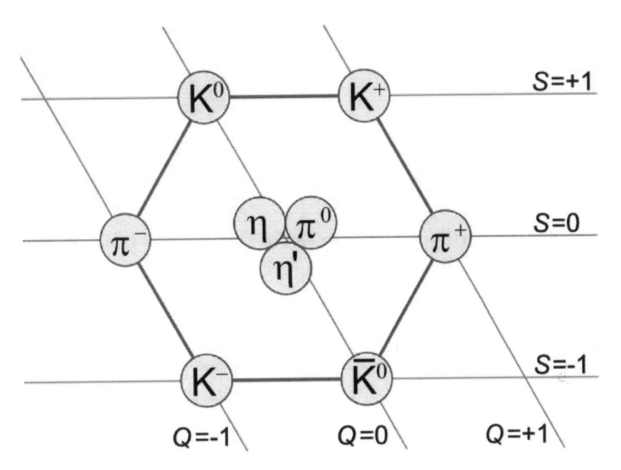

Mesons of spin 0 form a nonet

6.2.4 Molecules

Molecules are the smallest particles into which a non-elemental substance can be divided while maintaining the physical properties of the substance. Each type of molecule corresponds to a specific chemical compound. Molecules are a composite of two or more atoms. See list of compounds for a list of molecules.

6.3 Condensed matter

The field equations of condensed matter physics are remarkably similar to those of high energy particle physics. As a result, much of the theory of particle physics applies to condensed matter physics as well; in particular, there are a selection of field excitations, called quasi-particles, that can be created and explored. These include:

- Phonons are vibrational modes in a crystal lattice.

- Excitons are bound states of an electron and a hole.

- Plasmons are coherent excitations of a plasma.

- Polaritons are mixtures of photons with other quasi-particles.

- Polarons are moving, charged (quasi-) particles that are surrounded by ions in a material.

- Magnons are coherent excitations of electron spins in a material.

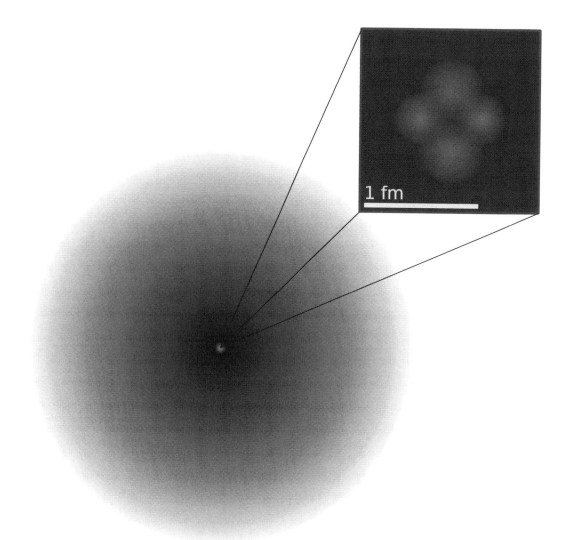

1 Å = 100,000 fm

A semi-accurate depiction of the helium atom. In the nucleus, the protons are in red and neutrons are in purple. In reality, the nucleus is also spherically symmetrical.

6.4 Other

- An anyon is a generalization of fermion and boson in two-dimensional systems like sheets of graphene that obeys braid statistics.

- A plekton is a theoretical kind of particle discussed as a generalization of the braid statistics of the anyon to dimension > 2.

- A WIMP (weakly interacting massive particle) is any one of a number of particles that might explain dark matter (such as the neutralino or the axion).

- The pomeron, used to explain the elastic scattering of hadrons and the location of Regge poles in Regge theory.

- The skyrmion, a topological solution of the pion field, used to model the low-energy properties of the nucleon, such as the axial vector current coupling and the mass.

- A genon is a particle existing in a closed timelike world line where spacetime is curled as in a Frank Tipler or Ronald Mallett time machine.

- A goldstone boson is a massless excitation of a field that has been spontaneously broken. The pions are quasi-goldstone bosons (quasi- because they are not exactly massless) of the broken chiral isospin symmetry of quantum chromodynamics.

- A goldstino is a goldstone fermion produced by the spontaneous breaking of supersymmetry.

- An instanton is a field configuration which is a local minimum of the Euclidean action. Instantons are used in nonperturbative calculations of tunneling rates.

- A dyon is a hypothetical particle with both electric and magnetic charges.

- A geon is an electromagnetic or gravitational wave which is held together in a confined region by the gravitational attraction of its own field energy.

- An inflaton is the generic name for an unidentified scalar particle responsible for the cosmic inflation.

- A spurion is the name given to a "particle" inserted mathematically into an isospin-violating decay in order to analyze it as though it conserved isospin.

- What is called "true muonium", a bound state of a muon and an antimuon, is a theoretical exotic atom which has never been observed.

- A diphoton A resonance particle formed from two identical photons.

6.5 Classification by speed

- A tardyon or bradyon travels slower than light and has a non-zero rest mass.

- A luxon travels at the speed of light and has no rest mass.

- A tachyon (mentioned above) is a hypothetical particle that travels faster than the speed of light and has an imaginary rest mass.

6.6 See also

- Acceleron

- List of baryons

- List of compounds for a list of molecules.

- List of fictional elements, materials, isotopes and atomic particles

- List of mesons

- Periodic table for an overview of atoms.

- Standard Model for the current theory of these particles.

- Table of nuclides

- Timeline of particle discoveries

6.7 References

[1] Observation of a new boson at a mass of 125 GeV with the CMS experiment at the LHC (2013). *arXiv:1207.7235.*

[2] Observation of a new particle in the search for the Standard Model Higgs boson with the ATLAS detector at the LHC (2012). *arXiv:1207.7214.*

[3] B. Kayser, *Two Questions About Neutrinos*, arXiv:1012.4469v1 [hep-ph] (2010).

[4] R. Maartens (2004). *Brane-World Gravity* (PDF). *Living Reviews in Relativity* **7**. p. 7. Also available in web format at http://www.livingreviews.org/lrr-2004-7.

- C. Amsler *et al.* (Particle Data Group) (2008). "Review of Particle Physics". *Physics Letters B* **667** (1–5): 1. Bibcode:2008PhLB..667....1P. doi:10.1016/j.physletb.2008.07.018. *(All information on this list, and more, can be found in the extensive, biannually-updated review by the Particle Data Group)*

Chapter 7

Elementary particle

This article is about the physics concept. For the novel, see The Elementary Particles.

In particle physics, an **elementary particle** or **fundamental particle** is a particle whose substructure is unknown, thus

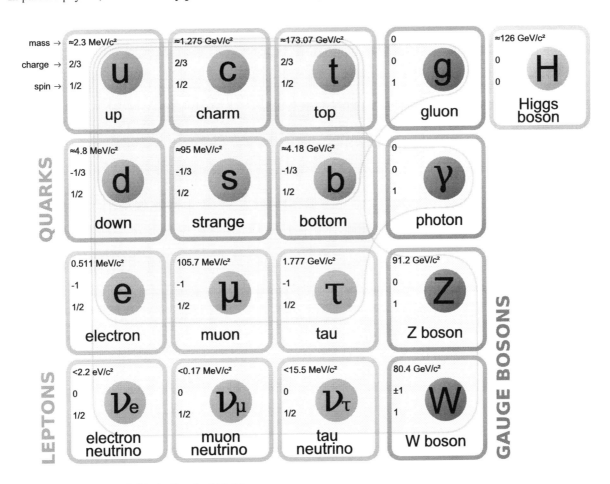

Elementary particles included in the Standard Model

it is unknown whether it is composed of other particles.[1] Known elementary particles include the fundamental fermions (quarks, leptons, antiquarks, and antileptons), which generally are "matter particles" and "antimatter particles", as well as the fundamental bosons (gauge bosons and the Higgs boson), which generally are "force particles" that mediate interactions among fermions.[1] A particle containing two or more elementary particles is a *composite particle*.

Everyday matter is composed of atoms, once presumed to be matter's elementary particles—*atom* meaning "indivisible" in Greek—although the atom's existence remained controversial until about 1910, as some leading physicists regarded molecules as mathematical illusions, and matter as ultimately composed of energy.[1][2] Soon, subatomic constituents of the atom were identified. As the 1930s opened, the electron and the proton had been observed, along with the photon, the particle of electromagnetic radiation.[1] At that time, the recent advent of quantum mechanics was radically altering the conception of particles, as a single particle could seemingly span a field as would a wave, a paradox still eluding satisfactory explanation.[3][4][5]

Via quantum theory, protons and neutrons were found to contain quarks—up quarks and down quarks—now considered elementary particles.[1] And within a molecule, the electron's three degrees of freedom (charge, spin, orbital) can separate via wavefunction into three quasiparticles (holon, spinon, orbiton).[6] Yet a free electron—which, not orbiting an atomic nucleus, lacks orbital motion—appears unsplittable and remains regarded as an elementary particle.[6]

Around 1980, an elementary particle's status as indeed elementary—an *ultimate constituent* of substance—was mostly discarded for a more practical outlook,[1] embodied in particle physics' Standard Model, science's most experimentally successful theory.[5][7] Many elaborations upon and theories beyond the Standard Model, including the extremely popular supersymmetry, double the number of elementary particles by hypothesizing that each known particle associates with a "shadow" partner far more massive,[8][9] although all such superpartners remain undiscovered.[7][10] Meanwhile, an elementary boson mediating gravitation—the graviton—remains hypothetical.[1]

7.1   Overview

Main article: Standard Model
See also: Physics beyond the Standard Model
All elementary particles are—depending on their *spin*—either bosons or fermions. These are differentiated via the spin–

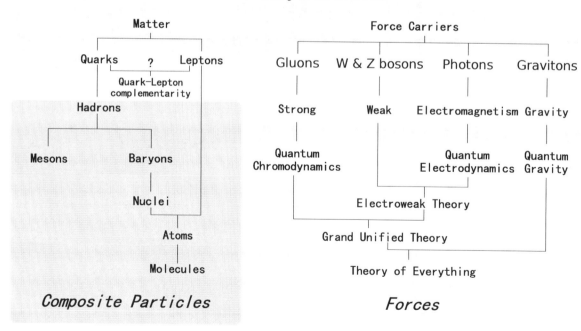

An overview of the various families of elementary and composite particles, and the theories describing their interactions

statistics theorem of quantum statistics. Particles of *half-integer* spin exhibit Fermi–Dirac statistics and are fermions.[1] Particles of *integer* spin, in other words full-integer, exhibit Bose–Einstein statistics and are bosons.[1]

Elementary fermions:

- Matter particles

  - Quarks:

    - up, down
    - charm, strange
    - top, bottom

  - Leptons:

    - electron, electron neutrino (a.k.a., "neutrino")
    - muon, muon neutrino
    - tau, tau neutrino

- Antimatter particles

  - Antiquarks
  - Antileptons

Elementary bosons:

- Force particles (gauge bosons):

  - photon
  - gluon (numbering eight)[1]
  - W^+, W^-, and Z^0 bosons
  - graviton (hypothetical)[1]

- Scalar boson

  - Higgs boson

A particle's mass is quantified in units of energy versus the electron's (electronvolts). Through conversion of energy into mass, any particle can be produced through collision of other particles at high energy,[1][11] although the output particle might not contain the input particles, for instance matter creation from colliding photons. Likewise, the composite fermions protons were collided at nearly light speed to produce the relatively more massive Higgs boson.[11] The most massive elementary particle, the top quark, rapidly decays, but apparently does not contain, lighter particles.

When probed at energies available in experiments, particles exhibit spherical sizes. In operating particle physics' Standard Model, elementary particles are usually represented for predictive utility as point particles, which, as zero-dimensional, lack spatial extension. Though extremely successful, the Standard Model is limited to the microcosm by its omission of gravitation, and has some parameters arbitrarily added but unexplained.[12] Seeking to resolve those shortcomings, string theory posits that elementary particles are ultimately composed of one-dimensional energy strings whose absolute minimum size is the Planck length.

7.2   Common elementary particles

Main article: cosmic abundance of elements

According to the current models of big bang nucleosynthesis, the primordial composition of visible matter of the universe should be about 75% hydrogen and 25% helium-4 (in mass). Neutrons are made up of one up and two down quark, while protons are made of two up and one down quark. Since the other common elementary particles (such as electrons, neutrinos, or weak bosons) are so light or so rare when compared to atomic nuclei, we can neglect their mass contribution to the observable universe's total mass. Therefore, one can conclude that most of the visible mass of the universe consists of protons and neutrons, which, like all baryons, in turn consist of up quarks and down quarks.

Some estimates imply that there are roughly 10^{80} baryons (almost entirely protons and neutrons) in the observable universe.[13][14][15]

The number of protons in the observable universe is called the Eddington number.

In terms of number of particles, some estimates imply that nearly all the matter, excluding dark matter, occurs in neutrinos, and that roughly 10^{86} elementary particles of matter exist in the visible universe, mostly neutrinos.[15] Other estimates imply that roughly 10^{97} elementary particles exist in the visible universe (not including dark matter), mostly photons, gravitons, and other massless force carriers.[15]

7.3 Standard Model

Main article: Standard Model
The Standard Model of particle physics contains 12 flavors of elementary fermions, plus their corresponding antiparticles,

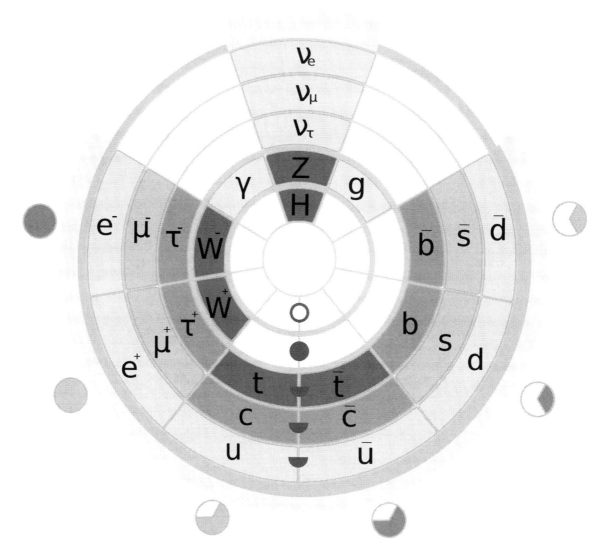

Graphic representation of the standard model. Spin, charge, mass and participation in different force interactions are shown. Click on the image to see the full description

as well as elementary bosons that mediate the forces and the Higgs boson, which was reported on July 4, 2012, as having been likely detected by the two main experiments at the LHC (ATLAS and CMS). However, the Standard Model is widely considered to be a provisional theory rather than a truly fundamental one, since it is not known if it is compatible with Einstein's general relativity. There may be hypothetical elementary particles not described by the Standard Model, such as the graviton, the particle that would carry the gravitational force, and sparticles, supersymmetric partners of the ordinary particles.

7.3.1 Fundamental fermions

Main article: Fermion

The 12 fundamental fermionic flavours are divided into three generations of four particles each. Six of the particles are quarks. The remaining six are leptons, three of which are neutrinos, and the remaining three of which have an electric charge of −1: the electron and its two cousins, the muon and the tau.

Antiparticles

Main article: Antimatter

There are also 12 fundamental fermionic antiparticles that correspond to these 12 particles. For example, the antielectron (positron) $e+$ is the electron's antiparticle and has an electric charge of +1.

Quarks

Main article: Quark

Isolated quarks and antiquarks have never been detected, a fact explained by confinement. Every quark carries one of three color charges of the strong interaction; antiquarks similarly carry anticolor. Color-charged particles interact via gluon exchange in the same way that charged particles interact via photon exchange. However, gluons are themselves color-charged, resulting in an amplification of the strong force as color-charged particles are separated. Unlike the electromagnetic force, which diminishes as charged particles separate, color-charged particles feel increasing force.

However, color-charged particles may combine to form color neutral composite particles called hadrons. A quark may pair up with an antiquark: the quark has a color and the antiquark has the corresponding anticolor. The color and anticolor cancel out, forming a color neutral meson. Alternatively, three quarks can exist together, one quark being "red", another "blue", another "green". These three colored quarks together form a color-neutral baryon. Symmetrically, three antiquarks with the colors "antired", "antiblue" and "antigreen" can form a color-neutral antibaryon.

Quarks also carry fractional electric charges, but, since they are confined within hadrons whose charges are all integral, fractional charges have never been isolated. Note that quarks have electric charges of either +2/3 or −1/3, whereas antiquarks have corresponding electric charges of either −2/3 or +1/3.

Evidence for the existence of quarks comes from deep inelastic scattering: firing electrons at nuclei to determine the distribution of charge within nucleons (which are baryons). If the charge is uniform, the electric field around the proton should be uniform and the electron should scatter elastically. Low-energy electrons do scatter in this way, but, above a particular energy, the protons deflect some electrons through large angles. The recoiling electron has much less energy and a jet of particles is emitted. This inelastic scattering suggests that the charge in the proton is not uniform but split among smaller charged particles: quarks.

7.3.2 Fundamental bosons

Main article: Boson

In the Standard Model, vector (spin–1) bosons (gluons, photons, and the W and Z bosons) mediate forces, whereas the Higgs boson (spin-0) is responsible for the intrinsic mass of particles. Bosons differ from fermions in the fact that multiple bosons can occupy the same quantum state (Pauli exclusion principle). Also, bosons can be either elementary, like photons, or a combination, like mesons. The spin of bosons are integers instead of half integers.

Gluons

Main article: Gluon

Gluons mediate the strong interaction, which join quarks and thereby form hadrons, which are either baryons (three quarks) or mesons (one quark and one antiquark). Protons and neutrons are baryons, joined by gluons to form the atomic nucleus. Like quarks, gluons exhibit colour and anticolour—unrelated to the concept of visual color—sometimes in combinations, altogether eight variations of gluons.

Electroweak bosons

Main articles: W and Z bosons and Photon

There are three weak gauge bosons: W^+, W^-, and Z^0; these mediate the weak interaction. The W bosons are known for their mediation in nuclear decay. The W^- converts a neutron into a proton then decay into an electron and electron antineutrino pair. The Z^0 does not convert charge but rather changes momentum and is the only mechanism for elastically scattering neutrinos. The weak gauge bosons were discovered due to momentum change in electrons from neutrino-Z exchange. The massless photon mediates the electromagnetic interaction. These four gauge bosons form the electroweak interaction among elementary particles.

Higgs boson

Main article: Higgs boson

Although the weak and electromagnetic forces appear quite different to us at everyday energies, the two forces are theorized to unify as a single electroweak force at high energies. This prediction was clearly confirmed by measurements of cross-sections for high-energy electron-proton scattering at the HERA collider at DESY. The differences at low energies is a consequence of the high masses of the W and Z bosons, which in turn are a consequence of the Higgs mechanism. Through the process of spontaneous symmetry breaking, the Higgs selects a special direction in electroweak space that causes three electroweak particles to become very heavy (the weak bosons) and one to remain massless (the photon). On 4 July 2012, after many years of experimentally searching for evidence of its existence, the Higgs boson was announced to have been observed at CERN's Large Hadron Collider. Peter Higgs who first posited the existence of the Higgs boson was present at the announcement.[16] The Higgs boson is believed to have a mass of approximately 125 GeV.[17] The statistical significance of this discovery was reported as 5-sigma, which implies a certainty of roughly 99.99994%. In particle physics, this is the level of significance required to officially label experimental observations as a discovery. Research into the properties of the newly discovered particle continues.

Graviton

Main article: Graviton

The graviton is hypothesized to mediate gravitation, but remains undiscovered and yet is sometimes included in tables of elementary particles.[1] Its spin would be two—thus a boson—and it would lack charge or mass. Besides mediating an extremely feeble force, the graviton would have its own antiparticle and rapidly annihilate, rendering its detection extremely difficult even if it exists.

7.4   Beyond the Standard Model

Although experimental evidence overwhelmingly confirms the predictions derived from the Standard Model, some of its parameters were added arbitrarily, not determined by a particular explanation, which remain mysteries, for instance the hierarchy problem. Theories beyond the Standard Model attempt to resolve these shortcomings.

7.4.1   Grand unification

Main article: Grand Unified Theory

One extension of the Standard Model attempts to combine the electroweak interaction with the strong interaction into a single 'grand unified theory' (GUT). Such a force would be spontaneously broken into the three forces by a Higgs-like mechanism. The most dramatic prediction of grand unification is the existence of X and Y bosons, which cause proton decay. However, the non-observation of proton decay at the Super-Kamiokande neutrino observatory rules out the simplest GUTs, including SU(5) and SO(10).

7.4.2   Supersymmetry

Main article: Supersymmetry

Supersymmetry extends the Standard Model by adding another class of symmetries to the Lagrangian. These symmetries exchange fermionic particles with bosonic ones. Such a symmetry predicts the existence of supersymmetric particles, abbreviated as *sparticles*, which include the sleptons, squarks, neutralinos, and charginos. Each particle in the Standard Model would have a superpartner whose spin differs by 1/2 from the ordinary particle. Due to the breaking of supersymmetry, the sparticles are much heavier than their ordinary counterparts; they are so heavy that existing particle colliders would not be powerful enough to produce them. However, some physicists believe that sparticles will be detected by the Large Hadron Collider at CERN.

7.4.3   String theory

Main article: String theory

String theory is a model of physics where all "particles" that make up matter are composed of strings (measuring at the Planck length) that exist in an 11-dimensional (according to M-theory, the leading version) universe. These strings vibrate at different frequencies that determine mass, electric charge, color charge, and spin. A string can be open (a line) or closed in a loop (a one-dimensional sphere, like a circle). As a string moves through space it sweeps out something called a *world sheet*. String theory predicts 1- to 10-branes (a 1-brane being a string and a 10-brane being a 10-dimensional object) that prevent tears in the "fabric" of space using the uncertainty principle (E.g., the electron orbiting a hydrogen atom has the probability, albeit small, that it could be anywhere else in the universe at any given moment).

String theory proposes that our universe is merely a 4-brane, inside which exist the 3 space dimensions and the 1 time dimension that we observe. The remaining 6 theoretical dimensions either are very tiny and curled up (and too small to be macroscopically accessible) or simply do not/cannot exist in our universe (because they exist in a grander scheme called the "multiverse" outside our known universe).

Some predictions of the string theory include existence of extremely massive counterparts of ordinary particles due to vibrational excitations of the fundamental string and existence of a massless spin-2 particle behaving like the graviton.

7.4.4 Technicolor

Main article: Technicolor (physics)

Technicolor theories try to modify the Standard Model in a minimal way by introducing a new QCD-like interaction. This means one adds a new theory of so-called Techniquarks, interacting via so called Technigluons. The main idea is that the Higgs-Boson is not an elementary particle but a bound state of these objects.

7.4.5 Preon theory

Main article: Preon

According to preon theory there are one or more orders of particles more fundamental than those (or most of those) found in the Standard Model. The most fundamental of these are normally called preons, which is derived from "pre-quarks". In essence, preon theory tries to do for the Standard Model what the Standard Model did for the particle zoo that came before it. Most models assume that almost everything in the Standard Model can be explained in terms of three to half a dozen more fundamental particles and the rules that govern their interactions. Interest in preons has waned since the simplest models were experimentally ruled out in the 1980s.

7.4.6 Acceleron theory

Accelerons are the hypothetical subatomic particles that integrally link the newfound mass of the neutrino and to the dark energy conjectured to be accelerating the expansion of the universe.[18]

In theory, neutrinos are influenced by a new force resulting from their interactions with accelerons. Dark energy results as the universe tries to pull neutrinos apart.[18]

7.5 See also

- Asymptotic freedom

- List of particles

- Physical ontology

- Quantum field theory

- Quantum gravity

- Quantum triviality

- UV fixed point

7.6 Notes

[1] Sylvie Braibant; Giorgio Giacomelli; Maurizio Spurio (2012). *Particles and Fundamental Interactions: An Introduction to Particle Physics* (2nd ed.). Springer. pp. 1–3. ISBN 978-94-007-2463-1.

[2] Ronald Newburgh; Joseph Peidle; Wolfgang Rueckner (2006). "Einstein, Perrin, and the reality of atoms: 1905 revisited" (PDF). *American Journal of Physics*. **74** (6): 478–481. Bibcode:2006AmJPh..74..478N. doi:10.1119/1.2188962.

[3] Friedel Weinert (2004). *The Scientist as Philosopher: Philosophical Consequences of Great Scientific Discoveries*. Springer. p. 43. ISBN 978-3-540-20580-7.

[4] Friedel Weinert (2004). *The Scientist as Philosopher: Philosophical Consequences of Great Scientific Discoveries*. Springer. pp. 57–59. ISBN 978-3-540-20580-7.

[5] Meinard Kuhlmann (24 Jul 2013). "Physicists debate whether the world is made of particles or fields—or something else entirely". *Scientific American*.

[6] Zeeya Merali (18 Apr 2012). "Not-quite-so elementary, my dear electron: Fundamental particle 'splits' into quasiparticles, including the new 'orbiton'". *Nature*. doi:10.1038/nature.2012.10471.

[7] Ian O'Neill (24 Jul 2013). "LHC discovery maims supersymmetry, again". *Discovery News*. Retrieved 2013-08-28.

[8] Particle Data Group. "Unsolved mysteries—supersymmetry". *The Particle Adventure*. Berkeley Lab. Retrieved 2013-08-28.

[9] National Research Council (2006). *Revealing the Hidden Nature of Space and Time: Charting the Course for Elementary Particle Physics*. National Academies Press. p. 68. ISBN 978-0-309-66039-6.

[10] "CERN latest data shows no sign of supersymmetry—yet". *Phys.Org*. 25 Jul 2013. Retrieved 2013-08-28.

[11] Ryan Avent (19 Jul 2012). "The Q&A: Brian Greene—Life after the Higgs". *The Economist*. Retrieved 2013-08-28.

[12] Sylvie Braibant; Giorgio Giacomelli; Maurizio Spurio (2012). *Particles and Fundamental Interactions: An Introduction to Particle Physics* (2nd ed.). Springer. p. 384. ISBN 978-94-007-2463-1.

[13] Frank Heile. "Is the Total Number of Particles in the Universe Stable Over Long Periods of Time?". 2014.

[14] Jared Brooks. "Galaxies and Cosmology". 2014. p. 4, equation 16.

[15] Robert Munafo (24 Jul 2013). "Notable Properties of Specific Numbers". Retrieved 2013-08-28.

[16] Lizzy Davies (4 July 2014). "Higgs boson announcement live: CERN scientists discover subatomic particle". *The Guardian*. Retrieved 2012-07-06.

[17] Lucas Taylor (4 Jul 2014). "Observation of a new particle with a mass of 125 GeV". CMS. Retrieved 2012-07-06.

[18] "New theory links neutrino's slight mass to accelerating Universe expansion". *ScienceDaily*. 28 Jul 2004. Retrieved 2008-06-05.

7.7 Further reading

7.7.1 General readers

- Feynman, R.P. & Weinberg, S. (1987) *Elementary Particles and the Laws of Physics: The 1986 Dirac Memorial Lectures*. Cambridge Univ. Press.

- Ford, Kenneth W. (2005) *The Quantum World*. Harvard Univ. Press.

- Brian Greene (1999). *The Elegant Universe*. W.W.Norton & Company. ISBN 0-393-05858-1.

- John Gribbin (2000) *Q is for Quantum – An Encyclopedia of Particle Physics*. Simon & Schuster. ISBN 0-684-85578-X.

- Oerter, Robert (2006) *The Theory of Almost Everything: The Standard Model, the Unsung Triumph of Modern Physics*. Plume.

- Schumm, Bruce A. (2004) *Deep Down Things: The Breathtaking Beauty of Particle Physics*. Johns Hopkins University Press. ISBN 0-8018-7971-X.

- Martinus Veltman (2003). *Facts and Mysteries in Elementary Particle Physics*. World Scientific. ISBN 981-238-149-X.

- Frank Close (2004). *Particle Physics: A Very Short Introduction*. Oxford: Oxford University Press. ISBN 0-19-280434-0.

- Seiden, Abraham (2005). *Particle Physics – A Comprehensive Introduction*. Addison Wesley. ISBN 0-8053-8736-6.

7.7.2 Textbooks

- Bettini, Alessandro (2008) *Introduction to Elementary Particle Physics*. Cambridge Univ. Press. ISBN 978-0-521-88021-3

- Coughlan, G. D., J. E. Dodd, and B. M. Gripaios (2006) *The Ideas of Particle Physics: An Introduction for Scientists*, 3rd ed. Cambridge Univ. Press. An undergraduate text for those not majoring in physics.

- Griffiths, David J. (1987) *Introduction to Elementary Particles*. John Wiley & Sons. ISBN 0-471-60386-4.

- Kane, Gordon L. (1987). *Modern Elementary Particle Physics*. Perseus Books. ISBN 0-201-11749-5.

- Perkins, Donald H. (2000) *Introduction to High Energy Physics*, 4th ed. Cambridge Univ. Press.

7.8 External links

The most important address about the current experimental and theoretical knowledge about elementary particle physics is the Particle Data Group, where different international institutions collect all experimental data and give short reviews over the contemporary theoretical understanding.

- Particle Data Group

other pages are:

- Greene, Brian, "*Elementary particles*", The Elegant Universe, NOVA (PBS)

- particleadventure.org, a well-made introduction also for non physicists

- CERNCourier: Season of Higgs and melodrama

- Pentaquark information page

- Interactions.org, particle physics news

- Symmetry Magazine, a joint Fermilab/SLAC publication

- "Sized Matter: perception of the extreme unseen", Michigan University project for artistic visualisation of sub-atomic particles

- Elementary Particles made thinkable, an interactive visualisation allowing physical properties to be compared

Chapter 8

Spin–statistics theorem

In quantum mechanics, the **spin–statistics theorem** relates the spin of a particle to the particle statistics it obeys. The spin of a particle is its intrinsic angular momentum (that is, the contribution to the total angular momentum that is not due to the orbital motion of the particle). All particles have either integer spin or half-integer spin (in units of the reduced Planck constant $\hbar$).[1][2]

The theorem states that:

- The wave function of a system of identical integer-spin particles has the same value when the positions of any two particles are swapped. Particles with wave functions symmetric under exchange are called *bosons*.

- The wave function of a system of identical half-integer spin particles changes sign when two particles are swapped. Particles with wave functions antisymmetric under exchange are called *fermions*.

In other words, the spin–statistics theorem states that integer-spin particles are bosons, while half-integer–spin particles are fermions.

The spin–statistics relation was first formulated in 1939 by Markus Fierz[3] and was rederived in a more systematic way by Wolfgang Pauli.[4] Fierz and Pauli argued their result by enumerating all free field theories subject to the requirement that there be quadratic forms for locally commuting observables including a positive-definite energy density. A more conceptual argument was provided by Julian Schwinger in 1950. Richard Feynman gave a demonstration by demanding unitarity for scattering as an external potential is varied,[5] which when translated to field language is a condition on the quadratic operator that couples to the potential.[6]

8.1 General discussion

In a given system, two indistinguishable particles, occupying two separate points, have only one state, not two. This means that if we exchange the positions of the particles, we do not get a new state, but rather the same physical state. In fact, one cannot tell which particle is in which position.

A physical state is described by a wavefunction, or – more generally – by a vector, which is also called a "state"; if interactions with other particles are ignored, then two different wavefunctions are physically equivalent if their absolute value is equal. So, while the physical state does not change under the exchange of the particles' positions, the wavefunction may get a minus sign.

Bosons are particles whose wavefunction is symmetric under such an exchange, so if we swap the particles the wavefunction does not change. Fermions are particles whose wavefunction is antisymmetric, so under such a swap the wavefunction gets a minus sign, meaning that the amplitude for two identical fermions to occupy the same state must be zero. This is the Pauli exclusion principle: two identical fermions cannot occupy the same state. This rule does not hold for bosons.

In quantum field theory, a state or a wavefunction is described by field operators operating on some basic state called the *vacuum*. In order for the operators to project out the symmetric or antisymmetric component of the creating wavefunction, they must have the appropriate commutation law. The operator

$$\iint \psi(x,y)\phi(x)\phi(y)\,dx\,dy$$

(with ϕ an operator and $\psi(x,y)$ a numerical function) creates a two-particle state with wavefunction $\psi(x,y)$, and depending on the commutation properties of the fields, either only the antisymmetric parts or the symmetric parts matter.

Let us assume that $x \neq y$ and the two operators take place at the same time; more generally, they may have spacelike separation, as is explained hereafter.

If the fields **commute**, meaning that the following holds:

$$\phi(x)\phi(y) = \phi(y)\phi(x)$$

then only the symmetric part of ψ contributes, so that $\psi(x,y) = \psi(y,x)$, and the field will create bosonic particles.

On the other hand, if the fields **anti-commute**, meaning that ϕ has the property that

$$\phi(x)\phi(y) = -\phi(y)\phi(x),$$

then only the antisymmetric part of ψ contributes, so that $\psi(x,y) = -\psi(y,x)$, and the particles will be fermionic.

Naively, neither has anything to do with the spin, which determines the rotation properties of the particles, not the exchange properties.

8.2 A suggestive bogus argument

Consider the two-field operator product

$$R(\pi)\phi(x)\phi(-x),$$

where R is the matrix that rotates the spin polarization of the field by 180 degrees when one does a 180-degree rotation around some particular axis. The components of ϕ are not shown in this notation, ϕ has many components, and the matrix R mixes them up with one another.

In a non-relativistic theory, this product can be interpreted as annihilating two particles at positions x and $-x$ with polarizations that are rotated by π relative to each other. Now rotate this configuration by π around the origin. Under this rotation, the two points x and $-x$ switch places, and the two field polarizations are additionally rotated by a π. So we get

$$R(2\pi)\phi(-x)R(\pi)\phi(x),$$

which for integer spin is equal to

$$\phi(-x)R(\pi)\phi(x)$$

and for half-integer spin is equal to

$-\phi(-x)R(\pi)\phi(x)$

(proved here). Both the operators $\pm\phi(-x)R(\pi)\phi(x)$ still annihilate two particles at x and $-x$. Hence we claim to have shown that, with respect to particle states:

$$R(\pi)\phi(x)\phi(-x) = \begin{cases} \phi(-x)R(\pi)\phi(x) & \text{spins integral for ,} \\ -\phi(-x)R(\pi)\phi(x) & \text{spins half-integral for .} \end{cases}$$

So exchanging the order of two appropriately polarized operator insertions into the vacuum can be done by a rotation, at the cost of a sign in the half-integer case.

This argument by itself does not prove anything like the spin–statistics relation. To see why, consider a nonrelativistic spin-0 field described by a free Schrödinger equation. Such a field can be anticommuting or commuting. To see where it fails, consider that a nonrelativistic spin-0 field has no polarization, so that the product above is simply:

$\phi(-x)\phi(x).$

In the nonrelativistic theory, this product annihilates two particles at x and $-x$, and has zero expectation value in any state. In order to have a nonzero matrix element, this operator product must be between states with two more particles on the right than on the left:

$\langle 0|\phi(-x)\phi(x)|\psi\rangle.$

Performing the rotation, all that we learn is that rotating the 2-particle state $|\psi\rangle$ gives the same sign as changing the operator order. This gives no additional information, so this argument does not prove anything.

8.3   Why the bogus argument fails

To prove spin–statistics theorem, it is necessary to use relativity, as is obvious from the consistency of the nonrelativistic spinless fermion, and the nonrelativistic spinning bosons. There are claims in the literature of proofs of spin–statistics theorem that do not require relativity,[7][8] but they are not proofs of a theorem, as the counterexamples show, rather they are arguments for why spin–statistics is "natural", while wrong-statistics is "unnatural". In relativity, the connection is required.

In relativity, there are no local fields that are pure creation operators or annihilation operators. Every local field both creates particles and annihilates the corresponding antiparticle. This means that in relativity, the product of the free real spin-0 field has a *nonzero* vacuum expectation value, because in addition to creating particles and annihilating particles, it also includes a part that creates and then annihilates a particle:

$G(x) = \langle 0|\phi(-x)\phi(x)|0\rangle.$

And now the heuristic argument can be used to see that $G(x)$ is equal to $G(-x)$, which tells us that the fields cannot be anti-commuting.

8.4   Proof

The essential ingredient in proving the spin/statistics relation is relativity, that the physical laws do not change under Lorentz transformations. The field operators transform under Lorentz transformations according to the spin of the particle that they create, by definition.

Additionally, the assumption (known as microcausality) that spacelike separated fields either commute or anticommute can be made only for relativistic theories with a time direction. Otherwise, the notion of being spacelike is meaningless. However, the proof involves looking at a Euclidean version of spacetime, in which the time direction is treated as a spatial one, as will be now explained.

Lorentz transformations include 3-dimensional rotations as well as boosts. A boost transfers to a frame of reference with a different velocity, and is mathematically like a rotation into time. By analytic continuation of the correlation functions of a quantum field theory, the time coordinate may become imaginary, and then boosts become rotations. The new "spacetime" has only spatial directions and is termed *Euclidean*.

A π rotation in the Euclidean xt plane can be used to rotate vacuum expectation values of the field product of the previous section. The *time rotation* turns the argument of the previous section into the spin–statistics theorem.

The proof requires the following assumptions:

1. The theory has a Lorentz-invariant Lagrangian.

2. The vacuum is Lorentz-invariant.

3. The particle is a localized excitation. Microscopically, it is not attached to a string or domain wall.

4. The particle is propagating, meaning that it has a finite, not infinite, mass.

5. The particle is a real excitation, meaning that states containing this particle have a positive-definite norm.

These assumptions are for the most part necessary, as the following examples show:

1. The spinless anticommuting field shows that spinless fermions are nonrelativistically consistent. Likewise, the theory of a spinor commuting field shows that spinning bosons are too.

2. This assumption may be weakened.

3. In 2+1 dimensions, sources for the Chern–Simons theory can have exotic spins, despite the fact that the three-dimensional rotation group has only integer and half-integer spin representations.

4. An ultralocal field can have either statistics independently of its spin. This is related to Lorentz invariance, since an infinitely massive particle is always nonrelativistic, and the spin decouples from the dynamics. Although colored quarks are attached to a QCD string and have infinite mass, the spin-statistics relation for quarks can be proved in the short distance limit.

5. Gauge ghosts are spinless fermions, but they include states of negative norm.

Assumptions 1 and 2 imply that the theory is described by a path integral, and assumption 3 implies that there is a local field which creates the particle.

The rotation plane includes time, and a rotation in a plane involving time in the Euclidean theory defines a CPT transformation in the Minkowski theory. If the theory is described by a path integral, a CPT transformation takes states to their conjugates, so that the correlation function

$$\langle 0 | R\phi(x)\phi(-x) | 0 \rangle$$

must be positive definite at x=0 by assumption 5, the particle states have positive norm. The assumption of finite mass implies that this correlation function is nonzero for x spacelike. Lorentz invariance now allows the fields to be rotated inside the correlation function in the manner of the argument of the previous section:

$$\langle 0 | RR\phi(x)R\phi(-x) | 0 \rangle = \pm \langle 0 | \phi(-x)R\phi(x) | 0 \rangle$$

Where the sign depends on the spin, as before. The CPT invariance, or Euclidean rotational invariance, of the correlation function guarantees that this is equal to G(x). So

$$\langle 0|(R\phi(x)\phi(y) - \phi(y)R\phi(x))|0\rangle = 0$$

for integer spin fields and

$$\langle 0|R\phi(x)\phi(y) + \phi(y)R\phi(x)|0\rangle = 0$$

for half-integer spin fields.

Since the operators are spacelike separated, a different order can only create states that differ by a phase. The argument fixes the phase to be −1 or 1 according to the spin. Since it is possible to rotate the space-like separated polarizations independently by local perturbations, the phase should not depend on the polarization in appropriately chosen field coordinates.

This argument is due to Julian Schwinger.[9]

8.5 Consequences

The spin-statistics theorem implies that half-integer spin particles are subject to the Pauli exclusion principle, while integer-spin particles are not. Only one fermion can occupy a given quantum state at any time, while the number of bosons that can occupy a quantum state is not restricted. The basic building blocks of matter such as protons, neutrons, and electrons are fermions. Particles such as the photon, which mediate forces between matter particles, are bosons.

There are a couple of interesting phenomena arising from the two types of statistics. The Bose–Einstein distribution which describes bosons leads to Bose–Einstein condensation. Below a certain temperature, most of the particles in a bosonic system will occupy the ground state (the state of lowest energy). Unusual properties such as superfluidity can result. The Fermi–Dirac distribution describing fermions also leads to interesting properties. Since only one fermion can occupy a given quantum state, the lowest single-particle energy level for spin-1/2 fermions contains at most two particles, with the spins of the particles oppositely aligned. Thus, even at absolute zero, the system still has a significant amount of energy. As a result, a fermionic system exerts an outward pressure. Even at non-zero temperatures, such a pressure can exist. This degeneracy pressure is responsible for keeping certain massive stars from collapsing due to gravity. See white dwarf, neutron star, and black hole.

Ghost fields do not obey the spin-statistics relation. See Klein transformation on how to patch up a loophole in the theorem.

8.6 Relation to representation theory of the Lorentz group

The Lorentz group has no non-trivial unitary representations of finite dimension. Thus it seems impossible to construct a Hilbert space in which all states have finite, non-zero spin and positive, Lorentz-invariant norm. This problem is overcome in different ways depending on particle spin-statistics.

For a state of integer spin the negative norm states (known as "unphysical polarization") are set to zero, which makes the use of gauge symmetry necessary.

For a state of half-integer spin the argument can be circumvented by having fermionic statistics.[10]

8.7 Literature

- Markus Fierz: *Über die relativistische Theorie kräftefreier Teilchen mit beliebigem Spin.* Helv. Phys. Acta **12**, 3–17 (1939)

- Wolfgang Pauli: *The connection between spin and statistics.* Phys. Rev. **58**, 716–722 (1940)

- Ray F. Streater and Arthur S. Wightman: *PCT, Spin & Statistics, and All That.* 5th edition: Princeton University Press, Princeton (2000)

- Ian Duck and Ennackel Chandy George Sudarshan: *Pauli and the Spin-Statistics Theorem.* World Scientific, Singapore (1997)

- Arthur S Wightman: *Pauli and the Spin-Statistics Theorem* (book review). Am. J. Phys. **67** (8), 742–746 (1999)

- Arthur Jabs: *Connecting spin and statistics in quantum mechanics.* http://arXiv.org/abs/0810.2399 (Found. Phys. **40**, 776–792, 793–794 (2010))

8.8 Notes

[1] Dirac, Paul Adrien Maurice (1981-01-01). *The Principles of Quantum Mechanics.* Clarendon Press. p. 149. ISBN978019852

[2] Pauli, Wolfgang (1980-01-01). *General principles of quantum mechanics.* Springer-Verlag. ISBN 9783540098423.

[3] Markus Fierz (1939). "Über die relativistische Theorie kräftefreier Teilchen mit beliebigem Spin". *Helvetica Physica Acta* **12** (1): 3–37. doi:10.5169/seals-110930.

[4] Wolfgang Pauli (15 October 1940). "The Connection Between Spin and Statistics" (PDF). *Physical Review* **58** (8): 716–722. doi:10.1103/PhysRev.58.716.

[5] Richard Feynman (1961). *Quantum Electrodynamics.* Basic Books. ISBN 978-0-201-36075-2.

[6] Wolfgang Pauli (1950). "On the Connection Between Spin and Statistics". *Progress of Theoretical Physics* **5** (4): 526–543. doi:10.1143/ptp/5.4.526.

[7] Jabs, Arthur (5 April 2002). "Connecting Spin and Statistics in Quantum Mechanics". *Foundations of Physics.* Foundations of Physics **40** (7): 776–792. arXiv:0810.2399. Bibcode:2010FoPh...40..776J. doi:10.1007/s10701-009-9351-4. Retrieved May 29, 2011.

[8] Horowitz, Joshua (14 April 2009). "From Path Integrals to Fractional Quantum Statistics" (PDF).

[9] Julian Schwinger (June 15, 1951). "The Quantum Theory of Fields I". *Physical Review* **82** (6): 914–917. doi:10.1103/PhysRev.82.914..The only difference between the argument in this paper and the argument presented here is that the operator"R"in Schwinger's paper is a pure time reversal,instead of a CPT operation,but this is the same for CP invariant freefield theories which were all that Schwinger considered.

[10] Peskin, Michael E.; Schroeder, Daniel V. (1995). *An Introduction to Quantum Field Theory.* Addison-Wesley. ISBN 0-201-50397-2.

8.9 See also

- Parastatistics

- Anyonic statistics

- Braid statistics

8.10   References

- Paul O'Hara, Rotational Invariance and the Spin-Statistics Theorem, Foun. Phys. 33, 1349–1368(2003).

- Ian Duck and E. C. G. Sudarshan, Toward an understanding of the spin-statistics theorem, Am. J. Phys. 66 (4), 284–303 April 1998. Archived from the original on 2009-01-02.

8.11   External links

- A nice nearly-proof at John Baez's home page

- Animation of the Dirac belt trick with a double belt, showing that belts behave as spin 1/2 particles

- Animation of a Dirac belt trick variant showing that spin 1/2 particles are fermions

Chapter 9

Quantum state

In quantum physics, **quantum state** refers to the state of a quantum system.

A quantum state can be either **pure** or **mixed**. A pure quantum state corresponds[1][3] to a vector, called a **state vector**, in a Hilbert space. For example, when dealing with the energy spectrum of the electron in a hydrogen atom, the relevant state vectors are identified by the principal quantum number, written $\{n\}$. For a more complicated case, consider Bohm's formulation of the EPR experiment, where the state vector

$$|\psi\rangle = \frac{1}{\sqrt{2}}\left(|\uparrow\downarrow\rangle - |\downarrow\uparrow\rangle\right)$$

involves superposition of joint spin states for two particles. Mathematically, a pure quantum state corresponds to a state vector in a Hilbert space over complex numbers, which is a generalization of our more usual three-dimensional space.[4]:93–96 If this Hilbert space is represented as a function space, then its elements are called wave functions.

A mixed quantum state corresponds to a probabilistic mixture of pure states; however, different distributions of pure states can generate equivalent (i.e., physically indistinguishable) mixed states. Mixed states are described by so-called density matrices. A pure state can also be recast as a density matrix; in this way, pure states can be represented as a subset of the more general mixed states.

For example, if the spin of an electron is measured in any direction, e.g. with a Stern–Gerlach experiment, there are two possible results: up or down. The Hilbert space for the electron's spin is therefore two-dimensional. A pure state here is represented by a two-dimensional complex vector (α, β), with a length of one; that is, with

$$|\alpha|^2 + |\beta|^2 = 1,$$

where $|\alpha|$ and $|\beta|$ are the absolute values of α and β. A mixed state, in this case, is a 2×2 matrix that is Hermitian, positive-definite, and has trace 1.

Before a particular measurement is performed on a quantum system, the theory usually gives only a probability distribution for the outcome, and the form that this distribution takes is completely determined by the quantum state and the observable describing the measurement. These probability distributions arise for both mixed states and pure states: it is impossible in quantum mechanics (unlike classical mechanics) to prepare a state in which all properties of the system are fixed and certain. This is exemplified by the uncertainty principle, and reflects a core difference between classical and quantum physics. Even in quantum theory, however, for every observable there are some states that have an exact and determined value for that observable.[4]:4–5[5]

9.1 Conceptual description

9.1.1 Pure states

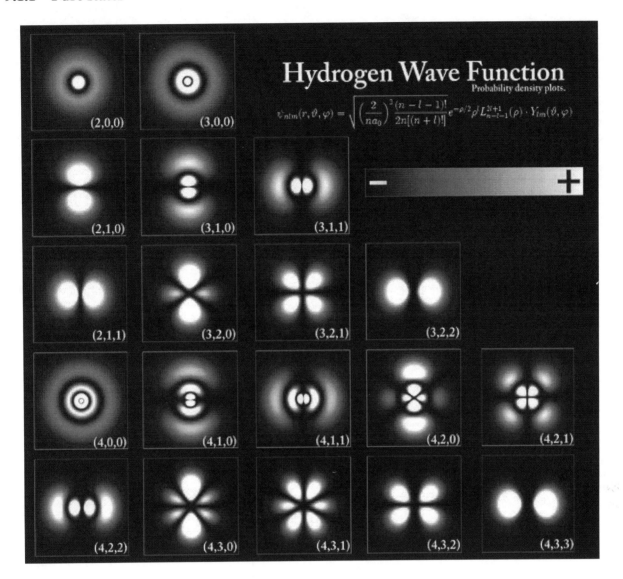

Probability densities for the electron of a hydrogen atom in different quantum states.

In the mathematical formulation of quantum mechanics, pure quantum states correspond to vectors in a Hilbert space, while each observable quantity (such as the energy or momentum of a particle) is associated with a mathematical operator. The operator serves as a linear function which acts on the states of the system. The eigenvalues of the operator correspond to the possible values of the observable, i.e. it is possible to observe a particle with a momentum of 1 kg·m/s if and only if one of the eigenvalues of the momentum operator is 1 kg·m/s. The corresponding eigenvector (which physicists call an **eigenstate**) with eigenvalue 1 kg·m/s would be a quantum state with a definite, well-defined value of momentum of 1 kg·m/s, with no quantum uncertainty. If its momentum were measured, the result is guaranteed to be 1 kg·m/s.

On the other hand, a system in a linear combination of multiple different eigenstates *does* in general have quantum uncertainty for the given observable. We can represent this linear combination of eigenstates as:

$$|\Psi(t)\rangle = \sum_n C_n(t)|\Phi_n\rangle$$

The coefficient which corresponds to a particular state in the linear combination is complex thus allowing interference effects between states. The coefficients are time dependent. How a quantum system changes in time is governed by the time evolution operator. The symbols $|$ and $\rangle$ [lower-alpha 1] surrounding the Ψ are part of bra–ket notation.

Statistical mixtures of states are different from a linear combination. A statistical mixture of states is a statistical ensemble of independent systems. Statistical mixtures represent the degree of knowledge whilst the uncertainty within quantum mechanics is fundamental. Mathematically, a statistical mixture is not a combination using complex coefficients, but rather a combination using real-valued, positive probabilities of different states Φ_n. A number P_n represents the probability of a randomly selected system being in the state Φ_n. Unlike the linear combination case each system is in a definite eigenstate.[6][7]

The expectation value $\langle A \rangle_\sigma$ of an observable A is a statistical mean of measured values of the observable. It is this mean, and the distribution of probabilities, that is predicted by physical theories.

There is no state which is simultaneously an eigenstate for *all* observables. For example, we cannot prepare a state such that both the position measurement $Q(t)$ and the momentum measurement $P(t)$ (at the same time t) are known exactly; at least one of them will have a range of possible values.[lower-alpha 2] This is the content of the Heisenberg uncertainty relation.

Moreover, in contrast to classical mechanics, it is unavoidable that *performing a measurement on the system generally changes its state*.[8][9] More precisely: After measuring an observable A, the system will be in an eigenstate of A; thus the state has changed, unless the system was already in that eigenstate. This expresses a kind of logical consistency: If we measure A twice in the same run of the experiment, the measurements being directly consecutive in time,[lower-alpha 3] then they will produce the same results. This has some strange consequences, however, as follows.

Consider two incompatible observables, A and B, where A corresponds to a measurement earlier in time than B.[lower-alpha 4] Suppose that the system is in an eigenstate of B at the experiment's begin. If we measure only B, all runs of the experiment will yield the same result. If we measure first A and then B in the same run of the experiment, the system will transfer to an eigenstate of A after the first measurement, and we will generally notice that the results of B are statistical. Thus: *Quantum mechanical measurements influence one another*, and it is important in which order they are performed.

Another feature of quantum states becomes relevant if we consider a physical system that consists of multiple subsystems; for example, an experiment with two particles rather than one. Quantum physics allows for certain states, called *entangled states*, that show certain statistical correlations between measurements on the two particles which cannot be explained by classical theory. For details, see entanglement. These entangled states lead to experimentally testable properties (Bell's theorem) that allow us to distinguish between quantum theory and alternative classical (non-quantum) models.

9.1.2 Schrödinger picture vs. Heisenberg picture

One can take the observables to be dependent on time, while the state σ was fixed once at the beginning of the experiment. This approach is called the Heisenberg picture. (This approach was taken in the later part of the discussion above, with time-varying observables $P(t)$, $Q(t)$.) One can, equivalently, treat the observables as fixed, while the state of the system depends on time; that is known as the Schrödinger picture. (This approach was taken in the earlier part of the discussion above, with a time-varying state $|\Psi(t)\rangle = \sum_n C_n(t)|\Phi_n\rangle$.) Conceptually (and mathematically), the two approaches are equivalent; choosing one of them is a matter of convention.

Both viewpoints are used in quantum theory. While non-relativistic quantum mechanics is usually formulated in terms of the Schrödinger picture, the Heisenberg picture is often preferred in a relativistic context, that is, for quantum field theory. Compare with Dirac picture.[10]:65

9.2 Formalism in quantum physics

See also: Mathematical formulation of quantum mechanics

9.2.1 Pure states as rays in a Hilbert space

Quantum physics is most commonly formulated in terms of linear algebra, as follows. Any given system is identified with some finite- or infinite-dimensional Hilbert space. The pure states correspond to vectors of norm 1. Thus the set of all pure states corresponds to the unit sphere in the Hilbert space.

Multiplying a pure state by a scalar is physically inconsequential (as long as the state is considered by itself). If one vector is obtained from the other by multiplying by a scalar of unit magnitude, the two vectors are said to correspond to the same "ray" in Hilbert space[11] and also to the same point in the projective Hilbert space.

9.2.2 Bra–ket notation

Main article: Bra–ket notation

Calculations in quantum mechanics make frequent use of linear operators, scalar products, dual spaces and Hermitian conjugation. In order to make such calculations flow smoothly, and to obviate the need (in some contexts) to fully understand the underlying linear algebra, Paul Dirac invented a notation to describe quantum states, known as *bra-ket notation*. Although the details of this are beyond the scope of this article (see the article bra–ket notation), some consequences of this are:

- The expression used to denote a state vector (which corresponds to a pure quantum state) takes the form $|\psi\rangle$ (where the " ψ " can be replaced by any other symbols, letters, numbers, or even words). This can be contrasted with the usual *mathematical* notation, where vectors are usually bold, lower-case letters, or letters with arrows on top.

- Dirac defined two kinds of vector, *bra* and *ket*, dual to each other.[12]

- Each ket $|\psi\rangle$ is uniquely associated with a so-called *bra*, denoted $\langle\psi|$, which corresponds to the same physical quantum state. Technically, the bra is the adjoint of the ket. It is an element of the dual space, and related to the ket by the Riesz representation theorem. In a finite-dimensional space with a chosen basis, writing $|\psi\rangle$ as a column vector, $\langle\psi|$ is a row vector; to obtain it just take the transpose and entry-wise complex conjugate of $|\psi\rangle$.

- Scalar products[13][14] (also called *brackets*) are written so as to look like a bra and ket next to each other: $\langle\psi_1|\psi_2\rangle$. (The phrase "bra-ket" is supposed to resemble "bracket".)

9.2.3 Spin

Main article: Mathematical formulation of quantum mechanics § Spin

The angular momentum has the same dimension as the Planck constant and, at quantum scale, behaves as a *discrete* degree of freedom. Most particles possess a kind of intrinsic angular momentum that does not appear at all in classical mechanics and arises from Dirac's relativistic generalization of the theory. Mathematically it is described with spinors. In non-relativistic quantum mechanics the group representations of the Lie group SU(2) are used to describe this additional freedom. For a given particle, the choice of representation (and hence the range of possible values of the spin observable) is specified by a non-negative number S that, in units of Planck's reduced constant $\hbar$, is either an integer (0, 1, 2 ...) or a half-integer (1/2, 3/2, 5/2 ...). For a massive particle with spin S, its spin quantum number m always assumes one of the $2S + 1$ possible values in the set

$$\{-S, -S+1, \ldots +S-1, +S\}$$

As a consequence, the quantum state of a particle with spin is described by a vector-valued wave function with values in $\mathbf{C}^{2S+1}$. Equivalently, it is represented by a complex-valued function of four variables: one discrete quantum number variable (for the spin) is added to the usual three continuous variables (for the position in space).

9.2.4 Many-body states and particle statistics

Further information: Particle statistics

The quantum state of a system of N particles, each potentially with spin, is described by a complex-valued function with four variables per particle, e.g.

$$|\psi(\mathbf{r}_1, m_1; \ldots; \mathbf{r}_N, m_N)\rangle.$$

Here, the spin variables $m\nu$ assume values from the set

$$\{-S_\nu, -S_\nu + 1, \ldots + S_\nu - 1, +S_\nu\}$$

where S_ν is the spin of νth particle. $S_\nu = 0$ for a particle that does not exhibit spin.

The treatment of identical particles is very different for bosons (particles with integer spin) versus fermions (particles with half-integer spin). The above N-particle function must either be symmetrized (in the bosonic case) or anti-symmetrized (in the fermionic case) with respect to the particle numbers. If not all N particles are identical, but some of them are, then the function must be (anti)symmetrized separately over the variables corresponding to each group of identical variables, according to its statistics (bosonic or fermionic).

Electrons are fermions with $S = 1/2$, photons (quanta of light) are bosons with $S = 1$ (although in the vacuum they are massless and can't be described with Schrödingerian mechanics).

When symmetrization or anti-symmetrization is unnecessary, N-particle spaces of states can be obtained simply by tensor products of one-particle spaces, to which we will return later.

9.2.5 Basis states of one-particle systems

As with any Hilbert space, if a basis is chosen for the Hilbert space of a system, then any ket can be expanded as a linear combination of those basis elements. Symbolically, given basis kets $|k_i\rangle$, any ket $|\psi\rangle$ can be written

$$|\psi\rangle = \sum_i c_i |k_i\rangle$$

where ci are complex numbers. In physical terms, this is described by saying that $|\psi\rangle$ has been expressed as a *quantum superposition* of the states $|k_i\rangle$. If the basis kets are chosen to be orthonormal (as is often the case), then $c_i = \langle k_i | \psi \rangle$.

One property worth noting is that the *normalized* states $|\psi\rangle$ are characterized by

$$\langle \psi | \psi \rangle = 1,$$

and for orthonormal basis this translates to

$$\sum_i |c_i|^2 = 1.$$

Expansions of this sort play an important role in measurement in quantum mechanics. In particular, if the $|k_i\rangle$ are eigenstates (with eigenvalues ki) of an observable, and that observable is measured on the normalized state $|\psi\rangle$, then the probability that the result of the measurement is ki is $|ci|^2$. (The normalization condition above mandates that the total sum of probabilities is equal to one.)

A particularly important example is the *position basis*, which is the basis consisting of eigenstates $|\mathbf{r}\rangle$ with eigenvalues $\mathbf{r}$ of the observable which corresponds to measuring position.[lower-alpha 5] If these eigenstates are nondegenerate (for example, if the system is a single, spinless particle), then any ket $|\psi\rangle$ is associated with a complex-valued function of three-dimensional space

$$\psi(\mathbf{r}) \equiv \langle \mathbf{r}|\psi\rangle.\text{ [lower-alpha 6]}$$

This function is called the **wave function** corresponding to $|\psi\rangle$. Similarly to the discrete case above, the probability *density* of the particle being found at position $\mathbf{r}$ is $|\psi(\mathbf{r})|^2$ and the normalized states have

$$\int \mathrm{d}^3\mathbf{r}|\psi(\mathbf{r})|^2 = 1$$

In terms of the continuous set of position basis $|\mathbf{r}\rangle$, the state $|\psi\rangle$ is:

$$|\psi\rangle \equiv \int \mathrm{d}^3\mathbf{r}\, \psi(\mathbf{r})|\mathbf{r}\rangle$$

9.2.6   Superposition of pure states

Main article: Quantum superposition

One aspect of quantum states, mentioned above, is that superpositions of them can be formed. If $|\alpha\rangle$ and $|\beta\rangle$ are two kets corresponding to quantum states, the ket

$$c_\alpha|\alpha\rangle + c_\beta|\beta\rangle$$

is a different quantum state (possibly not normalized). Note that *which* quantum state it is depends on both the amplitudes and phases (arguments) of c_α and c_β . In other words, for example, even though $|\psi\rangle$ and $e^{i\theta}|\psi\rangle$ (for real θ) correspond to the same physical quantum state, they are *not interchangeable*, since for example $|\phi\rangle + |\psi\rangle$ and $|\phi\rangle + e^{i\theta}|\psi\rangle$ do *not* (in general) correspond to the same physical state. However, $|\phi\rangle + |\psi\rangle$ and $e^{i\theta}(|\phi\rangle + |\psi\rangle)$ *do* correspond to the same physical state. This is sometimes described by saying that "global" phase factors are unphysical, but "relative" phase factors are physical and important.

One example of a quantum interference phenomenon that arises from superposition is the double-slit experiment. The photon state is a superposition of two different states, one of which corresponds to the photon having passed through the left slit, and the other corresponding to passage through the right slit. The relative phase of those two states has a value which depends on the distance from each of the two slits. Depending on what that phase is, the interference is constructive at some locations and destructive in others, creating the interference pattern. By the analogy with coherence in other wave phenomena, a superposed state can be referred to as a *coherent superposition*.

Another example of the importance of relative phase in quantum superposition is Rabi oscillations, where the relative phase of two states varies in time due to the Schrödinger equation. The resulting superposition ends up oscillating back and forth between two different states.

9.2.7   Mixed states

See also: Density matrix

A *pure quantum state* is a state which can be described by a single ket vector, as described above. A *mixed quantum state* is a statistical ensemble of pure states (see quantum statistical mechanics). Mixed states inevitably arise from pure states when, for a composite quantum system $H_1 \otimes H_2$ with an entangled state on it, the part H_2 is inaccessible to the observer. The state of the part H_1 is expressed then as the partial trace over H_2 .

A mixed state *cannot* be described as a ket vector. Instead, it is described by its associated *density matrix* (or *density operator*), usually denoted ϱ. Note that density matrices can describe both mixed *and* pure states, treating them on the same footing. Moreover, a mixed quantum state on a given quantum system described by a Hilbert space H can be always represented as the partial trace of a pure quantum state (called a purification) on a larger bipartite system $H \otimes K$ for a sufficiently large Hilbert space K .

The density matrix describing a mixed state is defined to be an operator of the form

$$\rho = \sum_s p_s |\psi_s\rangle\langle\psi_s|$$

where p_s is the fraction of the ensemble in each pure state $|\psi_s\rangle$. The density matrix can be thought of as a way of using the one-particle formalism to describe the behavior of many similar particles by giving a probability distribution (or ensemble) of states that these particles can be found in.

A simple criterion for checking whether a density matrix is describing a pure or mixed state is that the trace of ϱ^2 is equal to 1 if the state is pure, and less than 1 if the state is mixed.[lower-alpha 7][17] Another, equivalent, criterion is that the von Neumann entropy is 0 for a pure state, and strictly positive for a mixed state.

The rules for measurement in quantum mechanics are particularly simple to state in terms of density matrices. For example, the ensemble average (expectation value) of a measurement corresponding to an observable A is given by

$$\langle A \rangle = \sum_s p_s \langle\psi_s|A|\psi_s\rangle = \sum_s \sum_i p_s a_i |\langle\alpha_i|\psi_s\rangle|^2 = \mathrm{tr}(\rho A)$$

where $|\alpha_i\rangle$, a_i are eigenkets and eigenvalues, respectively, for the operator A, and "tr" denotes trace. It is important to note that two types of averaging are occurring, one being a weighted quantum superposition over the basis kets $|\psi_s\rangle$ of the pure states, and the other being a statistical (said *incoherent*) average with the probabilities ps of those states.

According to Wigner,[18] the concept of mixture was put forward by Landau.[19][16]:38–41

9.3 Interpretation

Main article: Interpretations of quantum mechanics

Although theoretically, for a given quantum system, a state vector provides the full information about its evolution, it is not easy to understand what information about the "real world" it carries. Due to the uncertainty principle, a state, even if it has the value of one observable exactly defined (i.e. the observable has this state as an eigenstate), cannot exactly define values of *all* observables.

For state vectors (pure states), probability amplitudes offer a probabilistic interpretation. It can be generalized for all states (including mixed), for instance, as expectation values mentioned above.

9.4 Mathematical generalizations

States can be formulated in terms of observables, rather than as vectors in a vector space. These are positive normalized linear functionals on a C*-algebra, or sometimes other classes of algebras of observables. See State on a C*-algebra and Gelfand–Naimark–Segal construction for more details.

9.5   See also

- Atomic electron transition

- Bloch sphere

- Ground state

- Introduction to quantum mechanics

- No-cloning theorem

- Orthonormal basis

- PBR theorem

- Quantum harmonic oscillator

- Qubit

- State vector reduction, for historical reasons called a *wave function collapse*

- Stationary state

- W state

9.6   Notes

[1] Sometimes written ">"; see angle brackets.

[2] To avoid misunderstandings: Here we mean that $Q(t)$ and $P(t)$ are measured in the same state, but *not* in the same run of the experiment.

[3] i.e. separated by a zero delay. One can think of it as stopping the time, then making the two measurements one after the other, then resuming the time. Thus, the measurements occured at the same time, but it is still possible to tell which was first.

[4] For concreteness' sake, suppose that $A = Q(t_1)$ and $B = P(t_2)$ in the above example, with $t_2 > t_1 > 0$.

[5] Note that a state $|\psi\rangle$ is a superposition of different basis states $|\mathbf{r}\rangle$, so $|\psi\rangle$ and $|\mathbf{r}\rangle$ are elements of the same Hilbert space. A particle in state $|\mathbf{r}\rangle$ is located precisely at position $\mathbf{r} = (x, y, z)$, while a particle in state $|\psi\rangle$ can be found at different positions with corresponding probabilities.

[6] In the continuous case, the basis kets $|\mathbf{r}\rangle$ are not unit kets (unlike the state $|\psi\rangle$): They are normalized according to $\int d^3\mathbf{r}' \langle \mathbf{r}|\mathbf{r}'\rangle = 1$, [15] i.e. $\langle \mathbf{r}|\mathbf{r}'\rangle = \delta(\mathbf{r}' - \mathbf{r})$ (a Dirac delta function), which means that $\langle \mathbf{r}|\mathbf{r}\rangle = \infty$.

[7] Note that this criterion works when the density matrix is normalized so that the trace of ρ is 1, as it is for the standard definition given in this section. Occasionally a density matrix will be normalized differently, in which case the criterion is $\text{Tr}(\rho^2) = (\text{Tr}\,\rho)^2$

9.7   References

[1] [2] p. 16: "each state of a dynamical system at a particular time corresponds to a ket vector."

[2] Dirac, P.A.M. (1958). *The Principles of Quantum Mechanics*, 4th edition, Oxford University Press, Oxford UK.

[3] Feynman, R.P., Leighton, R.B., Sands, M. (1963). *The Feynman Lectures on Physics*, Addison-Wesley, Reading MA, available at http://www.feynmanlectures.info/, Volume III, p. 8–2: "The states χ and φ correspond to the two vectors B and A."

[4] Griffiths, David J. (2004), *Introduction to Quantum Mechanics (2nd ed.)*, Prentice Hall, ISBN 0-13-111892-7

[5] Ballentine, L. E. (1970), "The Statistical Interpretation of Quantum Mechanics", *Reviews of Modern Physics* **42**: 358–381, Bibcode:1970RvMP...42..358B, doi:10.1103/RevModPhys.42.358

[6] Statistical Mixture of States

[7] http://electron6.phys.utk.edu/qm1/modules/m6/statistical.htm

[8] Heisenberg, W. (1927). Über den anschaulichen Inhalt der quantentheoretischen Kinematik und Mechanik, *Z. Phys.* **43**: 172–198. Translation as 'The actual content of quantum theoretical kinematics and mechanics'. Also translated as 'The physical content of quantum kinematics and mechanics' at pp. 62–84 by editors John Wheeler and Wojciech Zurek, in *Quantum Theory and Measurement* (1983), Princeton University Press, Princeton NJ.

[9] Bohr, N. (1927/1928). The quantum postulate and the recent development of atomic theory, *Nature* Supplement April 14 1928, **121**: 580–590.

[10] Gottfried, Kurt; Yan, Tung-Mow (2003). *Quantum Mechanics: Fundamentals* (2nd, illustrated ed.). Springer. ISBN 978038795

[11] Weinberg, Steven. "The Quantum Theory of Fields", Vol. 1. Cambridge University Press, 1995 p. 50.

[12] [2] p. 20: "The bra vectors, as they have been here introduced, are quite a different kind of vector from the kets, and so far there is no connexion between them except for the existence of a scalar product of a bra and a ket."

[13] [2] p. 19: "A scalar product $\langle B|A \rangle$ now appears as a complete bracket expression."

[14] [10] p. 31: "to define the scalar products as being between bras and kets."

[15] [16] p. 17: "$\int \Psi_{f'} \Psi_f^* \, dq = \delta(f' - f)$" (the left side corresponds to $\langle f|f' \rangle$), "$\int \delta(f' - f) \, df' = 1$".

[16] Lev Landau; Evgeny Lifshitz (1965). *Quantum Mechanics — Non-Relativistic Theory* (PDF). Course of Theoretical Physics **3** (2nd ed.). London: Pergamon Press.

[17] Blum, *Density matrix theory and applications*, page 39.

[18] Eugene Wigner (1962). "Remarks on the mind-body question" (PDF). In I.J. Good. *The Scientist Speculates*. London: Heinemann. pp. 284–302. Footnote 13 on p.180

[19] Lev Landau (1927). "Das Dämpfungsproblem in der Wellenmechanik (The Damping Problem in Wave Mechanics)". *Zeitschrift für Physik* **45** (5–6): 430–441. Bibcode:1927ZPhy...45..430L. doi:10.1007/bf01343064. English translation reprinted in: D. Ter Haar, ed. (1965). *Collected papers of L.D. Landau*. Oxford: Pergamon Press. p.8–18

9.8 Further reading

The concept of quantum states, in particular the content of the section Formalism in quantum physics above, is covered in most standard textbooks on quantum mechanics.

For a discussion of conceptual aspects and a comparison with classical states, see:

- Isham, Chris J (1995). *Lectures on Quantum Theory: Mathematical and Structural Foundations*. Imperial College Press. ISBN 978-1-86094-001-9.

For a more detailed coverage of mathematical aspects, see:

- Bratteli, Ola; Robinson, Derek W (1987). *Operator Algebras and Quantum Statistical Mechanics 1*. Springer. ISBN 978-3-540-17093-8. 2nd edition. In particular, see Sec. 2.3.

For a discussion of purifications of mixed quantum states, see Chapter 2 of John Preskill's lecture notes for Physics 219 at Caltech.

Chapter 10

Quantum field theory

"Relativistic quantum field theory" redirects here. For other uses, see Relativity.

In theoretical physics, **quantum field theory** (**QFT**) is a theoretical framework for constructing quantum mechanical models of subatomic particles in particle physics and quasiparticles in condensed matter physics. A QFT treats particles as excited states of an underlying physical field, so these are called field quanta.

In quantum field theory, quantum mechanical interactions between particles are described by interaction terms between the corresponding underlying quantum fields.

10.1 Definition

Quantum electrodynamics (QED) has one electron field and one photon field; quantum chromodynamics (QCD) has one field for each type of quark; and, in condensed matter, there is an atomic displacement field that gives rise to phonon particles. Edward Witten describes QFT as "by far" the most difficult theory in modern physics.[1]

10.1.1 Dynamics

See also: Relativistic dynamics

Ordinary quantum mechanical systems have a fixed number of particles, with each particle having a finite number of degrees of freedom. In contrast, the excited states of a QFT can represent any number of particles. This makes quantum field theories especially useful for describing systems where the particle count/number may change over time, a crucial feature of relativistic dynamics.

10.1.2 States

QFT interaction terms are similar in spirit to those between charges with electric and magnetic fields in Maxwell's equations. However, unlike the classical fields of Maxwell's theory, fields in QFT generally exist in quantum superpositions of states and are subject to the laws of quantum mechanics.

Because the fields are continuous quantities over space, there exist excited states with arbitrarily large numbers of particles in them, providing QFT systems with an effectively infinite number of degrees of freedom. Infinite degrees of freedom can easily lead to divergences of calculated quantities (e.g., the quantities become infinite). Techniques such as renormalization of QFT parameters or discretization of spacetime, as in lattice QCD, are often used to avoid such infinities so as to yield physically meaningful results.

10.1.3 Fields and radiation

The gravitational field and the electromagnetic field are the only two fundamental fields in nature that have infinite range and a corresponding classical low-energy limit, which greatly diminishes and hides their "particle-like" excitations. Albert Einstein in 1905, attributed "particle-like" and discrete exchanges of momenta and energy, characteristic of "field quanta", to the electromagnetic field. Originally, his principal motivation was to explain the thermodynamics of radiation. Although the photoelectric effect and Compton scattering strongly suggest the existence of the photon, it might alternately be explained by a mere quantization of emission; more definitive evidence of the quantum nature of radiation is now taken up into modern quantum optics as in the antibunching effect.[2]

10.2 Theories

There is currently no complete quantum theory of the remaining fundamental force, gravity. Many of the proposed theories to describe gravity as a QFT postulate the existence of a graviton particle that mediates the gravitational force. Presumably, the as yet unknown correct quantum field-theoretic treatment of the gravitational field will behave like Einstein's general theory of relativity in the low-energy limit. Quantum field theory of the fundamental forces itself has been postulated to be the low-energy effective field theory limit of a more fundamental theory such as superstring theory.

Most theories in standard particle physics are formulated as **relativistic quantum field theories**, such as QED, QCD, and the Standard Model. QED, the quantum field-theoretic description of the electromagnetic field, approximately reproduces Maxwell's theory of electrodynamics in the low-energy limit, with small non-linear corrections to the Maxwell equations required due to virtual electron–positron pairs.

In the perturbative approach to quantum field theory, the full field interaction terms are approximated as a perturbative expansion in the number of particles involved. Each term in the expansion can be thought of as forces between particles being mediated by other particles. In QED, the electromagnetic force between two electrons is caused by an exchange of photons. Similarly, intermediate vector bosons mediate the weak force and gluons mediate the strong force in QCD. The notion of a force-mediating particle comes from perturbation theory, and does not make sense in the context of non-perturbative approaches to QFT, such as with bound states.

10.3 History

Main article: History of quantum field theory

10.3.1 Foundations

The early development of the field involved Dirac, Fock, Pauli, Heisenberg and Bogolyubov. This phase of development culminated with the construction of the theory of quantum electrodynamics in the 1950s.

10.3.2 Gauge theory

Gauge theory was formulated and quantized, leading to the **unification of forces** embodied in the standard model of particle physics. This effort started in the 1950s with the work of Yang and Mills, was carried on by Martinus Veltman and a host of others during the 1960s and completed by the 1970s through the work of Gerard 't Hooft, Frank Wilczek, David Gross and David Politzer.

10.3.3   Grand synthesis

Parallel developments in the understanding of phase transitions in condensed matter physics led to the study of the renormalization group. This in turn led to the **grand synthesis** of theoretical physics, which unified theories of particle and condensed matter physics through quantum field theory. This involved the work of Michael Fisher and Leo Kadanoff in the 1970s, which led to the seminal reformulation of quantum field theory by Kenneth G. Wilson in 1975.

10.4   Principles

10.4.1   Classical and quantum fields

Main article: Classical field theory

A classical field is a function defined over some region of space and time.[3] Two physical phenomena which are described by classical fields are Newtonian gravitation, described by Newtonian gravitational field $\mathbf{g}(\mathbf{x}, t)$, and classical electromagnetism, described by the electric and magnetic fields $\mathbf{E}(\mathbf{x}, t)$ and $\mathbf{B}(\mathbf{x}, t)$. Because such fields can in principle take on distinct values at each point in space, they are said to have infinite degrees of freedom.[3]

Classical field theory does not, however, account for the quantum-mechanical aspects of such physical phenomena. For instance, it is known from quantum mechanics that certain aspects of electromagnetism involve discrete particles—photons—rather than continuous fields. The business of *quantum* field theory is to write down a field that is, like a classical field, a function defined over space and time, but which also accommodates the observations of quantum mechanics. This is a *quantum field*.

It is not immediately clear *how* to write down such a quantum field, since quantum mechanics has a structure very unlike a field theory. In its most general formulation, quantum mechanics is a theory of abstract operators (observables) acting on an abstract state space (Hilbert space), where the observables represent physically observable quantities and the state space represents the possible states of the system under study.[4] For instance, the fundamental observables associated with the motion of a single quantum mechanical particle are the position and momentum operators $\hat{x}$ and $\hat{p}$. Field theory, in contrast, treats x as a way to index the field rather than as an operator.[5]

There are two common ways of developing a quantum field: the path integral formalism and canonical quantization.[6] The latter of these is pursued in this article.

Lagrangian formalism

Quantum field theory frequently makes use of the Lagrangian formalism from classical field theory. This formalism is analogous to the Lagrangian formalism used in classical mechanics to solve for the motion of a particle under the influence of a field. In classical field theory, one writes down a Lagrangian density, $\mathcal{L}$, involving a field, $\varphi(\mathbf{x},t)$, and possibly its first derivatives ($\partial\varphi/\partial t$ and $\nabla\varphi$), and then applies a field-theoretic form of the Euler–Lagrange equation. Writing coordinates $(t, \mathbf{x}) = (x^0, x^1, x^2, x^3) = x^\mu$, this form of the Euler–Lagrange equation is[3]

$$\frac{\partial}{\partial x^\mu}\left[\frac{\partial\mathcal{L}}{\partial(\partial\phi/\partial x^\mu)}\right] - \frac{\partial\mathcal{L}}{\partial\phi} = 0,$$

where a sum over μ is performed according to the rules of Einstein notation.

By solving this equation, one arrives at the "equations of motion" of the field.[3] For example, if one begins with the Lagrangian density

$$\mathcal{L}(\phi, \nabla\phi) = -\rho(t, \mathbf{x})\,\phi(t, \mathbf{x}) - \frac{1}{8\pi G}|\nabla\phi|^2,$$

and then applies the Euler–Lagrange equation, one obtains the equation of motion

$$4\pi G \rho(t, \mathbf{x}) = \nabla^2 \phi.$$

This equation is Newton's law of universal gravitation, expressed in differential form in terms of the gravitational potential $\varphi(t, \mathbf{x})$ and the mass density $\rho(t, \mathbf{x})$. Despite the nomenclature, the "field" under study is the gravitational potential, φ, rather than the gravitational field, $\mathbf{g}$. Similarly, when classical field theory is used to study electromagnetism, the "field" of interest is the electromagnetic four-potential (V/c, $\mathbf{A}$), rather than the electric and magnetic fields $\mathbf{E}$ and $\mathbf{B}$.

Quantum field theory uses this same Lagrangian procedure to determine the equations of motion for quantum fields. These equations of motion are then supplemented by commutation relations derived from the canonical quantization procedure described below, thereby incorporating quantum mechanical effects into the behavior of the field.

10.4.2 Single- and many-particle quantum mechanics

Main articles: Quantum mechanics and First quantization

In quantum mechanics, a particle (such as an electron or proton) is described by a complex wavefunction, $\psi(x, t)$, whose time-evolution is governed by the Schrödinger equation:

$$-\frac{\hbar^2}{2m}\frac{\partial^2}{\partial x^2}\psi(x,t) + V(x)\psi(x,t) = i\hbar\frac{\partial}{\partial t}\psi(x,t).$$

Here m is the particle's mass and $V(x)$ is the applied potential. Physical information about the behavior of the particle is extracted from the wavefunction by constructing expected values for various quantities; for example, the expected value of the particle's position is given by integrating $\psi^*(x)$ x $\psi(x)$ over all space, and the expected value of the particle's momentum is found by integrating $-i\hbar\psi^*(x)\mathrm{d}\psi/\mathrm{d}x$. The quantity $\psi^*(x)\psi(x)$ is itself in the Copenhagen interpretation of quantum mechanics interpreted as a probability density function. This treatment of quantum mechanics, where a particle's wavefunction evolves against a classical background potential $V(x)$, is sometimes called *first quantization*.

This description of quantum mechanics can be extended to describe the behavior of multiple particles, so long as the number and the type of particles remain fixed. The particles are described by a wavefunction $\psi(x_1, x_2, ..., xN, t)$, which is governed by an extended version of the Schrödinger equation.

Often one is interested in the case where N particles are all of the same type (for example, the 18 electrons orbiting a neutral argon nucleus). As described in the article on identical particles, this implies that the state of the entire system must be either symmetric (bosons) or antisymmetric (fermions) when the coordinates of its constituent particles are exchanged. This is achieved by using a Slater determinant as the wavefunction of a fermionic system (and a Slater permanent for a bosonic system), which is equivalent to an element of the symmetric or antisymmetric subspace of a tensor product.

For example, the general quantum state of a system of N bosons is written as

$$|\phi_1 \cdots \phi_N\rangle = \sqrt{\frac{\prod_j N_j!}{N!}} \sum_{p \in S_N} |\phi_{p(1)}\rangle \otimes \cdots \otimes |\phi_{p(N)}\rangle,$$

where $|\phi_i\rangle$ are the single-particle states, Nj is the number of particles occupying state j, and the sum is taken over all possible permutations p acting on N elements. In general, this is a sum of $N!$ (N factorial) distinct terms. $\sqrt{\frac{\prod_j N_j!}{N!}}$ is a normalizing factor.

There are several shortcomings to the above description of quantum mechanics, which are addressed by quantum field theory. First, it is unclear how to extend quantum mechanics to include the effects of special relativity.[7] Attempted replacements for the Schrödinger equation, such as the Klein–Gordon equation or the Dirac equation, have many unsatisfactory qualities; for instance, they possess energy eigenvalues that extend to $-\infty$, so that there seems to be no easy definition of a ground state. It turns out that such inconsistencies arise from relativistic wavefunctions not having a well-defined probabilistic interpretation in position space, as probability conservation is not a relativistically covariant concept. The second shortcoming, related to the first, is that in quantum mechanics there is no mechanism to describe particle creation and annihilation;[8] this is crucial for describing phenomena such as pair production, which result from the conversion between mass and energy according to the relativistic relation $E = mc^2$.

10.4.3   Second quantization

Main article: Second quantization

In this section, we will describe a method for constructing a quantum field theory called **second quantization**. This basically involves choosing a way to index the quantum mechanical degrees of freedom in the space of multiple identical-particle states. It is based on the Hamiltonian formulation of quantum mechanics.

Several other approaches exist, such as the Feynman path integral,[9] which uses a Lagrangian formulation. For an overview of some of these approaches, see the article on quantization.

Bosons

For simplicity, we will first discuss second quantization for bosons, which form perfectly symmetric quantum states. Let us denote the mutually orthogonal single-particle states which are possible in the system by $|\phi_1\rangle, |\phi_2\rangle, |\phi_3\rangle$, and so on. For example, the 3-particle state with one particle in state $|\phi_1\rangle$ and two in state $|\phi_2\rangle$ is

$$\frac{1}{\sqrt{3}} \left[|\phi_1\rangle|\phi_2\rangle|\phi_2\rangle + |\phi_2\rangle|\phi_1\rangle|\phi_2\rangle + |\phi_2\rangle|\phi_2\rangle|\phi_1\rangle \right].$$

The first step in second quantization is to express such quantum states in terms of **occupation numbers**, by listing the number of particles occupying each of the single-particle states $|\phi_1\rangle, |\phi_2\rangle$, etc. This is simply another way of labelling the states. For instance, the above 3-particle state is denoted as

$$|1, 2, 0, 0, 0, \dots\rangle.$$

An N-particle state belongs to a space of states describing systems of N particles. The next step is to combine the individual N-particle state spaces into an extended state space, known as Fock space, which can describe systems of any number of particles. This is composed of the state space of a system with no particles (the so-called vacuum state, written as $|0\rangle$), plus the state space of a 1-particle system, plus the state space of a 2-particle system, and so forth. States describing a definite number of particles are known as Fock states: a general element of Fock space will be a linear combination of Fock states. There is a one-to-one correspondence between the occupation number representation and valid boson states in the Fock space.

At this point, the quantum mechanical system has become a quantum field in the sense we described above. The field's elementary degrees of freedom are the occupation numbers, and each occupation number is indexed by a number j indicating which of the single-particle states $|\phi_1\rangle, |\phi_2\rangle, \dots, |\phi_j\rangle, \dots$ it refers to:

$$|N_1, N_2, N_3, \dots, N_j, \dots\rangle.$$

The properties of this quantum field can be explored by defining creation and annihilation operators, which add and subtract particles. They are analogous to ladder operators in the quantum harmonic oscillator problem, which added and subtracted energy quanta. However, these operators literally create and annihilate particles of a given quantum state. The bosonic annihilation operator a_2 and creation operator $a_2^\dagger$ are easily defined in the occupation number representation as having the following effects:

$$a_2|N_1, N_2, N_3, \ldots\rangle = \sqrt{N_2} \mid N_1, (N_2 - 1), N_3, \ldots\rangle,$$

$$a_2^\dagger|N_1, N_2, N_3, \ldots\rangle = \sqrt{N_2 + 1} \mid N_1, (N_2 + 1), N_3, \ldots\rangle.$$

It can be shown that these are operators in the usual quantum mechanical sense, i.e. linear operators acting on the Fock space. Furthermore, they are indeed Hermitian conjugates, which justifies the way we have written them. They can be shown to obey the commutation relation

$$[a_i, a_j] = 0 \quad , \quad \left[a_i^\dagger, a_j^\dagger\right] = 0 \quad , \quad \left[a_i, a_j^\dagger\right] = \delta_{ij},$$

where δ stands for the Kronecker delta. These are precisely the relations obeyed by the ladder operators for an infinite set of independent quantum harmonic oscillators, one for each single-particle state. Adding or removing bosons from each state is therefore analogous to exciting or de-exciting a quantum of energy in a harmonic oscillator.

Applying an annihilation operator a_k followed by its corresponding creation operator $a_k^\dagger$ returns the number N_k of particles in the kth single-particle eigenstate:

$$a_k^\dagger a_k|\ldots, N_k, \ldots\rangle = N_k|\ldots, N_k, \ldots\rangle.$$

The combination of operators $a_k^\dagger a_k$ is known as the number operator for the kth eigenstate.

The Hamiltonian operator of the quantum field (which, through the Schrödinger equation, determines its dynamics) can be written in terms of creation and annihilation operators. For instance, for a field of free (non-interacting) bosons, the total energy of the field is found by summing the energies of the bosons in each energy eigenstate. If the kth single-particle energy eigenstate has energy E_k and there are N_k bosons in this state, then the total energy of these bosons is $E_k N_k$. The energy in the *entire* field is then a sum over k :

$$E_{\text{tot}} = \sum_k E_k N_k$$

This can be turned into the Hamiltonian operator of the field by replacing N_k with the corresponding number operator, $a_k^\dagger a_k$. This yields

$$H = \sum_k E_k \, a_k^\dagger \, a_k.$$

Fermions

It turns out that a different definition of creation and annihilation must be used for describing fermions. According to the Pauli exclusion principle, fermions cannot share quantum states, so their occupation numbers N_i can only take on the value 0 or 1. The fermionic annihilation operators c and creation operators $c^\dagger$ are defined by their actions on a Fock state thus

$$c_j|N_1, N_2, \ldots, N_j = 0, \ldots\rangle = 0$$

$$c_j|N_1, N_2, \ldots, N_j = 1, \ldots\rangle = (-1)^{(N_1 + \cdots + N_{j-1})}|N_1, N_2, \ldots, N_j = 0, \ldots\rangle$$

$$c_j^\dagger|N_1, N_2, \ldots, N_j = 0, \ldots\rangle = (-1)^{(N_1 + \cdots + N_{j-1})}|N_1, N_2, \ldots, N_j = 1, \ldots\rangle$$

$$c_j^\dagger |N_1, N_2, \ldots, N_j = 1, \ldots\rangle = 0.$$

These obey an anticommutation relation:

$$\{c_i, c_j\} = 0 \quad , \quad \{c_i^\dagger, c_j^\dagger\} = 0 \quad , \quad \{c_i, c_j^\dagger\} = \delta_{ij}.$$

One may notice from this that applying a fermionic creation operator twice gives zero, so it is impossible for the particles to share single-particle states, in accordance with the exclusion principle.

Field operators

We have previously mentioned that there can be more than one way of indexing the degrees of freedom in a quantum field. Second quantization indexes the field by enumerating the single-particle quantum states. However, as we have discussed, it is more natural to think about a "field", such as the electromagnetic field, as a set of degrees of freedom indexed by position.

To this end, we can define *field operators* that create or destroy a particle at a particular point in space. In particle physics, these operators turn out to be more convenient to work with, because they make it easier to formulate theories that satisfy the demands of relativity.

Single-particle states are usually enumerated in terms of their momenta (as in the particle in a box problem.) We can construct field operators by applying the Fourier transform to the creation and annihilation operators for these states. For example, the bosonic field annihilation operator $\phi(\mathbf{r})$ is

$$\phi(\mathbf{r}) \overset{\text{def}}{=} \sum_j e^{i\mathbf{k}_j \cdot \mathbf{r}} a_j.$$

The bosonic field operators obey the commutation relation

$$[\phi(\mathbf{r}), \phi(\mathbf{r}')] = 0 \quad , \quad [\phi^\dagger(\mathbf{r}), \phi^\dagger(\mathbf{r}')] = 0 \quad , \quad [\phi(\mathbf{r}), \phi^\dagger(\mathbf{r}')] = \delta^3(\mathbf{r} - \mathbf{r}')$$

where $\delta(x)$ stands for the Dirac delta function. As before, the fermionic relations are the same, with the commutators replaced by anticommutators.

The field operator is not the same thing as a single-particle wavefunction. The former is an operator acting on the Fock space, and the latter is a quantum-mechanical amplitude for finding a particle in some position. However, they are closely related, and are indeed commonly denoted with the same symbol. If we have a Hamiltonian with a space representation, say

$$H = -\frac{\hbar^2}{2m} \sum_i \nabla_i^2 + \sum_{i<j} U(|\mathbf{r}_i - \mathbf{r}_j|)$$

where the indices i and j run over all particles, then the field theory Hamiltonian (in the non-relativistic limit and for negligible self-interactions) is

$$H = -\frac{\hbar^2}{2m} \int d^3r\, \phi^\dagger(\mathbf{r})\nabla^2\phi(\mathbf{r}) + \frac{1}{2} \int d^3r \int d^3r'\, \phi^\dagger(\mathbf{r})\phi^\dagger(\mathbf{r}')U(|\mathbf{r} - \mathbf{r}'|)\phi(\mathbf{r}')\phi(\mathbf{r}).$$

This looks remarkably like an expression for the expectation value of the energy, with ϕ playing the role of the wavefunction. This relationship between the field operators and wavefunctions makes it very easy to formulate field theories starting from space-projected Hamiltonians.

10.4.4 Dynamics

Once the Hamiltonian operator is obtained as part of the canonical quantization process, the time dependence of the state is described with the Schrödinger equation, just as with other quantum theories. Alternatively, the Heisenberg picture can be used where the time dependence is in the operators rather than in the states.

10.4.5 Implications

Unification of fields and particles

The "second quantization" procedure that we have outlined in the previous section takes a set of single-particle quantum states as a starting point. Sometimes, it is impossible to define such single-particle states, and one must proceed directly to quantum field theory. For example, a quantum theory of the electromagnetic field *must* be a quantum field theory, because it is impossible (for various reasons) to define a wavefunction for a single photon.[10] In such situations, the quantum field theory can be constructed by examining the mechanical properties of the classical field and guessing the corresponding quantum theory. For free (non-interacting) quantum fields, the quantum field theories obtained in this way have the same properties as those obtained using second quantization, such as well-defined creation and annihilation operators obeying commutation or anticommutation relations.

Quantum field theory thus provides a unified framework for describing "field-like" objects (such as the electromagnetic field, whose excitations are photons) and "particle-like" objects (such as electrons, which are treated as excitations of an underlying electron field), so long as one can treat interactions as "perturbations" of free fields. There are still unsolved problems relating to the more general case of interacting fields that may or may not be adequately described by perturbation theory. For more on this topic, see Haag's theorem.

Physical meaning of particle indistinguishability

The second quantization procedure relies crucially on the particles being identical. We would not have been able to construct a quantum field theory from a distinguishable many-particle system, because there would have been no way of separating and indexing the degrees of freedom.

Many physicists prefer to take the converse interpretation, which is that *quantum field theory explains what identical particles are*. In ordinary quantum mechanics, there is not much theoretical motivation for using symmetric (bosonic) or antisymmetric (fermionic) states, and the need for such states is simply regarded as an empirical fact. From the point of view of quantum field theory, particles are identical if and only if they are excitations of the same underlying quantum field. Thus, the question "why are all electrons identical?" arises from mistakenly regarding individual electrons as fundamental objects, when in fact it is only the electron field that is fundamental.

Particle conservation and non-conservation

During second quantization, we started with a Hamiltonian and state space describing a fixed number of particles (N), and ended with a Hamiltonian and state space for an arbitrary number of particles. Of course, in many common situations N is an important and perfectly well-defined quantity, e.g. if we are describing a gas of atoms sealed in a box. From the point of view of quantum field theory, such situations are described by quantum states that are eigenstates of the number operator $\hat{N}$, which measures the total number of particles present. As with any quantum mechanical observable, $\hat{N}$ is conserved if it commutes with the Hamiltonian. In that case, the quantum state is trapped in the N-particle subspace of the total Fock space, and the situation could equally well be described by ordinary N-particle quantum mechanics. (Strictly speaking, this is only true in the noninteracting case or in the low energy density limit of renormalized quantum field theories)

For example, we can see that the free-boson Hamiltonian described above conserves particle number. Whenever the Hamiltonian operates on a state, each particle destroyed by an annihilation operator a_k is immediately put back by the creation operator $a_k^\dagger$.

On the other hand, it is possible, and indeed common, to encounter quantum states that are *not* eigenstates of $\hat{N}$, which do not have well-defined particle numbers. Such states are difficult or impossible to handle using ordinary quantum mechanics, but they can be easily described in quantum field theory as quantum superpositions of states having different values of N. For example, suppose we have a bosonic field whose particles can be created or destroyed by interactions with a fermionic field. The Hamiltonian of the combined system would be given by the Hamiltonians of the free boson and free fermion fields, plus a "potential energy" term such as

$$H_I = \sum_{k,q} V_q(a_q + a_{-q}^\dagger)c_{k+q}^\dagger c_k,$$

where $a_k^\dagger$ and a_k denotes the bosonic creation and annihilation operators, $c_k^\dagger$ and c_k denotes the fermionic creation and annihilation operators, and V_q is a parameter that describes the strength of the interaction. This "interaction term" describes processes in which a fermion in state k either absorbs or emits a boson, thereby being kicked into a different eigenstate $k + q$. (In fact, this type of Hamiltonian is used to describe interaction between conduction electrons and phonons in metals. The interaction between electrons and photons is treated in a similar way, but is a little more complicated because the role of spin must be taken into account.) One thing to notice here is that even if we start out with a fixed number of bosons, we will typically end up with a superposition of states with different numbers of bosons at later times. The number of fermions, however, is conserved in this case.

In condensed matter physics, states with ill-defined particle numbers are particularly important for describing the various superfluids. Many of the defining characteristics of a superfluid arise from the notion that its quantum state is a superposition of states with different particle numbers. In addition, the concept of a coherent state (used to model the laser and the BCS ground state) refers to a state with an ill-defined particle number but a well-defined phase.

10.4.6   Axiomatic approaches

The preceding description of quantum field theory follows the spirit in which most physicists approach the subject. However, it is not mathematically rigorous. Over the past several decades, there have been many attempts to put quantum field theory on a firm mathematical footing by formulating a set of axioms for it. These attempts fall into two broad classes.

The first class of axioms, first proposed during the 1950s, include the Wightman, Osterwalder–Schrader, and Haag–Kastler systems. They attempted to formalize the physicists' notion of an "operator-valued field" within the context of functional analysis, and enjoyed limited success. It was possible to prove that any quantum field theory satisfying these axioms satisfied certain general theorems, such as the spin-statistics theorem and the CPT theorem. Unfortunately, it proved extraordinarily difficult to show that any realistic field theory, including the Standard Model, satisfied these axioms. Most of the theories that could be treated with these analytic axioms were physically trivial, being restricted to low-dimensions and lacking interesting dynamics. The construction of theories satisfying one of these sets of axioms falls in the field of constructive quantum field theory. Important work was done in this area in the 1970s by Segal, Glimm, Jaffe and others.

During the 1980s, a second set of axioms based on geometric ideas was proposed. This line of investigation, which restricts its attention to a particular class of quantum field theories known as topological quantum field theories, is associated most closely with Michael Atiyah and Graeme Segal, and was notably expanded upon by Edward Witten, Richard Borcherds, and Maxim Kontsevich. However, most of the physically relevant quantum field theories, such as the Standard Model, are not topological quantum field theories; the quantum field theory of the fractional quantum Hall effect is a notable exception. The main impact of axiomatic topological quantum field theory has been on mathematics, with important applications in representation theory, algebraic topology, and differential geometry.

Finding the proper axioms for quantum field theory is still an open and difficult problem in mathematics. One of the Millennium Prize Problems—proving the existence of a mass gap in Yang–Mills theory—is linked to this issue.

10.5 Associated phenomena

In the previous part of the article, we described the most general features of quantum field theories. Some of the quantum field theories studied in various fields of theoretical physics involve additional special ideas, such as renormalizability, gauge symmetry, and supersymmetry. These are described in the following sections.

10.5.1 Renormalization

Main article: Renormalization

Early in the history of quantum field theory, it was found that many seemingly innocuous calculations, such as the perturbative shift in the energy of an electron due to the presence of the electromagnetic field, give infinite results. The reason is that the perturbation theory for the shift in an energy involves a sum over all other energy levels, and there are infinitely many levels at short distances that each give a finite contribution which results in a divergent series.

Many of these problems are related to failures in classical electrodynamics that were identified but unsolved in the 19th century, and they basically stem from the fact that many of the supposedly "intrinsic" properties of an electron are tied to the electromagnetic field that it carries around with it. The energy carried by a single electron—its self energy—is not simply the bare value, but also includes the energy contained in its electromagnetic field, its attendant cloud of photons. The energy in a field of a spherical source diverges in both classical and quantum mechanics, but as discovered by Weisskopf with help from Furry, in quantum mechanics the divergence is much milder, going only as the logarithm of the radius of the sphere.

The solution to the problem, presciently suggested by Stueckelberg, independently by Bethe after the crucial experiment by Lamb, implemented at one loop by Schwinger, and systematically extended to all loops by Feynman and Dyson, with converging work by Tomonaga in isolated postwar Japan, comes from recognizing that all the infinities in the interactions of photons and electrons can be isolated into redefining a finite number of quantities in the equations by replacing them with the observed values: specifically the electron's mass and charge: this is called renormalization. The technique of renormalization recognizes that the problem is essentially purely mathematical, that extremely short distances are at fault. In order to define a theory on a continuum, first place a cutoff on the fields, by postulating that quanta cannot have energies above some extremely high value. This has the effect of replacing continuous space by a structure where very short wavelengths do not exist, as on a lattice. Lattices break rotational symmetry, and one of the crucial contributions made by Feynman, Pauli and Villars, and modernized by 't Hooft and Veltman, is a symmetry-preserving cutoff for perturbation theory (this process is called regularization). There is no known symmetrical cutoff outside of perturbation theory, so for rigorous or numerical work people often use an actual lattice.

On a lattice, every quantity is finite but depends on the spacing. When taking the limit of zero spacing, we make sure that the physically observable quantities like the observed electron mass stay fixed, which means that the constants in the Lagrangian defining the theory depend on the spacing. Hopefully, by allowing the constants to vary with the lattice spacing, all the results at long distances become insensitive to the lattice, defining a continuum limit.

The renormalization procedure only works for a certain class of quantum field theories, called **renormalizable quantum field theories**. A theory is **perturbatively renormalizable** when the constants in the Lagrangian only diverge at worst as logarithms of the lattice spacing for very short spacings. The continuum limit is then well defined in perturbation theory, and even if it is not fully well defined non-perturbatively, the problems only show up at distance scales that are exponentially small in the inverse coupling for weak couplings. The Standard Model of particle physics is perturbatively renormalizable, and so are its component theories (quantum electrodynamics/electroweak theory and quantum chromodynamics). Of the three components, quantum electrodynamics is believed to not have a continuum limit, while the asymptotically free SU(2) and SU(3) weak hypercharge and strong color interactions are nonperturbatively well defined.

The renormalization group describes how renormalizable theories emerge as the long distance low-energy effective field theory for any given high-energy theory. Because of this, renormalizable theories are insensitive to the precise nature of the underlying high-energy short-distance phenomena. This is a blessing because it allows physicists to formulate low energy theories without knowing the details of high energy phenomenon. It is also a curse, because once a renormalizable theory like the standard model is found to work, it gives very few clues to higher energy processes. The only way high energy processes can be seen in the standard model is when they allow otherwise forbidden events, or if they predict quantitative relations between the coupling constants.

10.5.2   Haag's theorem

See also: Haag's theorem

From a mathematically rigorous perspective, there exists no interaction picture in a Lorentz-covariant quantum field theory. This implies that the perturbative approach of Feynman diagrams in QFT is not strictly justified, despite producing vastly precise predictions validated by experiment. This is called Haag's theorem, but most particle physicists relying on QFT largely shrug it off.

10.5.3   Gauge freedom

A gauge theory is a theory that admits a symmetry with a local parameter. For example, in every quantum theory the global phase of the wave function is arbitrary and does not represent something physical. Consequently, the theory is invariant under a global change of phases (adding a constant to the phase of all wave functions, everywhere); this is a global symmetry. In quantum electrodynamics, the theory is also invariant under a *local* change of phase, that is – one may shift the phase of all wave functions so that the shift may be different at every point in space-time. This is a *local* symmetry. However, in order for a well-defined derivative operator to exist, one must introduce a new field, the gauge field, which also transforms in order for the local change of variables (the phase in our example) not to affect the derivative. In quantum electrodynamics this gauge field is the electromagnetic field. The change of local gauge of variables is termed gauge transformation. It is worth noting that by Noether's theorem, for every such symmetry there exists an associated conserved current. The aforementioned symmetry of the wavefunction under global phase changes implies the conservation of electric charge.

In quantum field theory the excitations of fields represent particles. The particle associated with excitations of the gauge field is the gauge boson, which is the photon in the case of quantum electrodynamics.

The degrees of freedom in quantum field theory are local fluctuations of the fields. The existence of a gauge symmetry reduces the number of degrees of freedom, simply because some fluctuations of the fields can be transformed to zero by gauge transformations, so they are equivalent to having no fluctuations at all, and they therefore have no physical meaning. Such fluctuations are usually called "non-physical degrees of freedom" or *gauge artifacts*; usually some of them have a negative norm, making them inadequate for a consistent theory. Therefore, if a classical field theory has a gauge symmetry, then its quantized version (i.e. the corresponding quantum field theory) will have this symmetry as well. In other words, a gauge symmetry cannot have a quantum anomaly. If a gauge symmetry is anomalous (i.e. not kept in the quantum theory) then the theory is non-consistent: for example, in quantum electrodynamics, had there been a gauge anomaly, this would require the appearance of photons with longitudinal polarization and polarization in the time direction, the latter having a negative norm, rendering the theory inconsistent; another possibility would be for these photons to appear only in intermediate processes but not in the final products of any interaction, making the theory non-unitary and again inconsistent (see optical theorem).

In general, the gauge transformations of a theory consist of several different transformations, which may not be commutative. These transformations are together described by a mathematical object known as a gauge group. Infinitesimal gauge transformations are the gauge group generators. Therefore, the number of gauge bosons is the group dimension (i.e. number of generators forming a basis).

All the fundamental interactions in nature are described by gauge theories. These are:

- Quantum chromodynamics, whose gauge group is $\mathbf{SU}(3)$. The gauge bosons are eight gluons.

- The electroweak theory, whose gauge group is $\mathbf{U}(1) \times \mathbf{SU}(2)$, (a direct product of $\mathbf{U}(1)$ and $\mathbf{SU}(2)$).

- Gravity, whose classical theory is general relativity, admits the equivalence principle, which is a form of gauge symmetry. However, it is explicitly non-renormalizable.

10.5.4 Multivalued gauge transformations

The gauge transformations which leave the theory invariant involve, by definition, only single-valued gauge functions $\Lambda(x_i)$ which satisfy the Schwarz integrability criterion

$$\partial_{x_i x_j}\Lambda = \partial_{x_j x_i}\Lambda.$$

An interesting extension of gauge transformations arises if the gauge functions $\Lambda(x_i)$ are allowed to be multivalued functions which violate the integrability criterion. These are capable of changing the physical field strengths and are therefore not proper symmetry transformations. Nevertheless, the transformed field equations describe correctly the physical laws in the presence of the newly generated field strengths. See the textbook by H. Kleinert cited below for the applications to phenomena in physics.

10.5.5 Supersymmetry

Main article: Supersymmetry

Supersymmetry assumes that every fundamental fermion has a superpartner that is a boson and vice versa. It was introduced in order to solve the so-called Hierarchy Problem, that is, to explain why particles not protected by any symmetry (like the Higgs boson) do not receive radiative corrections to its mass driving it to the larger scales (GUT, Planck...). It was soon realized that supersymmetry has other interesting properties: its gauged version is an extension of general relativity (Supergravity), and it is a key ingredient for the consistency of string theory.

The way supersymmetry protects the hierarchies is the following: since for every particle there is a superpartner with the same mass, any loop in a radiative correction is cancelled by the loop corresponding to its superpartner, rendering the theory UV finite.

Since no superpartners have yet been observed, if supersymmetry exists it must be broken (through a so-called soft term, which breaks supersymmetry without ruining its helpful features). The simplest models of this breaking require that the energy of the superpartners not be too high; in these cases, supersymmetry is expected to be observed by experiments at the Large Hadron Collider. The Higgs particle has been detected at the LHC, and no such superparticles have been discovered.

10.6 See also

- Abraham–Lorentz force
- Basic concepts of quantum mechanics
- Common integrals in quantum field theory
- Einstein–Maxwell–Dirac equations
- Form factor (quantum field theory)
- Green–Kubo relations
- Green's function (many-body theory)
- Invariance mechanics
- List of quantum field theories
- Quantization of a field

- Quantum electrodynamics

- Quantum field theory in curved spacetime

- Quantum flavordynamics

- Quantum hydrodynamics

- Quantum triviality

- Relation between Schrödinger's equation and the path integral formulation of quantum mechanics

- Relationship between string theory and quantum field theory

- Schwinger–Dyson equation

- Static forces and virtual-particle exchange

- Symmetry in quantum mechanics

- Theoretical and experimental justification for the Schrödinger equation

- Ward–Takahashi identity

- Wheeler–Feynman absorber theory

- Wigner's classification

- Wigner's theorem

10.7   Notes

10.8   References

[1] "Beautiful Minds, Vol. 20: Ed Witten". la Repubblica. 2010. Retrieved 22 June 2012. See here.

[2] J. J. Thorn et al. (2004). Observing the quantum behavior of light in an undergraduate laboratory. . J. J. Thorn, M. S. Neel, V. W. Donato, G. S. Bergreen, R. E. Davies, and M. Beck. American Association of Physics Teachers, 2004.DOI: 10.1119/1.1737397.

[3] David Tong, *Lectures on Quantum Field Theory*, chapter 1.

[4] Srednicki, Mark. *Quantum Field Theory* (1st ed.). p. 19.

[5] Srednicki, Mark. *Quantum Field Theory* (1st ed.). pp. 25–6.

[6] Zee, Anthony. *Quantum Field Theory in a Nutshell* (2nd ed.). p. 61.

[7] David Tong, *Lectures on Quantum Field Theory*, Introduction.

[8] Zee, Anthony. *Quantum Field Theory in a Nutshell* (2nd ed.). p. 3.

[9] Abraham Pais, *Inward Bound: Of Matter and Forces in the Physical World* ISBN 0-19-851997-4. Pais recounts how his astonishment at the rapidity with which Feynman could calculate using his method. Feynman's method is now part of the standard methods for physicists.

[10] Newton, T.D.; Wigner, E.P. (1949). "Localized states for elementary systems". *Reviews of Modern Physics* **21** (3): 400–406. Bibcode:1949RvMP...21..400N. doi:10.1103/RevModPhys.21.400.

10.9 Further reading

General readers

- Feynman, R.P. (2001) [1964]. *The Character of Physical Law*. MIT Press. ISBN 0-262-56003-8.

- Feynman, R.P. (2006) [1985]. *QED: The Strange Theory of Light and Matter*. Princeton University Press. ISBN 0-691-12575-9.

- Gribbin, J. (1998). *Q is for Quantum: Particle Physics from A to Z*. Weidenfeld & Nicolson. ISBN 0-297-81752-3.

- Schumm, Bruce A. (2004) *Deep Down Things*. Johns Hopkins Univ. Press. Chpt. 4.

Introductory texts

- McMahon, D. (2008). *Quantum Field Theory*. McGraw-Hill. ISBN 978-0-07-154382-8.

- Bogoliubov, N.; Shirkov, D. (1982). *Quantum Fields*. Benjamin-Cummings. ISBN 0-8053-0983-7.

- Frampton, P.H. (2000). *Gauge Field Theories. Frontiers in Physics (2nd ed.). Wiley*.

- Greiner, W; Müller, B. (2000). *Gauge Theory of Weak Interactions*. Springer. ISBN 3-540-67672-4.

- Itzykson, C.; Zuber, J.-B. (1980). *Quantum Field Theory*. McGraw-Hill. ISBN 0-07-032071-3.

- Kane, G.L. (1987). *Modern Elementary Particle Physics*. Perseus Books. ISBN 0-201-11749-5.

- Kleinert, H.; Schulte-Frohlinde, Verena (2001). *Critical Properties of φ^4-Theories*. World Scientific. ISBN 981-02-4658-7.

- Kleinert, H. (2008). *Multivalued Fields in Condensed Matter, Electrodynamics, and Gravitation* (PDF). World Scientific. ISBN 978-981-279-170-2.

- Loudon, R (1983). *The Quantum Theory of Light*. Oxford University Press. ISBN 0-19-851155-8.

- Mandl, F.; Shaw, G. (1993). *Quantum Field Theory*. John Wiley & Sons. ISBN 978-0-471-94186-6.

- Peskin, M.; Schroeder, D. (1995). *An Introduction to Quantum Field Theory*. Westview Press. ISBN 0-201-50397-2.

- Ryder, L.H. (1985). *Quantum Field Theory*. Cambridge University Press. ISBN 0-521-33859-X.

- Schwartz, M.D. (2014). *Quantum Field Theory and the Standard Model*. Cambridge University Press. ISBN 978-1107034730.

- Srednicki, Mark (2007) *Quantum Field Theory*. Cambridge Univ. Press.

- Ynduráin, F.J. (1996). *Relativistic Quantum Mechanics and Introduction to Field Theory* (1st ed.). Springer. ISBN 978-3-540-60453-2.

- Zee, A. (2003). *Quantum Field Theory in a Nutshell*. Princeton University Press. ISBN 0-691-01019-6.

Advanced texts

- Brown, Lowell S. (1994). *Quantum Field Theory*. Cambridge University Press. ISBN 978-0-521-46946-3.

- Bogoliubov, N.; Logunov, A.A.; Oksak, A.I.; Todorov, I.T. (1990). *General Principles of Quantum Field Theory*. Kluwer Academic Publishers. ISBN 978-0-7923-0540-8.

- Weinberg, S. (1995). *The Quantum Theory of Fields* **1–3**. Cambridge University Press.

Articles:

- Gerard 't Hooft (2007) "The Conceptual Basis of Quantum Field Theory" in Butterfield, J., and John Earman, eds., *Philosophy of Physics, Part A*. Elsevier: 661–730.

- Frank Wilczek (1999) "Quantum field theory", *Reviews of Modern Physics* 71: S83–S95. Also doi=10.1103/Rev. Mod. Phys. 71.

10.10   External links

- Hazewinkel, Michiel, ed. (2001), "Quantum field theory", *Encyclopedia of Mathematics*, Springer, ISBN 978-1-55608-010-4

- Stanford Encyclopedia of Philosophy: "Quantum Field Theory", by Meinard Kuhlmann.

- Siegel, Warren, 2005. *Fields*. A free text, also available from arXiv:hep-th/9912205.

- Quantum Field Theory by P. J. Mulders

Chapter 11

Spin (physics)

This article is about spin in quantum mechanics. For rotation in classical mechanics, see angular momentum.

In quantum mechanics and particle physics, **spin** is an intrinsic form of angular momentum carried by elementary particles, composite particles (hadrons), and atomic nuclei.[1][2]

Spin is one of two types of angular momentum in quantum mechanics, the other being *orbital angular momentum*. The orbital angular momentum operator is the quantum-mechanical counterpart to the classical angular momentum of orbital revolution: it arises when a particle executes a rotating or twisting trajectory (such as when an electron orbits a nucleus).[3][4] The existence of spin angular momentum is inferred from experiments, such as the Stern–Gerlach experiment, in which particles are observed to possess angular momentum that cannot be accounted for by orbital angular momentum alone.[5]

In some ways, spin is like a vector quantity; it has a definite magnitude, and it has a "direction" (but quantization makes this "direction" different from the direction of an ordinary vector). All elementary particles of a given kind have the same magnitude of spin angular momentum, which is indicated by assigning the particle a *spin quantum number*.[2]

The SI unit of spin is the joule-second, just as with classical angular momentum. In practice, however, it is written as a multiple of the reduced Planck constant $\hbar$, usually in natural units, where the $\hbar$ is omitted, resulting in a unitless number. Spin quantum numbers are unitless numbers by definition.

When combined with the spin-statistics theorem, the spin of electrons results in the Pauli exclusion principle, which in turn underlies the periodic table of chemical elements.

Wolfgang Pauli was the first to propose the concept of spin, but he did not name it. In 1925, Ralph Kronig, George Uhlenbeck and Samuel Goudsmit at Leiden University suggested a physical interpretation of particles spinning around their own axis. The mathematical theory was worked out in depth by Pauli in 1927. When Paul Dirac derived his relativistic quantum mechanics in 1928, electron spin was an essential part of it.

11.1 Quantum number

Main article: Spin quantum number

As the name suggests, spin was originally conceived as the rotation of a particle around some axis. This picture is correct so far as spin obeys the same mathematical laws as quantized angular momenta do. On the other hand, spin has some peculiar properties that distinguish it from orbital angular momenta:

- Spin quantum numbers may take half-integer values.

- Although the direction of its spin can be changed, an elementary particle cannot be made to spin faster or slower.

- The spin of a charged particle is associated with a magnetic dipole moment with a g-factor differing from 1. This could only occur classically if the internal charge of the particle were distributed differently from its mass.

The conventional definition of the **spin quantum number**, s, is $s = n/2$, where n can be any non-negative integer. Hence the allowed values of s are 0, 1/2, 1, 3/2, 2, etc. The value of s for an elementary particle depends only on the type of particle, and cannot be altered in any known way (in contrast to the *spin direction* described below). The spin angular momentum, S, of any physical system is quantized. The allowed values of S are:

$$S = \frac{h}{2\pi}\sqrt{s(s+1)} = \frac{h}{4\pi}\sqrt{n(n+2)},$$

where h is the Planck constant. In contrast, orbital angular momentum can only take on integer values of s; i.e., even-numbered values of n.

11.1.1 Fermions and bosons

Those particles with half-integer spins, such as 1/2, 3/2, 5/2, are known as fermions, while those particles with integer spins, such as 0, 1, 2, are known as bosons. The two families of particles obey different rules and *broadly* have different roles in the world around us. A key distinction between the two families is that fermions obey the Pauli exclusion principle; that is, there cannot be two identical fermions simultaneously having the same quantum numbers (meaning, roughly, having the same position, velocity and spin direction). In contrast, bosons obey the rules of Bose–Einstein statistics and have no such restriction, so they may "bunch together" even if in identical states. Also, composite particles can have spins different from the particles which comprise them. For example, a helium atom can have spin 0 and therefore can behave like a boson even though the quarks and electrons which make it up are all fermions.

This has profound practical applications:

- Quarks and leptons (including electrons and neutrinos), which make up what is classically known as matter, are all fermions with spin 1/2. The common idea that "matter takes up space" actually comes from the Pauli exclusion principle acting on these particles to prevent the fermions that make up matter from being in the same quantum state. Further compaction would require electrons to occupy the same energy states, and therefore a kind of pressure (sometimes known as degeneracy pressure of electrons) acts to resist the fermions being overly close. It is also this pressure which prevents stars collapsing inwardly, and which, when it finally gives way under immense gravitational pressure in a dying massive star, triggers inward collapse and the dramatic explosion into a supernova.

  Elementary fermions with other spins (3/2, 5/2 etc.) are not known to exist, as of 2014.

- Elementary particles which are thought of as carrying forces are all bosons with spin 1. They include the photon which carries the electromagnetic force, the gluon (strong force), and the W and Z bosons (weak force). The ability of bosons to occupy the same quantum state is used in the laser, which aligns many photons having the same quantum number (the same direction and frequency), superfluid liquid helium resulting from helium-4 atoms being bosons, and superconductivity where pairs of electrons (which individually are fermions) act as single composite bosons.

  Elementary bosons with other spins (0, 2, 3 etc.) were not historically known to exist, although they have received considerable theoretical treatment and are well established within their respective mainstream theories. In particular theoreticians have proposed the graviton (predicted to exist by some quantum gravity theories) with spin 2, and the Higgs boson (explaining electroweak symmetry breaking) with spin 0. Since 2013 the Higgs boson with spin 0 has been considered proven to exist. It is the first scalar particle (spin 0) known to exist in nature.

Theoretical and experimental studies have shown that the spin possessed by elementary particles cannot be explained by postulating that they are made up of even smaller particles rotating about a common center of mass analogous to a classical electron radius; as far as can be presently determined, these elementary particles have no inner structure. The spin of an elementary particle is therefore seen as a truly intrinsic physical property, akin to the particle's electric charge and rest mass.

11.1.2 Spin-statistics theorem

The proof that particles with half-integer spin (fermions) obey Fermi–Dirac statistics and the Pauli Exclusion Principle, and particles with integer spin (bosons) obey Bose–Einstein statistics, occupy "symmetric states", and thus can share quantum states, is known as the spin-statistics theorem. The theorem relies on both quantum mechanics and the theory of special relativity, and this connection between spin and statistics has been called "one of the most important applications of the special relativity theory".[6]

11.2 Magnetic moments

Main article: Spin magnetic moment

Particles with spin can possess a magnetic dipole moment, just like a rotating electrically charged body in classical electrodynamics. These magnetic moments can be experimentally observed in several ways, e.g. by the deflection of particles by inhomogeneous magnetic fields in a Stern–Gerlach experiment, or by measuring the magnetic fields generated by the particles themselves.

The intrinsic magnetic moment $\boldsymbol{\mu}$ of a spin-1/2 particle with charge q, mass m, and spin angular momentum $\mathbf{S}$, is[7]

$$\boldsymbol{\mu} = \frac{g_s q}{2m}\mathbf{S}$$

where the dimensionless quantity g_s is called the spin g-factor. For exclusively orbital rotations it would be 1 (assuming that the mass and the charge occupy spheres of equal radius).

The electron, being a charged elementary particle, possesses a nonzero magnetic moment. One of the triumphs of the theory of quantum electrodynamics is its accurate prediction of the electron g-factor, which has been experimentally determined to have the value −2.0023193043622(15), with the digits in parentheses denoting measurement uncertainty in the last two digits at one standard deviation.[8] The value of 2 arises from the Dirac equation, a fundamental equation connecting the electron's spin with its electromagnetic properties, and the correction of 0.002319304... arises from the electron's interaction with the surrounding electromagnetic field, including its own field.[9] Composite particles also possess magnetic moments associated with their spin. In particular, the neutron possesses a non-zero magnetic moment despite being electrically neutral. This fact was an early indication that the neutron is not an elementary particle. In fact, it is made up of quarks, which are electrically charged particles. The magnetic moment of the neutron comes from the spins of the individual quarks and their orbital motions.

Neutrinos are both elementary and electrically neutral. The minimally extended Standard Model that takes into account non-zero neutrino masses predicts neutrino magnetic moments of:[10][11][12]

$$\mu_\nu \approx 3 \times 10^{-19}\mu_B \frac{m_\nu}{\text{eV}}$$

where the μ_ν are the neutrino magnetic moments, m_ν are the neutrino masses, and μ_B is the Bohr magneton. New physics above the electroweak scale could, however, lead to significantly higher neutrino magnetic moments. It can be shown in a model independent way that neutrino magnetic moments larger than about 10^{-14} μ_B are unnatural, because they would also lead to large radiative contributions to the neutrino mass. Since the neutrino masses cannot exceed about 1 eV, these radiative corrections must then be assumed to be fine tuned to cancel out to a large degree.[13]

The measurement of neutrino magnetic moments is an active area of research. As of 2001, the latest experimental results have put the neutrino magnetic moment at less than 1.2×10^{-10} times the electron's magnetic moment.

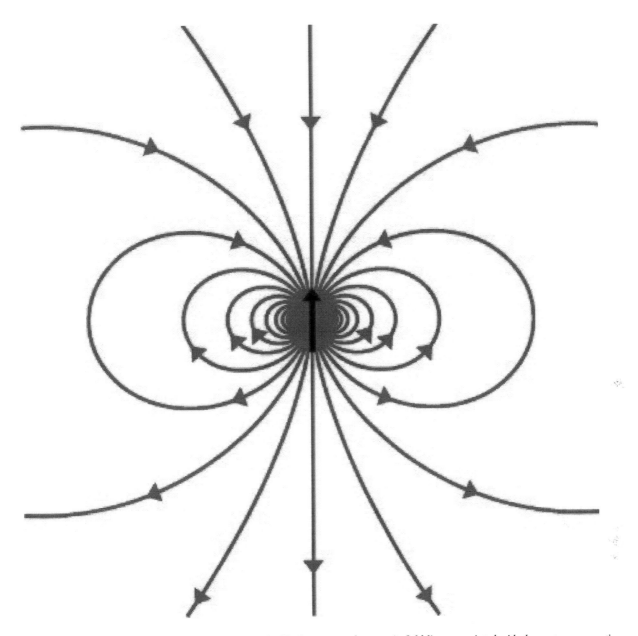

Schematic diagram depicting the spin of the neutron as the black arrow and magnetic field lines associated with the neutron magnetic moment. The neutron has a negative magnetic moment. While the spin of the neutron is upward in this diagram, the magnetic field lines at the center of the dipole are downward.

In ordinary materials, the magnetic dipole moments of individual atoms produce magnetic fields that cancel one another, because each dipole points in a random direction. Ferromagnetic materials below their Curie temperature, however, exhibit magnetic domains in which the atomic dipole moments are locally aligned, producing a macroscopic, non-zero magnetic field from the domain. These are the ordinary "magnets" with which we are all familiar.

In paramagnetic materials, the magnetic dipole moments of individual atoms spontaneously align with an externally applied magnetic field. In diamagnetic materials, on the other hand, the magnetic dipole moments of individual atoms spontaneously align oppositely to any externally applied magnetic field, even if it requires energy to do so.

The study of the behavior of such "spin models" is a thriving area of research in condensed matter physics. For instance, the Ising model describes spins (dipoles) that have only two possible states, up and down, whereas in the Heisenberg model the spin vector is allowed to point in any direction. These models have many interesting properties, which have led to interesting results in the theory of phase transitions.

11.3 Direction

Further information: Angular momentum operator

11.3.1 Spin projection quantum number and multiplicity

In classical mechanics, the angular momentum of a particle possesses not only a magnitude (how fast the body is rotating), but also a direction (either up or down on the axis of rotation of the particle). Quantum mechanical spin also contains information about direction, but in a more subtle form. Quantum mechanics states that the component of angular momentum measured along any direction can only take on the values [14]

$$S_i = \hbar s_i, \quad s_i \in \{-s, -(s-1), \ldots, s-1, s\}$$

where Si is the spin component along the i-axis (either x, y, or z), si is the spin projection quantum number along the i-axis, and s is the principal spin quantum number (discussed in the previous section). Conventionally the direction chosen is the z-axis:

$$S_z = \hbar s_z, \quad s_z \in \{-s, -(s-1), \ldots, s-1, s\}$$

where Sz is the spin component along the z-axis, sz is the spin projection quantum number along the z-axis.

One can see that there are $2s+1$ possible values of s_z. The number "$2s + 1$" is the multiplicity of the spin system. For example, there are only two possible values for a spin-1/2 particle: $s_z = +1/2$ and $s_z = -1/2$. These correspond to quantum states in which the spin is pointing in the +z or −z directions respectively, and are often referred to as "spin up" and "spin down". For a spin-3/2 particle, like a delta baryon, the possible values are +3/2, +1/2, −1/2, −3/2.

11.3.2 Vector

For a given quantum state, one could think of a spin vector $\langle S \rangle$ whose components are the expectation values of the spin components along each axis, i.e., $\langle S \rangle = [\langle S_x \rangle, \langle S_y \rangle, \langle S_z \rangle]$. This vector then would describe the "direction" in which the spin is pointing, corresponding to the classical concept of the axis of rotation. It turns out that the spin vector is not very useful in actual quantum mechanical calculations, because it cannot be measured directly: sx, sy and sz cannot possess simultaneous definite values, because of a quantum uncertainty relation between them. However, for statistically large collections of particles that have been placed in the same pure quantum state, such as through the use of a Stern–Gerlach apparatus, the spin vector does have a well-defined experimental meaning: It specifies the direction in ordinary space in which a subsequent detector must be oriented in order to achieve the maximum possible probability (100%) of detecting every particle in the collection. For spin-1/2 particles, this maximum probability drops off smoothly as the angle between the spin vector and the detector increases, until at an angle of 180 degrees—that is, for detectors oriented in the opposite direction to the spin vector—the expectation of detecting particles from the collection reaches a minimum of 0%.

As a qualitative concept, the spin vector is often handy because it is easy to picture classically. For instance, quantum mechanical spin can exhibit phenomena analogous to classical gyroscopic effects. For example, one can exert a kind of "torque" on an electron by putting it in a magnetic field (the field acts upon the electron's intrinsic magnetic dipole moment—see the following section). The result is that the spin vector undergoes precession, just like a classical gyroscope. This phenomenon is known as electron spin resonance (ESR). The equivalent behaviour of protons in atomic nuclei is used in nuclear magnetic resonance (NMR) spectroscopy and imaging.

Mathematically, quantum mechanical spin states are described by vector-like objects known as spinors. There are subtle differences between the behavior of spinors and vectors under coordinate rotations. For example, rotating a spin-1/2

A single point in space can spin continuously without becoming tangled. Notice that after a 360 degree rotation, the spiral flips between clockwise and counterclockwise orientations. It returns to its original configuration after spinning a full 720 degrees.

particle by 360 degrees does not bring it back to the same quantum state, but to the state with the opposite quantum phase; this is detectable, in principle, with interference experiments. To return the particle to its exact original state, one needs a 720 degree rotation. (The Plate trick and Mobius strip give non-quantum analogies.) A spin-zero particle can only have a single quantum state, even after torque is applied. Rotating a spin-2 particle 180 degrees can bring it back to the same quantum state and a spin-4 particle should be rotated 90 degrees to bring it back to the same quantum state. The spin 2 particle can be analogous to a straight stick that looks the same even after it is rotated 180 degrees and a spin 0 particle can be imagined as sphere which looks the same after whatever angle it is turned through.

11.4 Mathematical formulation

11.4.1 Operator

Spin obeys commutation relations analogous to those of the orbital angular momentum:

$$[S_i, S_j] = i\hbar\epsilon_{ijk}S_k$$

where ϵ_{ijk} is the Levi-Civita symbol. It follows (as with angular momentum) that the eigenvectors of S^2 and S_z (expressed as kets in the total S basis) are:

$$S^2|s, m\rangle = \hbar^2 s(s + 1)|s, m\rangle$$
$$S_z|s, m\rangle = \hbar m|s, m\rangle.$$

The spin raising and lowering operators acting on these eigenvectors give:

$$S_\pm|s, m\rangle = \hbar\sqrt{s(s + 1) - m(m \pm 1)}|s, m \pm 1\rangle \text{ , where } S_\pm = S_x \pm iS_y.$$

But unlike orbital angular momentum the eigenvectors are not spherical harmonics. They are not functions of θ and φ. There is also no reason to exclude half-integer values of s and m.

In addition to their other properties, all quantum mechanical particles possess an intrinsic spin (though it may have the intrinsic spin 0, too). The spin is quantized in units of the reduced Planck constant, such that the state function of the particle is, say, not $\psi = \psi(\mathbf{r})$, but $\psi = \psi(\mathbf{r}, \sigma)$ where σ is out of the following discrete set of values:

$$\sigma \in \{-s\hbar, -(s - 1)\hbar, \cdots, +(s - 1)\hbar, +s\hbar\}.$$

One distinguishes bosons (integer spin) and fermions (half-integer spin). The total angular momentum conserved in interaction processes is then the *sum* of the orbital angular momentum and the spin.

11.4.2 Pauli matrices

The quantum mechanical operators associated with spin-$\frac{1}{2}$ observables are:

$$\hat{\mathbf{S}} = \frac{\hbar}{2}\boldsymbol{\sigma}$$

where in Cartesian components:

$$S_x = \frac{\hbar}{2}\sigma_x, \quad S_y = \frac{\hbar}{2}\sigma_y, \quad S_z = \frac{\hbar}{2}\sigma_z \,.$$

For the special case of spin-1/2 particles, σx, σy and σz are the three Pauli matrices, given by:

$$\sigma_x = \begin{pmatrix} 0 & 1 \\ 1 & 0 \end{pmatrix} \quad \sigma_y = \begin{pmatrix} 0 & -i \\ i & 0 \end{pmatrix} \quad \sigma_z = \begin{pmatrix} 1 & 0 \\ 0 & -1 \end{pmatrix}.$$

11.4.3 Pauli exclusion principle

For systems of N identical particles this is related to the Pauli exclusion principle, which states that by interchanges of any two of the N particles one must have

$$\psi(\cdots \mathbf{r}_i, \sigma_i \cdots \mathbf{r}_j, \sigma_j \cdots) = (-1)^{2s} \psi(\cdots \mathbf{r}_j, \sigma_j \cdots \mathbf{r}_i, \sigma_i \cdots).$$

Thus, for bosons the prefactor $(-1)^{2s}$ will reduce to $+1$, for fermions to -1. In quantum mechanics all particles are either bosons or fermions. In some speculative relativistic quantum field theories "supersymmetric" particles also exist, where linear combinations of bosonic and fermionic components appear. In two dimensions, the prefactor $(-1)^{2s}$ can be replaced by any complex number of magnitude 1 such as in the Anyon.

The above permutation postulate for N-particle state functions has most-important consequences in daily life, e.g. the periodic table of the chemists or biologists.

11.4.4   Rotations

See also: symmetries in quantum mechanics

As described above, quantum mechanics states that components of angular momentum measured along any direction can only take a number of discrete values. The most convenient quantum mechanical description of particle's spin is therefore with a set of complex numbers corresponding to amplitudes of finding a given value of projection of its intrinsic angular momentum on a given axis. For instance, for a spin 1/2 particle, we would need two numbers $a_{\pm 1/2}$, giving amplitudes of finding it with projection of angular momentum equal to $\hbar/2$ and $-\hbar/2$, satisfying the requirement

$$\left| a_{\frac{1}{2}} \right|^2 + \left| a_{-\frac{1}{2}} \right|^2 = 1.$$

For a generic particle with spin s, we would need $2s + 1$ such parameters. Since these numbers depend on the choice of the axis, they transform into each other non-trivially when this axis is rotated. It's clear that the transformation law must be linear, so we can represent it by associating a matrix with each rotation, and the product of two transformation matrices corresponding to rotations A and B must be equal (up to phase) to the matrix representing rotation AB. Further, rotations preserve the quantum mechanical inner product, and so should our transformation matrices:

$$\sum_{m=-j}^{j} a_m^* b_m = \sum_{m=-j}^{j} \left(\sum_{n=-j}^{j} U_{nm} a_n \right)^* \left(\sum_{k=-j}^{j} U_{km} b_k \right)$$

$$\sum_{n=-j}^{j} \sum_{k=-j}^{j} U_{np}^* U_{kq} = \delta_{pq}.$$

Mathematically speaking, these matrices furnish a unitary projective representation of the rotation group SO(3). Each such representation corresponds to a representation of the covering group of SO(3), which is SU(2).[15] There is one n-dimensional irreducible representation of SU(2) for each dimension, though this representation is n-dimensional real for odd n and n-dimensional complex for even n (hence of real dimension $2n$). For a rotation by angle θ in the plane with normal vector $\hat{\boldsymbol{\theta}}$, U can be written

$$U = e^{-\frac{i}{\hbar} \boldsymbol{\theta} \cdot \mathbf{S}},$$

where $\boldsymbol{\theta} = \theta \hat{\boldsymbol{\theta}}$ is a and $\mathbf{S}$ is the vector of spin operators.

(Click "show" at right to see a proof or "hide" to hide it.)

Working in the coordinate system where $\hat{\theta} = \hat{z}$, we would like to show that S_x and S_y are rotated into each other by the angle θ. Starting with S_x. Using units where $\hbar = 1$:

$$S_x \rightarrow U^\dagger S_x U = e^{i\theta S_z} S_x e^{-i\theta S_z}$$

$$= S_x + (i\theta)[S_z, S_x] + \left(\frac{1}{2!}\right)(i\theta)^2[S_z,[S_z,S_x]] + \left(\frac{1}{3!}\right)(i\theta)^3[S_z,[S_z,[S_z,S_x]]] + \cdots$$

Using the spin operator commutation relations, we see that the commutators evaluate to iS_y for the odd terms in the series, and to S_x for all of the even terms. Thus:

$$U^\dagger S_x U = S_x \left[1 - \frac{\theta^2}{2!} + ...\right] - S_y \left[\theta - \frac{\theta^3}{3!}\cdots\right]$$

$$= S_x \cos\theta - S_y \sin\theta$$

as expected. Note that since we only relied on the spin operator commutation relations, this proof holds for any dimension (i.e., for any principal spin quantum number s).[16]

A generic rotation in 3-dimensional space can be built by compounding operators of this type using Euler angles:

$$\mathcal{R}(\alpha, \beta, \gamma) = e^{-i\alpha S_x} e^{-i\beta S_y} e^{-i\gamma S_z}$$

An irreducible representation of this group of operators is furnished by the Wigner D-matrix:

$$D^s_{m'm}(\alpha, \beta, \gamma) \equiv \langle sm'|\mathcal{R}(\alpha, \beta, \gamma)|sm\rangle = e^{-im'\alpha} d^s_{m'm}(\beta) e^{-im\gamma},$$

where

$$d^s_{m'm}(\beta) = \langle sm'|e^{-i\beta s_y}|sm\rangle$$

is Wigner's small d-matrix. Note that for $\gamma = 2\pi$ and $\alpha = \beta = 0$; i.e., a full rotation about the z-axis, the Wigner D-matrix elements become

$$D^s_{m'm}(0, 0, 2\pi) = d^s_{m'm}(0)e^{-im2\pi} = \delta_{m'm}(-1)^{2m}.$$

Recalling that a generic spin state can be written as a superposition of states with definite m, we see that if s is an integer, the values of m are all integers, and this matrix corresponds to the identity operator. However, if s is a half-integer, the values of m are also all half-integers, giving $(-1)^{2m} = -1$ for all m, and hence upon rotation by 2π the state picks up a minus sign. This fact is a crucial element of the proof of the spin-statistics theorem.

11.4.5 Lorentz transformations

We could try the same approach to determine the behavior of spin under general Lorentz transformations, but we would immediately discover a major obstacle. Unlike SO(3), the group of Lorentz transformations SO(3,1) is non-compact and therefore does not have any faithful, unitary, finite-dimensional representations.

In case of spin 1/2 particles, it is possible to find a construction that includes both a finite-dimensional representation and a scalar product that is preserved by this representation. We associate a 4-component Dirac spinor ψ with each particle. These spinors transform under Lorentz transformations according to the law

$$\psi' = \exp\left(\frac{1}{8}\omega_{\mu\nu}[\gamma_\mu, \gamma_\nu]\right)\psi$$

where γ_μ are gamma matrices and $\omega_{\mu\nu}$ is an antisymmetric 4×4 matrix parametrizing the transformation. It can be shown that the scalar product

$$\langle \psi | \phi \rangle = \bar{\psi}\phi = \psi^\dagger \gamma_0 \phi$$

is preserved. It is not, however, positive definite, so the representation is not unitary.

11.4.6   Metrology along the *x*, *y*, and *z* axes

Each of the (Hermitian) Pauli matrices has two eigenvalues, +1 and −1. The corresponding normalized eigenvectors are:

$$\psi_{x+} = \frac{1}{\sqrt{2}}\begin{pmatrix} 1 \\ 1 \end{pmatrix}, \quad \psi_{x-} = \frac{1}{\sqrt{2}}\begin{pmatrix} 1 \\ -1 \end{pmatrix},$$
$$\psi_{y+} = \frac{1}{\sqrt{2}}\begin{pmatrix} 1 \\ i \end{pmatrix}, \quad \psi_{y-} = \frac{1}{\sqrt{2}}\begin{pmatrix} 1 \\ -i \end{pmatrix},$$
$$\psi_{z+} = \begin{pmatrix} 1 \\ 0 \end{pmatrix}, \quad \psi_{z-} = \begin{pmatrix} 0 \\ 1 \end{pmatrix}.$$

By the postulates of quantum mechanics, an experiment designed to measure the electron spin on the *x*, *y* or *z* axis can only yield an eigenvalue of the corresponding spin operator (Sx, Sy or Sz) on that axis, i.e. $\hbar/2$ or $-\hbar/2$. The quantum state of a particle (with respect to spin), can be represented by a two component spinor:

$$\psi = \begin{pmatrix} a + bi \\ c + di \end{pmatrix}.$$

When the spin of this particle is measured with respect to a given axis (in this example, the *x*-axis), the probability that its spin will be measured as $\hbar/2$ is just $|\langle \psi_{x+} | \psi \rangle|^2$. Correspondingly, the probability that its spin will be measured as $-\hbar/2$ is just $|\langle \psi_{x-} | \psi \rangle|^2$. Following the measurement, the spin state of the particle will collapse into the corresponding eigenstate. As a result, if the particle's spin along a given axis has been measured to have a given eigenvalue, all measurements will yield the same eigenvalue (since $|\langle \psi_{x+} | \psi_{x+} \rangle|^2 = 1$, etc), provided that no measurements of the spin are made along other axes.

11.4.7   Metrology along an arbitrary axis

The operator to measure spin along an arbitrary axis direction is easily obtained from the Pauli spin matrices. Let $u = (ux, uy, uz)$ be an arbitrary unit vector. Then the operator for spin in this direction is simply

$$S_u = \frac{\hbar}{2}\left(u_x \sigma_x + u_y \sigma_y + u_z \sigma_z\right)$$

The operator Su has eigenvalues of $\pm\hbar/2$, just like the usual spin matrices. This method of finding the operator for spin in an arbitrary direction generalizes to higher spin states, one takes the dot product of the direction with a vector of the three operators for the three *x*, *y*, *z* axis directions.

A normalized spinor for spin-1/2 in the (ux, uy, uz) direction (which works for all spin states except spin down where it will give 0/0), is:

$$\frac{1}{\sqrt{2 + 2u_z}}\begin{pmatrix} 1 + u_z \\ u_x + iu_y \end{pmatrix}.$$

The above spinor is obtained in the usual way by diagonalizing the σ_u matrix and finding the eigenstates corresponding to the eigenvalues. In quantum mechanics, vectors are termed "normalized" when multiplied by a normalizing factor, which results in the vector having a length of unity.

11.4.8 Compatibility of metrology

Since the Pauli matrices do not commute, measurements of spin along the different axes are incompatible. This means that if, for example, we know the spin along the x-axis, and we then measure the spin along the y-axis, we have invalidated our previous knowledge of the x-axis spin. This can be seen from the property of the eigenvectors (i.e. eigenstates) of the Pauli matrices that:

$$| \langle \psi_{x\pm} | \psi_{y\pm} \rangle |^2 = | \langle \psi_{x\pm} | \psi_{z\pm} \rangle |^2 = | \langle \psi_{y\pm} | \psi_{z\pm} \rangle |^2 = \frac{1}{2}.$$

So when physicists measure the spin of a particle along the x-axis as, for example, $\hbar/2$, the particle's spin state collapses into the eigenstate $| \psi_{x+} \rangle$. When we then subsequently measure the particle's spin along the y-axis, the spin state will now collapse into either $| \psi_{y+} \rangle$ or $| \psi_{y-} \rangle$, each with probability 1/2. Let us say, in our example, that we measure $-\hbar/2$. When we now return to measure the particle's spin along the x-axis again, the probabilities that we will measure $\hbar/2$ or $-\hbar/2$ are each 1/2 (i.e. they are $| \langle \psi_{x+} | \psi_{y-} \rangle |^2$ and $| \langle \psi_{x-} | \psi_{y-} \rangle |^2$ respectively). This implies that the original measurement of the spin along the x-axis is no longer valid, since the spin along the x-axis will now be measured to have either eigenvalue with equal probability.

11.4.9 Higher spins

The spin-1/2 operator $\mathbf{S} = \hbar/2\boldsymbol{\sigma}$ form the fundamental representation of SU(2). By taking Kronecker products of this representation with itself repeatedly, one may construct all higher irreducible representations. That is, the resulting spin operators for higher spin systems in three spatial dimensions, for arbitrarily large s, can be calculated using this spin operator and ladder operators.

The resulting spin matrices for spin 1 are:

$$S_x = \frac{\hbar}{\sqrt{2}} \begin{pmatrix} 0 & 1 & 0 \\ 1 & 0 & 1 \\ 0 & 1 & 0 \end{pmatrix}$$

$$S_y = \frac{\hbar}{\sqrt{2}} \begin{pmatrix} 0 & -i & 0 \\ i & 0 & -i \\ 0 & i & 0 \end{pmatrix}$$

$$S_z = \hbar \begin{pmatrix} 1 & 0 & 0 \\ 0 & 0 & 0 \\ 0 & 0 & -1 \end{pmatrix}$$

for spin 3/2 they are

$$S_x = \frac{\hbar}{2} \begin{pmatrix} 0 & \sqrt{3} & 0 & 0 \\ \sqrt{3} & 0 & 2 & 0 \\ 0 & 2 & 0 & \sqrt{3} \\ 0 & 0 & \sqrt{3} & 0 \end{pmatrix}$$

$$S_y = \frac{\hbar}{2} \begin{pmatrix} 0 & -i\sqrt{3} & 0 & 0 \\ i\sqrt{3} & 0 & -2i & 0 \\ 0 & 2i & 0 & -i\sqrt{3} \\ 0 & 0 & i\sqrt{3} & 0 \end{pmatrix}$$

$$S_z = \frac{\hbar}{2} \begin{pmatrix} 3 & 0 & 0 & 0 \\ 0 & 1 & 0 & 0 \\ 0 & 0 & -1 & 0 \\ 0 & 0 & 0 & -3 \end{pmatrix}$$

and for spin 5/2 they are

$$S_x = \frac{\hbar}{2} \begin{pmatrix} 0 & \sqrt{5} & 0 & 0 & 0 & 0 \\ \sqrt{5} & 0 & 2\sqrt{2} & 0 & 0 & 0 \\ 0 & 2\sqrt{2} & 0 & 3 & 0 & 0 \\ 0 & 0 & 3 & 0 & 2\sqrt{2} & 0 \\ 0 & 0 & 0 & 2\sqrt{2} & 0 & \sqrt{5} \\ 0 & 0 & 0 & 0 & \sqrt{5} & 0 \end{pmatrix}$$

$$S_y = \frac{\hbar}{2} \begin{pmatrix} 0 & -i\sqrt{5} & 0 & 0 & 0 & 0 \\ i\sqrt{5} & 0 & -2i\sqrt{2} & 0 & 0 & 0 \\ 0 & 2i\sqrt{2} & 0 & -3i & 0 & 0 \\ 0 & 0 & 3i & 0 & -2i\sqrt{2} & 0 \\ 0 & 0 & 0 & 2i\sqrt{2} & 0 & -i\sqrt{5} \\ 0 & 0 & 0 & 0 & i\sqrt{5} & 0 \end{pmatrix}$$

$$S_z = \frac{\hbar}{2} \begin{pmatrix} 5 & 0 & 0 & 0 & 0 & 0 \\ 0 & 3 & 0 & 0 & 0 & 0 \\ 0 & 0 & 1 & 0 & 0 & 0 \\ 0 & 0 & 0 & -1 & 0 & 0 \\ 0 & 0 & 0 & 0 & -3 & 0 \\ 0 & 0 & 0 & 0 & 0 & -5 \end{pmatrix}.$$

The generalization of these matrices for arbitrary s is

$$(S_x)_{ab} = \frac{\hbar}{2}(\delta_{a,b+1} + \delta_{a+1,b})\sqrt{(s+1)(a+b-1) - ab}$$

$$(S_y)_{ab} = \frac{\hbar}{2i}(\delta_{a,b+1} - \delta_{a+1,b})\sqrt{(s+1)(a+b-1) - ab} \quad 1 \le a, b \le 2s+1$$

$$(S_z)_{ab} = \hbar(s+1-a)\delta_{a,b} = \hbar(s+1-b)\delta_{a,b}.$$

Also useful in the quantum mechanics of multiparticle systems, the general Pauli group *Gn* is defined to consist of all *n*-fold tensor products of Pauli matrices.

The analog formula of Euler's formula in terms of the Pauli matrices:

$$e^{i\theta(\hat{\mathbf{n}}\cdot\boldsymbol{\sigma})} = I\cos\theta + i(\hat{\mathbf{n}}\cdot\boldsymbol{\sigma})\sin\theta$$

for higher spins is tractable, but less simple.[17]

11.5   Parity

In tables of the spin quantum number *s* for nuclei or particles, the spin is often followed by a "+" or "−". This refers to the parity with "+" for even parity (wave function unchanged by spatial inversion) and "−" for odd parity (wave function negated by spatial inversion). For example, see the isotopes of bismuth.

11.6   Applications

Spin has important theoretical implications and practical applications. Well-established *direct* applications of spin include:

- Nuclear magnetic resonance (NMR) spectroscopy in chemistry;

- Electron spin resonance spectroscopy in chemistry and physics;

- Magnetic resonance imaging (MRI) in medicine, a type of applied NMR, which relies on proton spin density;

- Giant magnetoresistive (GMR) drive head technology in modern hard disks.

Electron spin plays an important role in magnetism, with applications for instance in computer memories. The manipulation of *nuclear spin* by radiofrequency waves (nuclear magnetic resonance) is important in chemical spectroscopy and medical imaging.

Spin-orbit coupling leads to the fine structure of atomic spectra, which is used in atomic clocks and in the modern definition of the second. Precise measurements of the g-factor of the electron have played an important role in the development and verification of quantum electrodynamics. *Photon spin* is associated with the polarization of light.

An emerging application of spin is as a binary information carrier in spin transistors. The original concept, proposed in 1990, is known as Datta-Das spin transistor.[18] Electronics based on spin transistors are referred to as spintronics. The manipulation of spin in dilute magnetic semiconductor materials, such as metal-doped ZnO or TiO_2 imparts a further degree of freedom and has the potential to facilitate the fabrication of more efficient electronics.[19]

There are many *indirect* applications and manifestations of spin and the associated Pauli exclusion principle, starting with the periodic table of chemistry.

11.7 History

Spin was first discovered in the context of the emission spectrum of alkali metals. In 1924 Wolfgang Pauli introduced what he called a "two-valued quantum degree of freedom" associated with the electron in the outermost shell. This allowed him to formulate the Pauli exclusion principle, stating that no two electrons can share the same quantum state at the same time.

The physical interpretation of Pauli's "degree of freedom" was initially unknown. Ralph Kronig, one of Landé's assistants, suggested in early 1925 that it was produced by the self-rotation of the electron. When Pauli heard about the idea, he criticized it severely, noting that the electron's hypothetical surface would have to be moving faster than the speed of light in order for it to rotate quickly enough to produce the necessary angular momentum. This would violate the theory of relativity. Largely due to Pauli's criticism, Kronig decided not to publish his idea.

In the autumn of 1925, the same thought came to two Dutch physicists, George Uhlenbeck and Samuel Goudsmit at Leiden University. Under the advice of Paul Ehrenfest, they published their results. It met a favorable response, especially after Llewellyn Thomas managed to resolve a factor-of-two discrepancy between experimental results and Uhlenbeck and Goudsmit's calculations (and Kronig's unpublished results). This discrepancy was due to the orientation of the electron's tangent frame, in addition to its position.

Mathematically speaking, a fiber bundle description is needed. The tangent bundle effect is additive and relativistic; that is, it vanishes if c goes to infinity. It is one half of the value obtained without regard for the tangent space orientation, but with opposite sign. Thus the combined effect differs from the latter by a factor two (Thomas precession).

Despite his initial objections, Pauli formalized the theory of spin in 1927, using the modern theory of quantum mechanics invented by Schrödinger and Heisenberg. He pioneered the use of Pauli matrices as a representation of the spin operators, and introduced a two-component spinor wave-function.

Pauli's theory of spin was non-relativistic. However, in 1928, Paul Dirac published the Dirac equation, which described the relativistic electron. In the Dirac equation, a four-component spinor (known as a "Dirac spinor") was used for the electron wave-function. In 1940, Pauli proved the *spin-statistics theorem*, which states that fermions have half-integer spin and bosons integer spin.

In retrospect, the first direct experimental evidence of the electron spin was the Stern–Gerlach experiment of 1922. However, the correct explanation of this experiment was only given in 1927.[20]

Wolfgang Pauli

11.8   See also

- Einstein–de Haas effect

- Spin-orbital

- Chirality (physics)

- Dynamic nuclear polarisation
- Helicity (particle physics)
- Pauli equation
- Pauli–Lubanski pseudovector
- Rarita–Schwinger equation
- Representation theory of SU(2)
- Spin-½
- Spin-flip
- Spin isomers of hydrogen
- Spin tensor
- Spin wave
- Spin engineering
- Yrast
- Zitterbewegung

11.9 References

[1] Merzbacher, Eugen (1998). *Quantum Mechanics* (3rd ed.). pp. 372–3.

[2] Griffiths, David (2005). *Introduction to Quantum Mechanics* (2nd ed.). pp. 183–4.

[3] "Angular Momentum Operator Algebra", class notes by Michael Fowler

[4] *A modern approach to quantum mechanics*, by Townsend, p. 31 and p. 80

[5] Eisberg, Robert; Resnick, Robert (1985). *Quantum Physics of Atoms, Molecules, Solids, Nuclei, and Particles* (2nd ed.). pp. 272–3.

[6] Pauli, Wolfgang(1940)."The Connection Between Spin and Statistics"(PDF).*Phys. Rev* **58**(8): 716–722.Bibcode:1940PhRv...doi:10.1103/PhysRev.58.716.

[7] Physics of Atoms and Molecules, B.H. Bransden, C.J.Joachain, Longman, 1983, ISBN 0-582-44401-2

[8] "CODATA Value: electron g factor". *The NIST Reference on Constants, Units, and Uncertainty*. NIST. 2006. Retrieved 2013-11-15.

[9] R.P. Feynman (1985). "Electrons and Their Interactions". *QED: The Strange Theory of Light and Matter*. Princeton, New Jersey: Princeton University Press. p. 115. ISBN 0-691-08388-6.

> "After some years, it was discovered that this value [$-g/2$] was not exactly 1, but slightly more—something like 1.00116. This correction was worked out for the first time in 1948 by Schwinger as $j*j$ divided by 2 pi [*sic*] [where j is the square root of the fine-structure constant], and was due to an alternative way the electron can go from place to place: instead of going directly from one point to another, the electron goes along for a while and suddenly emits a photon; then (horrors!) it absorbs its own photon."

[10] W.J. Marciano, A.I. Sanda (1977). "Exotic decays of the muon and heavy leptons in gauge theories". *Physics Letters* **B67** (3): 303–305. Bibcode:1977PhLB...67..303M. doi:10.1016/0370-2693(77)90377-X.

[11] B.W. Lee, R.E. Shrock (1977). "Natural suppression of symmetry violation in gauge theories: Muon- and electron-lepton-number nonconservation".*Physical Review***D16**(5): 1444–1473.Bibcode:1977PhRvD..16.1444L.doi:10.1103/PhysRevD.16.

[12] K. Fujikawa, R. E. Shrock (1980). "Magnetic Moment of a Massive Neutrino and Neutrino-Spin Rotation". *Physical Review Letters* **45** (12): 963–966. Bibcode:1980PhRvL..45..963F. doi:10.1103/PhysRevLett.45.963.

[13] N.F. Bell; Cirigliano, V.; Ramsey-Musolf, M.; Vogel, P.; Wise, Mark; et al. (2005). "How Magnetic is the Dirac Neutrino?". *Physical Review Letters* **95** (15): 151802. arXiv:hep-ph/0504134. Bibcode:2005PhRvL..95o1802B. doi:10.1103/PhysRevLett. PMID16241715.

[14] Quanta: A handbook of concepts, P.W. Atkins, Oxford University Press, 1974, ISBN 0-19-855493-1

[15] B.C. Hall (2013). *Quantum Theory for Mathematicians*. Springer. pp. 354–358.

[16] *Modern Quantum Mechanics*, by J. J. Sakurai, p159

[17] Curtright, T L; Fairlie, D B; Zachos, C K (2014). "A compact formula for rotations as spin matrix polynomials". *SIGMA* **10**: 084. doi:10.3842/SIGMA.2014.084.

[18] Datta. S and B. Das (1990). "Electronic analog of the electrooptic modulator". *Applied Physics Letters* **56** (7): 665–667. Bibcode:1990ApPhL..56..665D. doi:10.1063/1.102730.

[19] Assadi, M.H.N; Hanaor, D.A.H (2013). "Theoretical study on copper's energetics and magnetism in TiO_2 polymorphs" (PDF). *Journal of Applied Physics* **113** (23): 233913. doi:10.1063/1.4811539.

[20] B. Friedrich, D. Herschbach (2003). "Stern and Gerlach: How a Bad Cigar Helped Reorient Atomic Physics". *Physics Today* **56** (12): 53. Bibcode:2003PhT....56l..53F. doi:10.1063/1.1650229.

11.10   Further reading

- Cohen-Tannoudji, Claude; Diu, Bernard; Laloë, Franck (2006). *Quantum Mechanics* (2 volume set ed.). John Wiley & Sons. ISBN 978-0-471-56952-7.

- Condon, E. U.; Shortley, G. H. (1935). "Especially Chapter 3". *The Theory of Atomic Spectra*. Cambridge University Press. ISBN 0-521-09209-4.

- Hipple, J. A.; Sommer, H.; Thomas, H.A. (1949). *A precise method of determining the faraday by magnetic resonance*. doi:10.1103/PhysRev.76.1877.2.https://www.academia.edu/6483539/John_A._Hipple_1911-1985_technology_as_knowledge

- Edmonds, A. R. (1957). *Angular Momentum in Quantum Mechanics*. Princeton University Press. ISBN 0-691-07912-9.

- Jackson, John David (1998). *Classical Electrodynamics* (3rd ed.). John Wiley & Sons. ISBN 978-0-471-30932-1.

- Serway, Raymond A.; Jewett, John W. (2004). *Physics for Scientists and Engineers* (6th ed.). Brooks/Cole. ISBN 0-534-40842-7.

- Thompson, William J. (1994). *Angular Momentum: An Illustrated Guide to Rotational Symmetries for Physical Systems*. Wiley. ISBN 0-471-55264-X.

- Tipler, Paul (2004). *Physics for Scientists and Engineers: Mechanics, Oscillations and Waves, Thermodynamics* (5th ed.). W. H. Freeman. ISBN 0-7167-0809-4.

- Sin-Itiro Tomonaga, The Story of Spin, 1997

11.11 External links

- "Spintronics. Feature Article" in *Scientific American*, June 2002.

- Goudsmit on the discovery of electron spin.

- *Nature*: "Milestones in 'spin' since 1896."

- ECE 495N Lecture 36: Spin Online lecture by S. Datta

Chapter 12

Superfluidity

Superfluidity is a state of matter in which the matter behaves like a fluid with zero viscosity; where it appears to exhibit the ability to self-propel and travel in a way that defies the forces of gravity and surface tension. Superfluidity is found in astrophysics, high-energy physics, and theories of quantum gravity. The phenomenon is related to Bose–Einstein condensation, but neither is a specific type of the other: not all Bose-Einstein condensates can be regarded as superfluids, and not all superfluids are Bose–Einstein condensates.

12.1 Superfluidity of liquid helium

Main article: Superfluid helium-4

Superfluidity was originally discovered in liquid helium, by Pyotr Kapitsa and John F. Allen. It has since been described through phenomenology and microscopic theories. In liquid helium-4, the superfluidity occurs at far higher temperatures than it does in helium-3. Each atom of helium-4 is a boson particle, by virtue of its integer spin. A helium-3 atom is a fermion particle; it can form bosons only by pairing with itself at much lower temperatures. This process is similar to the electron pairing in superconductivity.

12.2 Ultracold atomic gases

Superfluidity in an ultracold fermionic gas was experimentally proven by Wolfgang Ketterle and his team who observed quantum vortices in ^{6}Li at a temperature of 50 nK at MIT in April 2005.[1][2] Such vortices had previously been observed in an ultracold bosonic gas using ^{87}Rb in 2000,[3] and more recently in two-dimensional gases.[4] As early as 1999 Lene Hau created such a condensate using sodium atoms[5] for the purpose of slowing light, and later stopping it completely.[6] Her team then subsequently used this system of compressed light[7] to generate the superfluid analogue of shock waves and tornadoes: "These dramatic excitations result in the formation of solitons that in turn decay into quantized vortices—created far out of equilibrium, in pairs of opposite circulation—revealing directly the process of superfluid breakdown in Bose-Einstein condensates. With a double light-roadblock setup, we can generate controlled collisions between shock waves resulting in completely unexpected, nonlinear excitations. We have observed hybrid structures consisting of vortex rings embedded in dark solitonic shells. The vortex rings act as 'phantom propellers' leading to very rich excitation dynamics."[8]

12.3 Superfluid in astrophysics

The idea that superfluidity exists inside neutron stars was first proposed by Arkady Migdal.[9][10] By analogy with electrons inside superconductors forming Cooper pairs due to electron-lattice interaction, it is expected that nucleons in a neutron star at sufficiently high density and low temperature can also form Cooper pairs due to the long-range attractive nuclear force and lead to superfluidity and superconductivity.[11]

12.4 Superfluidity in high-energy physics and quantum gravity

Main article: Superfluid vacuum theory

Superfluid vacuum theory (SVT) is an approach in theoretical physics and quantum mechanics where the physical vacuum is viewed as superfluid.

The ultimate goal of the approach is to develop scientific models that unify quantum mechanics (describing three of the four known fundamental interactions) with gravity. This makes SVT a candidate for the theory of quantum gravity and an extension of the Standard Model.

It is hoped that development of such theory would unify into a single consistent model of all fundamental interactions, and to describe all known interactions and elementary particles as different manifestations of the same entity, superfluid vacuum.

12.5 See also

- Boojum (superfluidity)

- Condensed matter physics

- Macroscopic quantum phenomena

- Quantum hydrodynamics

- Slow light

- Supersolid

12.6 References

[1] "MIT physicists create new form of matter". Retrieved November 22, 2010.

[2] Grimm, R. (2005). "Low-temperature physics: A quantum revolution". *Nature* **435** (7045): 1035–1036. doi:10.1038/4351035a. PMID 15973388.

[3] Madison, K.; Chevy, F.; Wohlleben, W.; Dalibard, J. (2000). "Vortex Formation in a Stirred Bose-Einstein Condensate". *Physical Review Letters* **84** (5): 806–809. arXiv:cond-mat/9912015. Bibcode:2000PhRvL..84..806M. doi:10.1103/PhysRevLett.84. 806.PMID11017378.

[4] Burnett, K. (2007). "Atomic physics: Cold gases venture into Flatland". *Nature Physics* **3** (9): 589. Bibcode:2007NatPh...3..589B. doi:10.1038/nphys704.

[5] Hau, L. V.; Harris, S. E.; Dutton, Z.; Behroozi, C. H. (1999). "Light speed reduction to 17 metres per second in an ultracold atomic gas". *Nature* **397** (6720): 594–598. doi:10.1038/17561.

[6] "Lene Hau". Physicscentral.com. Retrieved 2013-02-10.

[7] Lene Vestergaard Hau (2003). "Frozen Light" (PDF). *Scientific American*: 44–51.

[8] Shocking Bose-Einstein Condensates with Slow Light

[9] A. B. Migdal (1959). "Superfluidity and the moments of inertia of nuclei". *Nucl. Phys.* **13**(5): 655–674. Bibcode:1959NucPh.. doi:10.1016/0029-5582(59)90264-0.

[10] A. B. Migdal (1960). *Soviet Phys. JETP* **10**: 176. Missing or empty |title= (help)

[11] U. Lombardo & H.-J. Schulze (2001). "Superfluidity in Neutron Star Matter". *Physics of Neutron Star Interiors*. Lecture Notes in Physics **578**. Springer. pp. 30–53. arXiv:astro-ph/0012209. doi:10.1007/3-540-44578-1_2.

12.7   Further reading

- Guénault, Antony M. (2003). *Basic superfluids*. London: Taylor & Francis. ISBN 0-7484-0891-6.

- Annett, James F. (2005). *Superconductivity, superfluids, and condensates*. Oxford: Oxford Univ. Press. ISBN 978-0-19-850756-7.

- Volovik, G. E. (2003). *The Universe in a helium droplet*. Int. Ser. Monogr. Phys. **117**. pp. 1–507. ISBN 978-0198507826.

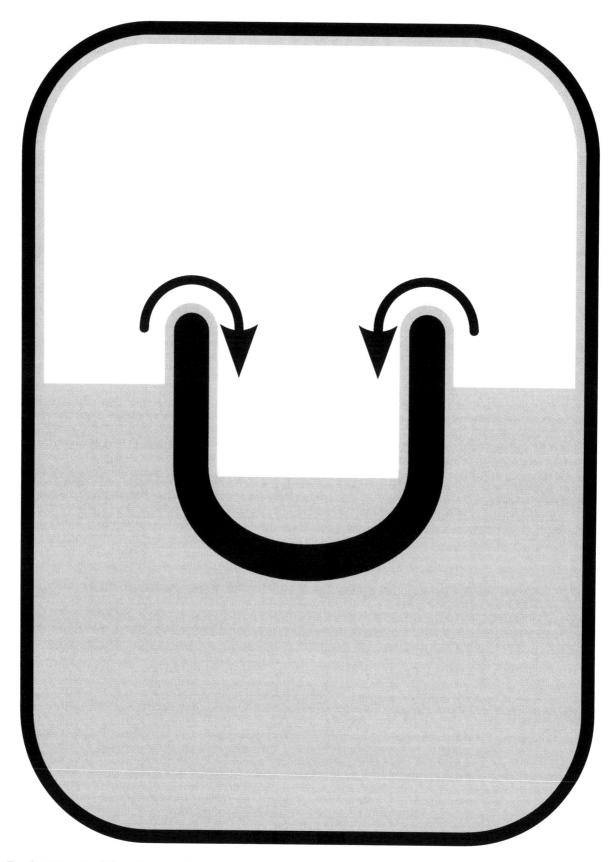

Fig. 1. Helium II will "creep" along surfaces in order to find its own level—after a short while, the levels in the two containers will equalize. The Rollin film also covers the interior of the larger container; if it were not sealed, the helium II would creep out and escape.

Fig. 2. *The liquid helium is in the superfluid phase. As long as it remains superfluid, it creeps up the wall of the cup as a thin film. It comes down on the outside, forming a drop which will fall into the liquid below. Another drop will form—and so on—until the cup is empty.*

Chapter 13

Superconductivity

A magnet levitating above a high-temperature superconductor, cooled with liquid nitrogen. Persistent electric current flows on the surface of the superconductor, acting to exclude the magnetic field of the magnet (Faraday's law of induction). This current effectively forms an electromagnet that repels the magnet.

Superconductivity is a phenomenon of exactly zero electrical resistance and expulsion of magnetic fields occurring in certain materials when cooled below a characteristic critical temperature. It was discovered by Dutch physicist Heike Kamerlingh Onnes on April 8, 1911 in Leiden. Like ferromagnetism and atomic spectral lines, superconductivity is a quantum mechanical phenomenon. It is characterized by the Meissner effect, the complete ejection of magnetic field lines from the interior of the superconductor as it transitions into the superconducting state. The occurrence of the Meissner effect indicates that superconductivity cannot be understood simply as the idealization of *perfect conductivity* in classical

111

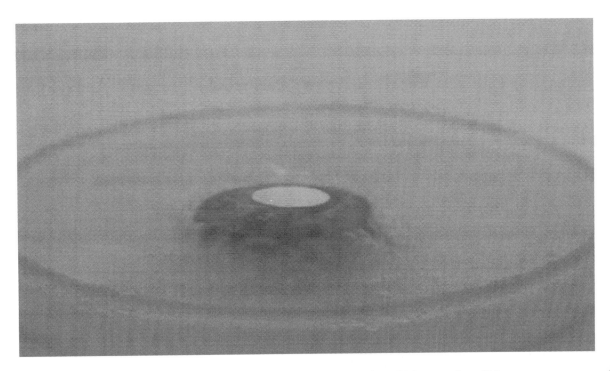

Video of a Meissner effect in a high temperature superconductor (black pellet) with a NdFeB magnet (metallic)

A high-temperature superconductor levitating above a magnet

physics.

The electrical resistivity of a metallic conductor decreases gradually as temperature is lowered. In ordinary conductors, such as copper or silver, this decrease is limited by impurities and other defects. Even near absolute zero, a real sample of a normal conductor shows some resistance. In a superconductor, the resistance drops abruptly to zero when the material is cooled below its critical temperature. An electric current flowing through a loop of superconducting wire can persist indefinitely with no power source.[1][2][3][4]

In 1986, it was discovered that some cuprate-perovskite ceramic materials have a critical temperature above 90 K (−183 °C).[5] Such a high transition temperature is theoretically impossible for a conventional superconductor, leading the materials to be termed high-temperature superconductors. Liquid nitrogen boils at 77 K, and superconduction at higher temperatures than this facilitates many experiments and applications that are less practical at lower temperatures.

13.1 Classification

Main article: Superconductor classification

There are many criteria by which superconductors are classified. The most common are:

- **Response to a magnetic field**: A superconductor can be *Type I*, meaning it has a single critical field, above which all superconductivity is lost; or *Type II*, meaning it has two critical fields, between which it allows partial penetration of the magnetic field.

- **By theory of operation**: It is *conventional* if it can be explained by the BCS theory or its derivatives, or *unconventional*, otherwise.

- **By critical temperature**: A superconductor is generally considered *high temperature* if it reaches a superconducting state when cooled using liquid nitrogen – that is, at only $T_c > 77$ K) – or *low temperature* if more aggressive cooling techniques are required to reach its critical temperature.

- **By material**: Superconductor material classes include chemical elements (e.g. mercury or lead), alloys (such as niobium-titanium, germanium-niobium, and niobium nitride), ceramics (YBCO and magnesium diboride), or organic superconductors (fullerenes and carbon nanotubes; though perhaps these examples should be included among the chemical elements, as they are composed entirely of carbon).

13.2 Elementary properties of superconductors

Most of the physical properties of superconductors vary from material to material, such as the heat capacity and the critical temperature, critical field, and critical current density at which superconductivity is destroyed.

On the other hand, there is a class of properties that are independent of the underlying material. For instance, all superconductors have *exactly* zero resistivity to low applied currents when there is no magnetic field present or if the applied field does not exceed a critical value. The existence of these "universal" properties implies that superconductivity is a thermodynamic phase, and thus possesses certain distinguishing properties which are largely independent of microscopic details.

13.2.1 Zero electrical DC resistance

The simplest method to measure the electrical resistance of a sample of some material is to place it in an electrical circuit in series with a current source I and measure the resulting voltage V across the sample. The resistance of the sample is given by Ohm's law as $R = V / I$. If the voltage is zero, this means that the resistance is zero.

Superconductors are also able to maintain a current with no applied voltage whatsoever, a property exploited in superconducting electromagnets such as those found inMRI machines.
Experiments have demonstrated that currents in superconducting

Electric cables for accelerators at CERN. Both the massive and slim cables are rated for 12,500 A. Top: *conventional cables for LEP;* bottom: *superconductor-based cables for the LHC*

coils can persist for years without any measurable degradation. Experimental evidence points to a current lifetime of at least 100,000 years. Theoretical estimates for the lifetime of a persistent current can exceed the estimated lifetime of the universe, depending on the wire geometry and the temperature.[3]

In a normal conductor, an electric current may be visualized as a fluid of electrons moving across a heavy ionic lattice. The electrons are constantly colliding with the ions in the lattice, and during each collision some of the energy carried by the current is absorbed by the lattice and converted into heat, which is essentially the vibrational kinetic energy of the lattice ions. As a result, the energy carried by the current is constantly being dissipated. This is the phenomenon of electrical resistance.

The situation is different in a superconductor. In a conventional superconductor, the electronic fluid cannot be resolved into individual electrons. Instead, it consists of bound *pairs* of electrons known as Cooper pairs. This pairing is caused by an attractive force between electrons from the exchange of phonons. Due to quantum mechanics, the energy spectrum of this Cooper pair fluid possesses an *energy gap*, meaning there is a minimum amount of energy ΔE that must be supplied in order to excite the fluid. Therefore, if ΔE is larger than the thermal energy of the lattice, given by kT, where k is Boltzmann's constant and T is the temperature, the fluid will not be scattered by the lattice. The Cooper pair fluid is thus a superfluid, meaning it can flow without energy dissipation.

In a class of superconductors known as type II superconductors, including all known high-temperature superconductors, an extremely small amount of resistivity appears at temperatures not too far below the nominal superconducting transition when an electric current is applied in conjunction with a strong magnetic field, which may be caused by the electric current. This is due to the motion of magnetic vortices in the electronic superfluid, which dissipates some of the energy carried by the current. If the current is sufficiently small, the vortices are stationary, and the resistivity vanishes. The resistance due to this effect is tiny compared with that of non-superconducting materials, but must be taken into account in sensitive experiments. However, as the temperature decreases far enough below the nominal superconducting transition, these vortices can become frozen into a disordered but stationary phase known as a "vortex glass". Below this vortex glass transition temperature, the resistance of the material becomes truly zero.

13.2.2 Superconducting phase transition

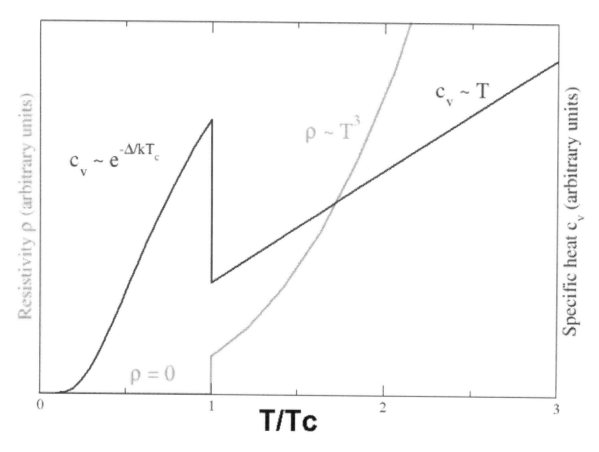

Behavior of heat capacity (cv, blue) and resistivity (ϱ, green) at the superconducting phase transition

In superconducting materials, the characteristics of superconductivity appear when the temperature T is lowered below a **critical temperature** Tc. The value of this critical temperature varies from material to material. Conventional superconductors usually have critical temperatures ranging from around 20 K to less than 1 K. Solid mercury, for example, has a critical temperature of 4.2 K. As of 2009, the highest critical temperature found for a conventional superconductor is 39 K for magnesium diboride (MgB_2),[6][7] although this material displays enough exotic properties that there is some doubt about classifying it as a "conventional" superconductor.[8] Cuprate superconductors can have much higher critical temperatures: $YBa_2Cu_3O_7$, one of the first cuprate superconductors to be discovered, has a critical temperature of 92 K, and mercury-based cuprates have been found with critical temperatures in excess of 130 K. The explanation for these high critical temperatures remains unknown. Electron pairing due to phonon exchanges explains superconductivity in conventional superconductors, but it does not explain superconductivity in the newer superconductors that have a very high critical temperature.

Similarly, at a fixed temperature below the critical temperature, superconducting materials cease to superconduct when an external magnetic field is applied which is greater than the *critical magnetic field*. This is because the Gibbs free energy of the superconducting phase increases quadratically with the magnetic field while the free energy of the normal phase is roughly independent of the magnetic field. If the material superconducts in the absence of a field, then the superconducting phase free energy is lower than that of the normal phase and so for some finite value of the magnetic field (proportional to the square root of the difference of the free energies at zero magnetic field) the two free energies will be equal and a phase transition to the normal phase will occur. More generally, a higher temperature and a stronger magnetic field lead to a smaller fraction of the electrons in the superconducting band and consequently a longer London penetration depth of external magnetic fields and currents. The penetration depth becomes infinite at the phase transition.

The onset of superconductivity is accompanied by abrupt changes in various physical properties, which is the hallmark of a phase transition. For example, the electronic heat capacity is proportional to the temperature in the normal (non-superconducting) regime. At the superconducting transition, it suffers a discontinuous jump and thereafter ceases to be linear. At low temperatures, it varies instead as $e^{-\alpha/T}$ for some constant, α. This exponential behavior is one of the pieces of evidence for the existence of the energy gap.

The order of the superconducting phase transition was long a matter of debate. Experiments indicate that the transition is second-order, meaning there is no latent heat. However, in the presence of an external magnetic field there is latent heat, because the superconducting phase has a lower entropy below the critical temperature than the normal phase. It has been experimentally demonstrated[9] that, as a consequence, when the magnetic field is increased beyond the critical field, the resulting phase transition leads to a decrease in the temperature of the superconducting material.

Calculations in the 1970s suggested that it may actually be weakly first-order due to the effect of long-range fluctuations in the electromagnetic field. In the 1980s it was shown theoretically with the help of a disorder field theory, in which the vortex lines of the superconductor play a major role, that the transition is of second order within the type II regime and of first order (i.e., latent heat) within the type I regime, and that the two regions are separated by a tricritical point.[10] The results were strongly supported by Monte Carlo computer simulations.[11]

13.2.3   Meissner effect

Main article: Meissner effect

When a superconductor is placed in a weak external magnetic field **H**, and cooled below its transition temperature, the magnetic field is ejected. The Meissner effect does not cause the field to be completely ejected but instead the field penetrates the superconductor but only to a very small distance, characterized by a parameter λ, called the London penetration depth, decaying exponentially to zero within the bulk of the material. The Meissner effect is a defining characteristic of superconductivity. For most superconductors, the London penetration depth is on the order of 100 nm.

The Meissner effect is sometimes confused with the kind of diamagnetism one would expect in a perfect electrical conductor: according to Lenz's law, when a *changing* magnetic field is applied to a conductor, it will induce an electric current in the conductor that creates an opposing magnetic field. In a perfect conductor, an arbitrarily large current can be induced, and the resulting magnetic field exactly cancels the applied field.

The Meissner effect is distinct from this—it is the spontaneous expulsion which occurs during transition to superconductivity. Suppose we have a material in its normal state, containing a constant internal magnetic field. When the material is cooled below the critical temperature, we would observe the abrupt expulsion of the internal magnetic field, which we would not expect based on Lenz's law.

The Meissner effect was given a phenomenological explanation by the brothers Fritz and Heinz London, who showed that the electromagnetic free energy in a superconductor is minimized provided

$$\nabla^2 \mathbf{H} = \lambda^{-2} \mathbf{H}$$

where **H** is the magnetic field and λ is the London penetration depth.

This equation, which is known as the London equation, predicts that the magnetic field in a superconductor decays exponentially from whatever value it possesses at the surface.

A superconductor with little or no magnetic field within it is said to be in the Meissner state. The Meissner state breaks down when the applied magnetic field is too large. Superconductors can be divided into two classes according to how this breakdown occurs. In Type I superconductors, superconductivity is abruptly destroyed when the strength of the applied field rises above a critical value *Hc*. Depending on the geometry of the sample, one may obtain an intermediate state[12] consisting of a baroque pattern[13] of regions of normal material carrying a magnetic field mixed with regions of superconducting material containing no field. In Type II superconductors, raising the applied field past a critical value Hc_1 leads to a mixed state (also known as the vortex state) in which an increasing amount of magnetic flux penetrates the material, but there remains no resistance to the flow of electric current as long as the current is not too large. At a second critical field strength Hc_2, superconductivity is destroyed. The mixed state is actually caused by vortices in the electronic superfluid, sometimes called fluxons because the flux carried by these vortices is quantized. Most pure elemental superconductors, except niobium and carbon nanotubes, are Type I, while almost all impure and compound superconductors are Type II.

13.2.4 London moment

Conversely, a spinning superconductor generates a magnetic field, precisely aligned with the spin axis. The effect, the London moment, was put to good use in Gravity Probe B. This experiment measured the magnetic fields of four superconducting gyroscopes to determine their spin axes. This was critical to the experiment since it is one of the few ways to accurately determine the spin axis of an otherwise featureless sphere.

13.3 History of superconductivity

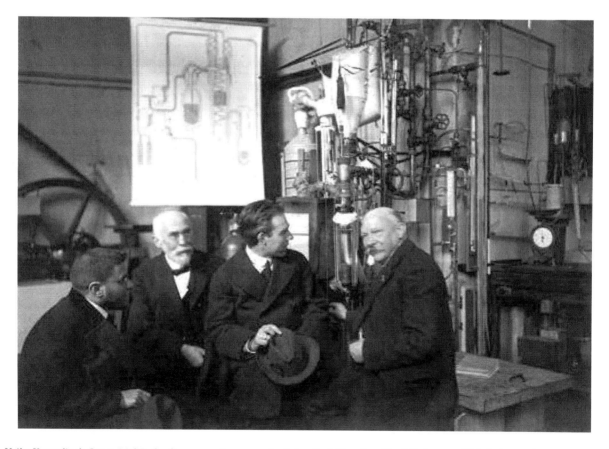

Heike Kamerlingh Onnes (right), the discoverer of superconductivity. Paul Ehrenfest, Hendrik Lorentz, Niels Bohr stand to his left.

Main article: History of superconductivity

Superconductivity was discovered on April 8, 1911 by Heike Kamerlingh Onnes, who was studying the resistance of solid mercury at cryogenic temperatures using the recently produced liquid helium as a refrigerant. At the temperature of

4.2 K, he observed that the resistance abruptly disappeared.[14] In the same experiment, he also observed the superfluid transition of helium at 2.2 K, without recognizing its significance. The precise date and circumstances of the discovery were only reconstructed a century later, when Onnes's notebook was found.[15] In subsequent decades, superconductivity was observed in several other materials. In 1913, lead was found to superconduct at 7 K, and in 1941 niobium nitride was found to superconduct at 16 K.

Great efforts have been devoted to finding out how and why superconductivity works; the important step occurred in 1933, when Meissner and Ochsenfeld discovered that superconductors expelled applied magnetic fields, a phenomenon which has come to be known as the Meissner effect.[16] In 1935, Fritz and Heinz London showed that the Meissner effect was a consequence of the minimization of the electromagnetic free energy carried by superconducting current.[17]

13.3.1   London theory

The first phenomenological theory of superconductivity was London theory. It was put forward by the brothers Fritz and Heinz London in 1935, shortly after the discovery that magnetic fields are expelled from superconductors. A major triumph of the equations of this theory is their ability to explain the Meissner effect,[18] wherein a material exponentially expels all internal magnetic fields as it crosses the superconducting threshold. By using the London equation, one can obtain the dependence of the magnetic field inside the superconductor on the distance to the surface.[19]

There are two London equations:

$$\frac{\partial \mathbf{j}_s}{\partial t} = \frac{n_s e^2}{m} \mathbf{E}, \qquad \nabla \times \mathbf{j}_s = -\frac{n_s e^2}{m} \mathbf{B}.$$

The first equation follows from Newton's second law for superconducting electrons.

13.3.2   Conventional theories (1950s)

During the 1950s, theoretical condensed matter physicists arrived at a solid understanding of "conventional" superconductivity, through a pair of remarkable and important theories: the phenomenological Ginzburg-Landau theory (1950) and the microscopic BCS theory (1957).[20][21]

In 1950, the phenomenological Ginzburg-Landau theory of superconductivity was devised by Landau and Ginzburg.[22] This theory, which combined Landau's theory of second-order phase transitions with a Schrödinger-like wave equation, had great success in explaining the macroscopic properties of superconductors. In particular, Abrikosov showed that Ginzburg-Landau theory predicts the division of superconductors into the two categories now referred to as Type I and Type II. Abrikosov and Ginzburg were awarded the 2003 Nobel Prize for their work (Landau had received the 1962 Nobel Prize for other work, and died in 1968). The four-dimensional extension of the Ginzburg-Landau theory, the Coleman-Weinberg model, is important in quantum field theory and cosmology.

Also in 1950, Maxwell and Reynolds *et al.* found that the critical temperature of a superconductor depends on the isotopic mass of the constituent element.[23][24] This important discovery pointed to the electron-phonon interaction as the microscopic mechanism responsible for superconductivity.

The complete microscopic theory of superconductivity was finally proposed in 1957 by Bardeen, Cooper and Schrieffer.[21] This BCS theory explained the superconducting current as a superfluid of Cooper pairs, pairs of electrons interacting through the exchange of phonons. For this work, the authors were awarded the Nobel Prize in 1972.

The BCS theory was set on a firmer footing in 1958, when N. N. Bogolyubov showed that the BCS wavefunction, which had originally been derived from a variational argument, could be obtained using a canonical transformation of the electronic Hamiltonian.[25] In 1959, Lev Gor'kov showed that the BCS theory reduced to the Ginzburg-Landau theory close to the critical temperature.[26][27]

Generalizations of BCS theory for conventional superconductors form the basis for understanding of the phenomenon of superfluidity, because they fall into the lambda transition universality class. The extent to which such generalizations can be applied to unconventional superconductors is still controversial.

13.3.3 Further history

The first practical application of superconductivity was developed in 1954 with Dudley Allen Buck's invention of the cryotron.[28] Two superconductors with greatly different values of critical magnetic field are combined to produce a fast, simple, switch for computer elements.

Soon after discovering superconductivity in 1911, Kamerlingh Onnes attempted to make an electromagnet with superconducting windings but found that relatively low magnetic fields destroyed superconductivity in the materials he investigated. Much later, in 1955, G.B. Yntema [29] succeeded in constructing a small 0.7-tesla iron-core electromagnet with superconducting niobium wire windings. Then, in 1961, J.E. Kunzler, E. Buehler, F.S.L. Hsu, and J.H. Wernick [30] made the startling discovery that, at 4.2 degrees kelvin, a compound consisting of three parts niobium and one part tin, was capable of supporting a current density of more than 100,000 amperes per square centimeter in a magnetic field of 8.8 tesla. Despite being brittle and difficult to fabricate, niobium-tin has since proved extremely useful in supermagnets generating magnetic fields as high as 20 tesla. In 1962 T.G. Berlincourt and R.R. Hake [31][32] discovered that alloys of niobium and titanium are suitable for applications up to 10 tesla. Promptly thereafter, commercial production of niobium-titanium supermagnet wire commenced at Westinghouse Electric Corporation and at Wah Chang Corporation. Although niobium-titanium boasts less-impressive superconducting properties than those of niobium-tin, niobium-titanium has, nevertheless, become the most widely-used "workhorse" supermagnet material, in large measure a consequence of its very-high ductility and ease of fabrication. However, both niobium-tin and niobium-titanium find wide application in MRI medical imagers, bending and focusing magnets for enormous high-energy-particle accelerators, and a host of other applications. Conectus, a European superconductivity consortium, estimated that in 2014, global economic activity for which superconductivity was indispensable amounted to about five billion euros, with MRI systems accounting for about 80% of that total.

In 1962, Josephson made the important theoretical prediction that a supercurrent can flow between two pieces of superconductor separated by a thin layer of insulator.[33] This phenomenon, now called the Josephson effect, is exploited by superconducting devices such as SQUIDs. It is used in the most accurate available measurements of the magnetic flux quantum $\Phi_0 = h/(2e)$, where h is the Planck constant. Coupled with the quantum Hall resistivity, this leads to a precise measurement of the Planck constant. Josephson was awarded the Nobel Prize for this work in 1973.

In 2008, it was proposed that the same mechanism that produces superconductivity could produce a superinsulator state in some materials, with almost infinite electrical resistance.[34]

13.4 High-temperature superconductivity

Main article: High-temperature superconductivity

Until 1986, physicists had believed that BCS theory forbade superconductivity at temperatures above about 30 K. In that year, Bednorz and Müller discovered superconductivity in a lanthanum-based cuprate perovskite material, which had a transition temperature of 35 K (Nobel Prize in Physics, 1987).[5] It was soon found that replacing the lanthanum with yttrium (i.e., making YBCO) raised the critical temperature to 92 K.[35]

This temperature jump is particularly significant, since it allows liquid nitrogen as a refrigerant, replacing liquid helium.[35] This can be important commercially because liquid nitrogen can be produced relatively cheaply, even on-site. Also, the higher temperatures help avoid some of the problems that arise at liquid helium temperatures, such as the formation of plugs of frozen air that can block cryogenic lines and cause unanticipated and potentially hazardous pressure buildup.[36][37]

Many other cuprate superconductors have since been discovered, and the theory of superconductivity in these materials is one of the major outstanding challenges of theoretical condensed matter physics.[38] There are currently two main hypotheses – the resonating-valence-bond theory, and spin fluctuation which has the most support in the research community.[39] The second hypothesis proposed that electron pairing in high-temperature superconductors is mediated by short-range spin waves known as paramagnons.[40][41]

Since about 1993, the highest temperature superconductor was a ceramic material consisting of mercury, barium, calcium, copper and oxygen ($HgBa_2Ca_2Cu_3O_{8+\delta}$) with T_c = 133–138 K.[42][43] The latter experiment (138 K) still awaits experimental confirmation, however.

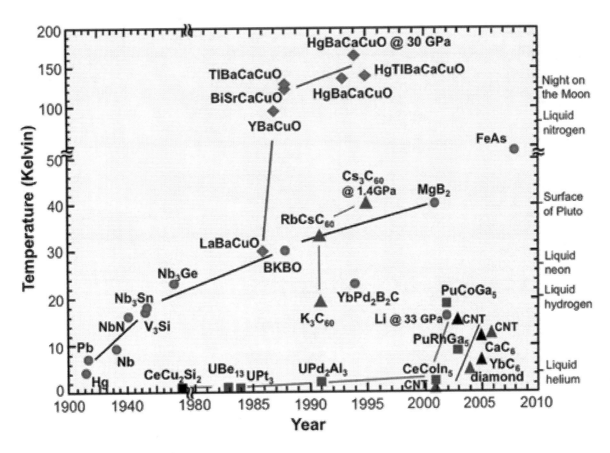

Timeline of superconducting materials

In February 2008, an iron-based family of high-temperature superconductors was discovered.[44][45] Hideo Hosono, of the Tokyo Institute of Technology, and colleagues found lanthanum oxygen fluorine iron arsenide ($LaO_{1-x}F_xFeAs$), an oxypnictide that superconducts below 26 K. Replacing the lanthanum in $LaO_{1-x}F_xFeAs$ with samarium leads to super-conductors that work at 55 K.[46]

In May 2014, hydrogen sulfide (H
2S) was predicted to be a high-temperature superconductor with a transition temperate of 80 at 160 gigapascals.[47] In 2015, H
2S has been observed to exhibit superconductivity at below 203 K but at extremely high pressures — around 150 gigapascals.[48]

13.5 Applications

Main article: Technological applications of superconductivity
Superconducting magnets are some of the most powerful electromagnets known. They are used in MRI/NMR machines, mass spectrometers, and the beam-steering magnets used in particle accelerators. They can also be used for magnetic separation, where weakly magnetic particles are extracted from a background of less or non-magnetic particles, as in the pigment industries.

In the 1950s and 1960s, superconductors were used to build experimental digital computers using cryotron switches. More recently, superconductors have been used to make digital circuits based on rapid single flux quantum technology and RF and microwave filters for mobile phone base stations.

Superconductors are used to build Josephson junctions which are the building blocks of SQUIDs (superconducting quantum interference devices), the most sensitive magnetometers known. SQUIDs are used in scanning SQUID microscopes

Video of superconducting levitation of YBCO

and magnetoencephalography. Series of Josephson devices are used to realize the SI volt. Depending on the particular mode of operation, a superconductor-insulator-superconductor Josephson junction can be used as a photon detector or as a mixer. The large resistance change at the transition from the normal- to the superconducting state is used to build thermometers in cryogenic micro-calorimeter photon detectors. The same effect is used in ultrasensitive bolometers made from superconducting materials.

Other early markets are arising where the relative efficiency, size and weight advantages of devices based on high-temperature superconductivity outweigh the additional costs involved. For example in wind turbines the lower weight and volume of superconducting generators could lead to savings in construction and tower costs, offsetting the higher costs for the generator and lowering the total LCOE.[49]

Promising future applications include high-performance smart grid, electric power transmission, transformers, power storage devices, electric motors (e.g. for vehicle propulsion, as in vactrains or maglev trains), magnetic levitation devices, fault current limiters, enhancing spintronic devices with superconducting materials,[50] and superconducting magnetic refrigeration. However, superconductivity is sensitive to moving magnetic fields so applications that use alternating current (e.g. transformers) will be more difficult to develop than those that rely upon direct current. Compared to to traditional power lines superconducting transmission lines are more efficient and require only a fraction of the space, which would not only lead to a better environmental performance but could also improve public acceptance for expansion of the electric grid.[51]

13.6   Nobel Prizes for superconductivity

- Heike Kamerlingh Onnes (1913), "for his investigations on the properties of matter at low temperatures which led, inter alia, to the production of liquid helium"

- John Bardeen, Leon N. Cooper, and J. Robert Schrieffer (1972), "for their jointly developed theory of superconductivity, usually called the BCS-theory"

- Leo Esaki, Ivar Giaever, and Brian D. Josephson (1973), "for their experimental discoveries regarding tunneling phenomena in semiconductors and superconductors, respectively," and "for his theoretical predictions of the properties of a supercurrent through a tunnel barrier, in particular those phenomena which are generally known as the Josephson effects"

- Georg Bednorz and K. Alex Müller (1987), "for their important break-through in the discovery of superconductivity in ceramic materials"

- Alexei A. Abrikosov, Vitaly L. Ginzburg, and Anthony J. Leggett (2003), "for pioneering contributions to the theory of superconductors and superfluids"[52]

13.7   See also

- Andreev reflection
- Charge transfer complex
- Color superconductivity in quarks
- Composite Reaction Texturing
- Conventional superconductor
- Covalent superconductors
- Flux pumping
- High-temperature superconductivity
- Homes's law
- Iron-based superconductor
- Kondo effect
- List of superconductors
- Little-Parks effect
- Magnetic levitation
- Macroscopic quantum phenomena
- Magnetic sail
- National Superconducting Cyclotron Laboratory
- Oxypnictide
- Persistent current
- Proximity effect
- Room-temperature superconductor
- Rutherford cable
- Spallation Neutron Source

- Superconducting RF

- Superconductor classification

- Superfluid film

- Superfluidity

- Superstripes

- Technological applications of superconductivity

- Timeline of low-temperature technology

- Type-I superconductor

- Type-II superconductor

- Unconventional superconductor

- BCS theory

- Bean's critical state model

13.8 References

[1] John Bardeen; Leon Cooper; J. R. Schriffer (December 1, 1957). "Theory of Superconductivity". *Physical Review* **8** (5): 1178. Bibcode:1957PhRv..108.1175B. doi:10.1103/physrev.108.1175. ISBN 9780677000800. Retrieved June 6, 2014. reprinted in Nikolaï Nikolaevich Bogoliubov (1963) *The Theory of Superconductivity, Vol. 4*, CRC Press, ISBN 0677000804, p. 73

[2] John Daintith (2009). *The Facts on File Dictionary of Physics* (4th ed.). Infobase Publishing. p. 238. ISBN 1438109490.

[3] John C. Gallop (1990). *SQUIDS, the Josephson Effects and Superconducting Electronics*. CRC Press. pp. 3, 20. ISBN 0-7503-0051-5.

[4] Durrant, Alan (2000). *Quantum Physics of Matter*. CRC Press. pp. 102–103. ISBN 0750307218.

[5] J. G. Bednorz & K. A. Müller (1986). "Possible high T_c superconductivity in the Ba–La–Cu–O system". *Z. Physik, B* **64** (1): 189–193. Bibcode:1986ZPhyB..64..189B. doi:10.1007/BF01303701.

[6] Jun Nagamatsu; Norimasa Nakagawa; Takahiro Muranaka; Yuji Zenitani; et al. (2001). "Superconductivity at 39 K in magnesium diboride". *Nature* **410** (6824): 63–4. Bibcode:2001Natur.410...63N. doi:10.1038/35065039. PMID 11242039.

[7] Paul Preuss (14 August 2002). "A most unusual superconductor and how it works: first-principles calculation explains the strange behavior of magnesium diboride". *Research News* (Lawrence Berkeley National Laboratory). Retrieved 2009-10-28.

[8] Hamish Johnston (17 February 2009). "Type-1.5 superconductor shows its stripes". *Physics World* (Institute of Physics). Retrieved 2009-10-28.

[9] R. L. Dolecek (1954). "Adiabatic Magnetization of a Superconducting Sphere".*Physical Review***96**(1): 25–28.Bibcode:1954P doi:10.1103/PhysRev.96.25.

[10] H. Kleinert (1982). "Disorder Version of the Abelian Higgs Model and the Order of the Superconductive Phase Transition" (PDF). *Lettere al Nuovo Cimento* **35** (13): 405–412. doi:10.1007/BF02754760.

[11] J. Hove; S. Mo; A. Sudbo (2002). "Vortex interactions and thermally induced crossover from type-I to type-II superconductivity" (PDF).*Physical Review B***66**(6): 064524.arXiv:cond-mat/0202215.Bibcode:2002PhRvB..66f4524H.doi:10.1103/PhysRev

[12] Lev D. Landau; Evgeny M. Lifschitz (1984). *Electrodynamics of Continuous Media*. Course of Theoretical Physics **8**. Oxford: Butterworth-Heinemann. ISBN 0-7506-2634-8.

[13] David J. E. Callaway (1990). "On the remarkable structure of the superconducting intermediate state". *Nuclear Physics B* **344** (3): 627–645. Bibcode:1990NuPhB.344..627C. doi:10.1016/0550-3213(90)90672-Z.

[14] H. K. Onnes (1911). "The resistance of pure mercury at helium temperatures". *Commun. Phys. Lab. Univ. Leiden* **12**: 120.

[15] Dirk vanDelft & Peter Kes (September 2010). "The Discovery of Superconductivity" (PDF). *Physics Today* (American Institute of Physics).

[16] W. Meissner & R. Ochsenfeld (1933). "Ein neuer Effekt bei Eintritt der Supraleitfähigkeit". *Naturwissenschaften* **21** (44): 787–788. Bibcode:1933NW.....21..787M. doi:10.1007/BF01504252.

[17] F. London & H. London (1935). "The Electromagnetic Equations of the Supraconductor". *Proceedings of the Royal Society of London A* **149** (866): 71–88. Bibcode:1935RSPSA.149...71L. doi:10.1098/rspa.1935.0048. JSTOR 96265.

[18] Meissner, W.; R. Ochsenfeld (1933). "Ein neuer Effekt bei Eintritt der Supraleitfähigkeit". *Naturwissenschaften* **21** (44): 787–788. Bibcode:1933NW.....21..787M. doi:10.1007/BF01504252.

[19] "The London equations". The Open University. Retrieved 2011-10-16.

[20] J. Bardeen; L. N. Cooper & J. R. Schrieffer (1957). "Microscopic Theory of Superconductivity". *Physical Review* **106** (1): 162–164. Bibcode:1957PhRv..106..162B. doi:10.1103/PhysRev.106.162.

[21] J. Bardeen; L. N. Cooper & J. R. Schrieffer (1957). "Theory of Superconductivity". *Physical Review* **108** (5): 1175–1205. Bibcode:1957PhRv..108.1175B. doi:10.1103/PhysRev.108.1175.

[22] V. L. Ginzburg & L.D. Landau (1950). "On the theory of superconductivity". *Zhurnal Eksperimental'noi i Teoreticheskoi Fiziki* **20**: 1064.

[23] E. Maxwell (1950). "Isotope Effect in the Superconductivity of Mercury". *Physical Review* **78** (4): 477. Bibcode:1950PhRv...78 doi:10.1103/PhysRev.78.477.

[24] C. A. Reynolds; B. Serin; W. H. Wright & L. B. Nesbitt (1950). "Superconductivity of Isotopes of Mercury". *Physical Review* **78** (4): 487. Bibcode:1950PhRv...78..487R. doi:10.1103/PhysRev.78.487.

[25] N. N. Bogoliubov (1958). "A new method in the theory of superconductivity". *Zhurnal Eksperimental'noi i Teoreticheskoi Fiziki* **34**: 58.

[26] L. P. Gor'kov (1959). "Microscopic derivation of the Ginzburg—Landau equations in the theory of superconductivity". *Zhurnal Eksperimental'noi i Teoreticheskoi Fiziki* **36**: 1364.

[27] M. Combescot; W.V. Pogosov and O. Betbeder-Matibet (2013). "BCS ansatz for superconductivity in the light of the Bogoliubov approach and the Richardson–Gaudin exact wave function". *Physica C: Superconductivity* **485**: 47–57. arXiv:1111.4781. Bibcode:2013PhyC..485...47C. doi:10.1016/j.physc.2012.10.011. Retrieved 11 August 2014.

[28] Buck, Dudley A. "The Cryotron - A Superconductive Computer Component" (PDF). Lincoln Laboratory, Massachusetts Institute of Technology. Retrieved 10 August 2014.

[29] G.B.Yntema (1955). "Superconducting Winding for Electromagnet". *Physical Review* **98** (4): 1197. Bibcode:1955PhRv...98.1 doi:10.1103/PhysRev.98.1144.

[30] J.E. Kunzler, E. Buehler, F.L.S. Hsu, and J.H. Wernick (1961). "Superconductivity in Nb3Sn at High Current Density in a Magnetic Field of 88 kgauss". *Physical Review Letters* **6** (3): 89–91. Bibcode:1961PhRvL...6...89K. doi:10.1103/PhysRevLett.6.89. line feed character in |title= at position 65 (help)

[31] T.G. Berlincourt and R.R. Hake (1962). "Pulsed-Magnetic-Field Studies of Superconducting Transition Metal Alloys at High and Low Current Densities". *Bulletin of the American Physical Society* **II–7**: 408. line feed character in |title= at position 60 (help)

[32] T.G. Berlincourt (1987). "Emergence of Nb-Ti as Supermagnet Material". *Cryogenics* **27** (6): 283–289. Bibcode:1987Cryo... doi:10.1016/0011-2275(87)90057-9.

[33] B. D. Josephson (1962). "Possible new effects in superconductive tunnelling". *Physics Letters* **1** (7): 251–253. Bibcode:1962Ph doi:10.1016/0031-9163(62)91369-0.

[34] "Newly discovered fundamental state of matter, a superinsulator, has been created.". Science Daily. April 9, 2008. Retrieved 2008-10-23.

[35] M. K. Wu; et al. (1987). "Superconductivity at 93 K in a New Mixed-Phase Y-Ba-Cu-O Compound System at Ambient Pressure". *Physical Review Letters* **58** (9): 908–910. Bibcode:1987PhRvL..58..908W. doi:10.1103/PhysRevLett.58.908. PMID 10035069.

[36] "Introduction to Liquid Helium". *"Cryogenics and Fluid Branch"*. Goddard Space Flight Center, NASA.

[37] "Section 4.1 "Air plug in the fill line"". *"Superconducting Rock Magnetometer Cryogenic System Manual"*. 2G Enterprises. Archived from the original on May 6, 2009. Retrieved 9 October 2012.

[38] Alexei A. Abrikosov (8 December 2003). "type II Superconductors and the Vortex Lattice". *Nobel Lecture*.

[39] Adam Mann (Jul 20, 2011). "High-temperature superconductivity at 25: Still in suspense". *Nature* **475** (7356): 280–2. Bibcode:2011Natur.475..280M. doi:10.1038/475280a. PMID 21776057.

[40] Pines, D. (2002), "The Spin Fluctuation Model for High Temperature Superconductivity: Progress and Prospects", *The Gap Symmetry and Fluctuations in High-Tc Superconductors*, NATO Science Series: B: **371**, New York: Kluwer Academic, pp. 111–142, doi:10.1007/0-306-47081-0_7, ISBN 0-306-45934-5

[41] P. Monthoux; A. V. Balatsky & D. Pines (1991). "Toward a theory of high-temperature superconductivity in the antiferromagnetically correlated cuprate oxides". *Phys. Rev. Lett.* **67** (24): 3448–3451. Bibcode:1991PhRvL..67.3448M. doi:10.1103/PhysRevLett.67.3448.PMID10044736.

[42] A. Schilling; et al. (1993). "Superconductivity above 130 K in the Hg–Ba–Ca–Cu–O system". *Nature* **363** (6424): 56–58. Bibcode:1993Natur.363..56C. doi:10.1038/363056a0.

[43] P. Dai; B. C. Chakoumakos; G. F. Sun; K. W. Wong; et al. (1995). "Synthesis and neutron powder diffraction study of the superconductor $HgBa_2Ca_2Cu_3O_{8+}\delta$ by Tl substitution". *Physica C* **243** (3–4): 201–206. Bibcode:1995PhyC..243..201D. doi:10.1016/0921-4534(94)02461-8.

[44] Hiroki Takahashi; Kazumi Igawa; Kazunobu Arii; Yoichi Kamihara; et al. (2008). "Superconductivity at 43 K in an iron-based layered compound $LaO_{1-x}F_xFeAs$". *Nature* **453** (7193): 376–378. Bibcode:2008Natur.453..376T. doi:10.1038/nature06972. PMID 18432191.

[45] Adrian Cho. "Second Family of High-Temperature Superconductors Discovered". ScienceNOW Daily News.

[46] Zhi-An Ren; et al. (2008). "Superconductivity and phase diagram in iron-based arsenic-oxides ReFeAsO1-d (Re = rare-earth metal) without fluorine doping". *EPL* **83**: 17002. arXiv:0804.2582. Bibcode:2008EL.....8317002R. doi:10.1209/0295-5075/83/17002.

[47] Li, Yinwei; Hao, Jian; Liu, Hanyu; Li, Yanling; Ma, Yanming (2014-05-07). "The metallization and superconductivity of dense hydrogen sulfide". *The Journal of Chemical Physics* **140** (17): 174712. doi:10.1063/1.4874158. ISSN 0021-9606.

[48] Drozdov, A. P.; Eremets, M. I.; Troyan, I. A.; Ksenofontov, V.; Shylin, S. I. (2015). "Conventional superconductivity at 203 kelvin at high pressures in the sulfur hydride system". *Nature* **525** (7567): 73–6. doi:10.1038/nature14964. ISSN 0028-0836. PMID 26280333.

[49] Islam et al, *A review of offshore wind turbine nacelle: Technical challenges, and research and developmental trends.* In: *Renewable and Sustainable Energy Reviews* 33, (2014), 161–176, doi:10.1016/j.rser.2014.01.085

[50] Linder, Jacob; Robinson, Jason W. A. (2 April 2015). "Superconducting spintronics". *Nature Physics* **11** (4): 307–315. doi:10.1038/nphys3242.

[51] Thomas et al, *Superconducting transmission lines – Sustainable electric energy transfer with higher public acceptance?* In: *Renewable and Sustainable Energy Reviews* 55, (2016), 59–72, doi:10.1016/j.rser.2015.10.041.

[52] "All Nobel Prizes in Physics". *Nobelprize.org*. Nobel Media AB 2014.

13.9 Further reading

- Hagen Kleinert (1989). "Superflow and Vortex Lines". *Gauge Fields in Condensed Matter* **1**. World Scientific. ISBN 9971-5-0210-0.

- Anatoly Larkin; Andrei Varlamov (2005). *Theory of Fluctuations in Superconductors*. Oxford University Press. ISBN 0-19-852815-9.

- A. G. Lebed (2008). *The Physics of Organic Superconductors and Conductors* **110** (1st ed.). Springer. ISBN 978-3-540-76667-4.

- Jean Matricon; Georges Waysand; Charles Glashausser (2003). *The Cold Wars: A History of Superconductivity*. Rutgers University Press. ISBN 0-8135-3295-7.

- "Physicist Discovers Exotic Superconductivity". ScienceDaily. 17 August 2006.

- Michael Tinkham (2004). *Introduction to Superconductivity* (2nd ed.). Dover Books. ISBN 0-486-43503-2.

- Terry Orlando; Kevin Delin (1991). *Foundations of Applied Superconductivity*. Prentice Hall. ISBN 978-0-201-18323-8.

- Paul Tipler; Ralph Llewellyn (2002). *Modern Physics* (4th ed.). W. H. Freeman. ISBN 0-7167-4345-0.

13.10 External links

- Everything about superconductivity: properties, research, applications with videos, animations, games

- Video about Type I Superconductors: R=0/transition temperatures/ B is a state variable/ Meissner effect/ Energy gap(Giaever)/ BCS model

- Superconductivity: Current in a Cape and Thermal Tights. An introduction to the topic for non-scientists National High Magnetic Field Laboratory

- Lectures on Superconductivity (series of videos, including interviews with leading experts)

- Superconductivity News Update

- Superconductor Week Newsletter – industry news, links, et cetera

- Superconducting Magnetic Levitation

- National Superconducting Cyclotron Laboratory at Michigan State University

- YouTube Video Levitating magnet

- International Workshop on superconductivity in Diamond and Related Materials (free download papers)

- New Diamond and Frontier Carbon Technology Volume 17, No.1 Special Issue on Superconductivity in CVD Diamond

- DoITPoMS Teaching and Learning Package – "Superconductivity"

- The Nobel Prize for Physics, 1901–2008

- folding hands-on activities about superconductivity

Chapter 14

Baryon

Not to be confused with Baryonyx.

A **baryon** is a composite subatomic particle made up of three quarks (as distinct from mesons, which are composed of one quark and one antiquark). Baryons and mesons belong to the hadron family of particles, which are the quark-based particles. The name "baryon" comes from the Greek word for "heavy" (βαρύς, *barys*), because, at the time of their naming, most known elementary particles had lower masses than the baryons.

As quark-based particles, baryons participate in the strong interaction, whereas leptons, which are not quark-based, do not. The most familiar baryons are the protons and neutrons that make up most of the mass of the visible matter in the universe. Electrons (the other major component of the atom) are leptons.

Each baryon has a corresponding antiparticle (antibaryon) where quarks are replaced by their corresponding antiquarks. For example, a proton is made of two up quarks and one down quark; and its corresponding antiparticle, the antiproton, is made of two up antiquarks and one down antiquark.

14.1   Background

Baryons are strongly interacting fermions that is, they experience the strong nuclear force and are described by Fermi–Dirac statistics, which apply to all particles obeying the Pauli exclusion principle. This is in contrast to the bosons, which do not obey the exclusion principle.

Baryons, along with mesons, are hadrons, meaning they are particles composed of quarks. Quarks have baryon numbers of $B = 1/3$ and antiquarks have baryon number of $B = -1/3$. The term "baryon" usually refers to *triquarks*—baryons made of three quarks ($B = 1/3 + 1/3 + 1/3 = 1$).

Other exotic baryons have been proposed, such as pentaquarks—baryons made of four quarks and one antiquark ($B = 1/3 + 1/3 + 1/3 + 1/3 - 1/3 = 1$), but their existence is not generally accepted. In theory, heptaquarks (5 quarks, 2 antiquarks), nonaquarks (6 quarks, 3 antiquarks), etc. could also exist. Until recently, it was believed that some experiments showed the existence of pentaquarks—baryons made of four quarks and one antiquark.[1][2] The particle physics community as a whole did not view their existence as likely in 2006,[3] and in 2008, considered evidence to be overwhelmingly against the existence of the reported pentaquarks.[4] However, in July 2015, the LHCb experiment observed two resonances consistent with pentaquark states in the Λ0
b → J/ψK−
p decay, with a combined statistical significance of 15σ.[5][6]

14.2 Baryonic matter

Nearly all matter that may be encountered or experienced in everyday life is baryonic matter, which includes atoms of any sort, and provides those with the quality of mass. Non-baryonic matter, as implied by the name, is any sort of matter that is not composed primarily of baryons. Those might include neutrinos or free electrons, dark matter, such as supersymmetric particles, axions, or black holes.

The very existence of baryons is also a significant issue in cosmology because it is assumed that the Big Bang produced a state with equal amounts of baryons and antibaryons. The process by which baryons came to outnumber their antiparticles is called baryogenesis.

14.3 Baryogenesis

Main article: Baryogenesis

Experiments are consistent with the number of quarks in the universe being a constant and, to be more specific, the number of baryons being a constant ; in technical language, the total baryon number appears to be *conserved*. Within the prevailing Standard Model of particle physics, the number of baryons may change in multiples of three due to the action of sphalerons, although this is rare and has not been observed under experiment. Some grand unified theories of particle physics also predict that a single proton can decay, changing the baryon number by one; however, this has not yet been observed under experiment. The excess of baryons over antibaryons in the present universe is thought to be due to non-conservation of baryon number in the very early universe, though this is not well understood.

14.4 Properties

14.4.1 Isospin and charge

Main article: Isospin

The concept of isospin was first proposed by Werner Heisenberg in 1932 to explain the similarities between protons and neutrons under the strong interaction.[7] Although they had different electric charges, their masses were so similar that physicists believed they were actually the same particle. The different electric charges were explained as being the result of some unknown excitation similar to spin. This unknown excitation was later dubbed *isospin* by Eugene Wigner in 1937.[8]

This belief lasted until Murray Gell-Mann proposed the quark model in 1964 (containing originally only the u, d, and s quarks).[9] The success of the isospin model is now understood to be the result of the similar masses of the u and d quarks. Since the u and d quarks have similar masses, particles made of the same number then also have similar masses. The exact specific u and d quark composition determines the charge, as u quarks carry charge +2/3 while d quarks carry charge −1/3. For example the four Deltas all have different charges (Δ++ (uuu), Δ+ (uud), Δ0 (udd), Δ− (ddd)), but have similar masses (~1,232 MeV/c^2) as they are each made of a combination of three u and d quarks. Under the isospin model, they were considered to be a single particle in different charged states.

The mathematics of isospin was modeled after that of spin. Isospin projections varied in increments of 1 just like those of spin, and to each projection was associated a "charged state". Since the "Delta particle" had four "charged states", it was said to be of isospin I = 3/2. Its "charged states" Δ++, Δ+, Δ0, and Δ−, corresponded to the isospin projections I_3 = +3/2, I_3 = +1/2, I_3 = −1/2, and I_3 = −3/2, respectively. Another example is the "nucleon particle". As there were two nucleon "charged states", it was said to be of isospin 1/2. The positive nucleon N+ (proton) was identified with I_3 = +1/2 and the neutral nucleon N0 (neutron) with I_3 = −1/2.[10] It was later noted that the isospin projections were related to the up and down quark content of particles by the relation:

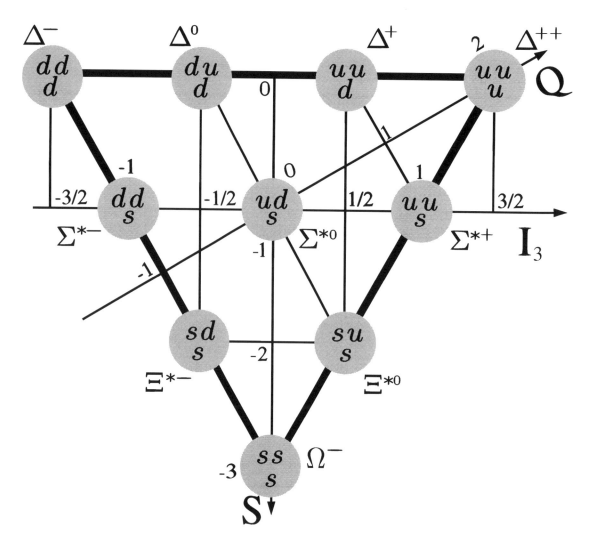

*Combinations of three **u, d** or **s** quarks forming baryons with a spin-3/2 form the* uds baryon decuplet

$$I_3 = \frac{1}{2}[(n_u - n_{\bar{u}}) - (n_d - n_{\bar{d}})],$$

where the n's are the number of up and down quarks and antiquarks.

In the "isospin picture", the four Deltas and the two nucleons were thought to be the different states of two particles. However in the quark model, Deltas are different states of nucleons (the N^{++} or N^- are forbidden by Pauli's exclusion principle). Isospin, although conveying an inaccurate picture of things, is still used to classify baryons, leading to unnatural and often confusing nomenclature.

14.4.2 Flavour quantum numbers

Main article: Flavour (particle physics) § Flavour quantum numbers

The strangeness flavour quantum number S (not to be confused with spin) was noticed to go up and down along with particle mass. The higher the mass, the lower the strangeness (the more s quarks). Particles could be described with

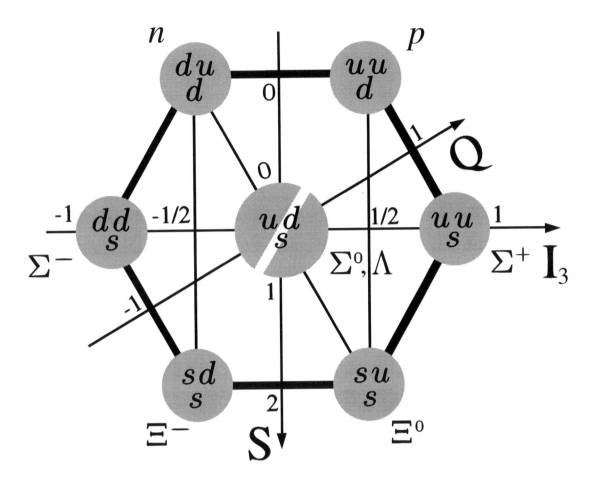

*Combinations of three **u**, **d** or s quarks forming baryons with a spin-1/2 form the* uds baryon octet

isospin projections (related to charge) and strangeness (mass) (see the uds octet and decuplet figures on the right). As other quarks were discovered, new quantum numbers were made to have similar description of udc and udb octets and decuplets. Since only the u and d mass are similar, this description of particle mass and charge in terms of isospin and flavour quantum numbers works well only for octet and decuplet made of one u, one d, and one other quark, and breaks down for the other octets and decuplets (for example, ucb octet and decuplet). If the quarks all had the same mass, their behaviour would be called *symmetric*, as they would all behave in exactly the same way with respect to the strong interaction. Since quarks do not have the same mass, they do not interact in the same way (exactly like an electron placed in an electric field will accelerate more than a proton placed in the same field because of its lighter mass), and the symmetry is said to be broken.

It was noted that charge (Q) was related to the isospin projection (I_3), the baryon number (B) and flavour quantum numbers (S, C, B', T) by the Gell-Mann–Nishijima formula:[10]

$$Q = I_3 + \frac{1}{2}(B + S + C + B' + T),$$

where S, C, B', and T represent the strangeness, charm, bottomness and topness flavour quantum numbers, respectively. They are related to the number of strange, charm, bottom, and top quarks and antiquark according to the relations:

$$S = -(n_{\rm s} - n_{\bar{\rm s}}),$$

$$C = +(n_{\rm c} - n_{\bar{\rm c}}),$$

$$B' = -(n_{\rm b} - n_{\bar{\rm b}}),$$

$$T = +(n_{\rm t} - n_{\bar{\rm t}}),$$

meaning that the Gell-Mann–Nishijima formula is equivalent to the expression of charge in terms of quark content:

$$Q = \frac{2}{3}[(n_{\rm u} - n_{\bar{\rm u}}) + (n_{\rm c} - n_{\bar{\rm c}}) + (n_{\rm t} - n_{\bar{\rm t}})] - \frac{1}{3}[(n_{\rm d} - n_{\bar{\rm d}}) + (n_{\rm s} - n_{\bar{\rm s}}) + (n_{\rm b} - n_{\bar{\rm b}})].$$

14.4.3 Spin, orbital angular momentum, and total angular momentum

Main articles: Spin (physics), Angular momentum operator, Quantum numbers and Clebsch–Gordan coefficients

Spin (quantum number S) is a vector quantity that represents the "intrinsic" angular momentum of a particle. It comes in increments of 1/2 ħ (pronounced "h-bar"). The ħ is often dropped because it is the "fundamental" unit of spin, and it is implied that "spin 1" means "spin 1 ħ". In some systems of natural units, ħ is chosen to be 1, and therefore does not appear anywhere.

Quarks are fermionic particles of spin 1/2 ($S = 1/2$). Because spin projections vary in increments of 1 (that is 1 ħ), a single quark has a spin vector of length 1/2, and has two spin projections ($S_z = +1/2$ and $S_z = -1/2$). Two quarks can have their spins aligned, in which case the two spin vectors add to make a vector of length $S = 1$ and three spin projections ($S_z = +1$, $S_z = 0$, and $S_z = -1$). If two quarks have unaligned spins, the spin vectors add up to make a vector of length $S = 0$ and has only one spin projection ($S_z = 0$), etc. Since baryons are made of three quarks, their spin vectors can add to make a vector of length $S = 3/2$, which has four spin projections ($S_z = +3/2$, $S_z = +1/2$, $S_z = -1/2$, and $S_z = -3/2$), or a vector of length $S = 1/2$ with two spin projections ($S_z = +1/2$, and $S_z = -1/2$).[11]

There is another quantity of angular momentum, called the orbital angular momentum, (azimuthal quantum number L), that comes in increments of 1 ħ, which represent the angular moment due to quarks orbiting around each other. The total angular momentum (total angular momentum quantum number J) of a particle is therefore the combination of intrinsic angular momentum (spin) and orbital angular momentum. It can take any value from $J = |L - S|$ to $J = |L + S|$, in increments of 1.

Particle physicists are most interested in baryons with no orbital angular momentum ($L = 0$), as they correspond to ground states—states of minimal energy. Therefore the two groups of baryons most studied are the $S = 1/2$; $L = 0$ and $S = 3/2$; $L = 0$, which corresponds to $J = 1/2^+$ and $J = 3/2^+$, respectively, although they are not the only ones. It is also possible to obtain $J = 3/2^+$ particles from $S = 1/2$ and $L = 2$, as well as $S = 3/2$ and $L = 2$. This phenomenon of having multiple particles in the same total angular momentum configuration is called *degeneracy*. How to distinguish between these degenerate baryons is an active area of research in baryon spectroscopy.[12][13]

14.4.4 Parity

Main article: Parity (physics)

If the universe were reflected in a mirror, most of the laws of physics would be identical—things would behave the same way regardless of what we call "left" and what we call "right". This concept of mirror reflection is called *intrinsic parity* or *parity* (*P*). Gravity, the electromagnetic force, and the strong interaction all behave in the same way regardless of whether or not the universe is reflected in a mirror, and thus are said to conserve parity (P-symmetry). However, the weak interaction *does* distinguish "left" from "right", a phenomenon called parity violation (P-violation).

Based on this, one might think that, if the wavefunction for each particle (in more precise terms, the quantum field for each particle type) were simultaneously mirror-reversed, then the new set of wavefunctions would perfectly satisfy the laws of physics (apart from the weak interaction). It turns out that this is not quite true: In order for the equations to be satisfied, the wavefunctions of certain types of particles have to be multiplied by −1, in addition to being mirror-reversed. Such particle types are said to have *negative* or *odd* parity ($P = -1$, or alternatively $P = -$), while the other particles are said to have *positive* or *even* parity ($P = +1$, or alternatively $P = +$).

For baryons, the parity is related to the orbital angular momentum by the relation:[14]

$$P = (-1)^L.$$

As a consequence, baryons with no orbital angular momentum ($L = 0$) all have even parity ($P = +$).

14.5   Nomenclature

Baryons are classified into groups according to their isospin (I) values and quark (q) content. There are six groups of baryons—nucleon (N), Delta (Δ), Lambda (Λ), Sigma (Σ), Xi (Ξ), and Omega (Ω). The rules for classification are defined by the Particle Data Group. These rules consider the up (u), down (d) and strange (s) quarks to be *light* and the charm (c), bottom (b), and top (t) quarks to be *heavy*. The rules cover all the particles that can be made from three of each of the six quarks, even though baryons made of t quarks are not expected to exist because of the t quark's short lifetime. The rules do not cover pentaquarks.[15]

- Baryons with three u and/or d quarks are N's ($I = 1/2$) or Δ's ($I = 3/2$).

- Baryons with two u and/or d quarks are Λ's ($I = 0$) or Σ's ($I = 1$). If the third quark is heavy, its identity is given by a subscript.

- Baryons with one u or d quark are Ξ's ($I = 1/2$). One or two subscripts are used if one or both of the remaining quarks are heavy.

- Baryons with no u or d quarks are Ω's ($I = 0$), and subscripts indicate any heavy quark content.

- Baryons that decay strongly have their masses as part of their names. For example, Σ^0 does not decay strongly, but $\Delta^{++}(1232)$ does.

It is also a widespread (but not universal) practice to follow some additional rules when distinguishing between some states that would otherwise have the same symbol.[10]

- Baryons in total angular momentum $J = 3/2$ configuration that have the same symbols as their $J = 1/2$ counterparts are denoted by an asterisk (*).

- Two baryons can be made of three different quarks in $J = 1/2$ configuration. In this case, a prime (′) is used to distinguish between them.

  - *Exception*: When two of the three quarks are one up and one down quark, one baryon is dubbed Λ while the other is dubbed Σ.

Quarks carry charge, so knowing the charge of a particle indirectly gives the quark content. For example, the rules above say that a Λ+
c contains a c quark and some combination of two u and/or d quarks. The c quark has a charge of ($Q = +2/3$), therefore the other two must be a u quark ($Q = +2/3$), and a d quark ($Q = -1/3$) to have the correct total charge ($Q = +1$).

14.6   See also

- Eightfold way

- List of baryons

- List of particles

- Meson

- Timeline of particle discoveries

14.7   Notes

[1] H. Muir (2003)

[2] K. Carter (2003)

[3] W.-M. Yao *et al.* (2006): Particle listings – Θ^+

[4] C. Amsler *et al.* (2008): Pentaquarks

[5] LHCb (14 July 2015). "Observation of particles composed of five quarks, pentaquark-charmonium states, seen in $\Lambda_b^0 \rightarrow J/\psi p K^-$ decays.". *CERN website*. Retrieved 2015-07-14.

[6] R. Aaij et al. (LHCb collaboration) (2015). "Observation of J/ψp resonances consistent with pentaquark states in Λ^0 b→J/ψK–
p decays". *Physical Review Letters* **115** (7). Bibcode:2015PhRvL.115g2001A. doi:10.1103/PhysRevLett.115.072001.

[7] W. Heisenberg (1932)

[8] E. Wigner (1937)

[9] M. Gell-Mann (1964)

[10] S.S.M. Wong (1998a)

[11] R. Shankar (1994)

[12] H. Garcilazo *et al.* (2007)

[13] D.M. Manley (2005)

[14] S.S.M. Wong (1998b)

[15] C. Amsler *et al.* (2008): Naming scheme for hadrons

14.8   References

- C. Amsler *et al.* (Particle Data Group) (2008). "Review of Particle Physics". *Physics Letters B* **667** (1): 1–1340. Bibcode:2008PhLB..667....1P. doi:10.1016/j.physletb.2008.07.018.

- H. Garcilazo, J. Vijande, and A. Valcarce (2007). "Faddeev study of heavy-baryon spectroscopy". *Journal of Physics G* **34** (5): 961–976. doi:10.1088/0954-3899/34/5/014.

- K. Carter (2006). "The rise and fall of the pentaquark". Fermilab and SLAC. Retrieved 2008-05-27.

- W.-M. Yao *et al.*(Particle Data Group) (2006). "Review of Particle Physics". *Journal of Physics G* **33**: 1–1232. arXiv:astro-ph/0601168. Bibcode:2006JPhG...33....1Y. doi:10.1088/0954-3899/33/1/001.

- D.M. Manley (2005). "Status of baryon spectroscopy". *Journal of Physics: Conference Series* **5**: 230–237. Bibcode:2005JPhCS...9..230M. doi:10.1088/1742-6596/9/1/043.

- H. Muir (2003). "Pentaquark discovery confounds sceptics". New Scientist. Retrieved 2008-05-27.

- S.S.M. Wong (1998a). "Chapter 2—Nucleon Structure". *Introductory Nuclear Physics* (2nd ed.). New York (NY): John Wiley & Sons. pp. 21–56. ISBN 0-471-23973-9.

- S.S.M. Wong (1998b). "Chapter 3—The Deuteron". *Introductory Nuclear Physics* (2nd ed.). New York (NY): John Wiley & Sons. pp. 57–104. ISBN 0-471-23973-9.

- R. Shankar (1994). *Principles of Quantum Mechanics* (2nd ed.). New York (NY): Plenum Press. ISBN 0-306-44790-8.

- E. Wigner (1937). "On the Consequences of the Symmetry of the Nuclear Hamiltonian on the Spectroscopy of Nuclei". *Physical Review* **51** (2): 106–119. Bibcode:1937PhRv...51..106W. doi:10.1103/PhysRev.51.106.

- M. Gell-Mann (1964). "A Schematic of Baryons and Mesons". *Physics Letters* **8**(3): 214–215. Bibcode:1964PhL.. doi:10.1016/S0031-9163(64)92001-3.

- W. Heisenberg (1932). "Über den Bau der Atomkerne I". *Zeitschrift für Physik* (in German) **77**: 1–11. Bibcode:1 doi:10.1007/BF01342433.

- W. Heisenberg (1932). "Über den Bau der Atomkerne II". *Zeitschrift für Physik* (in German) **78** (3–4): 156–164. Bibcode:1932ZPhy...78..156H. doi:10.1007/BF01337585.

- W. Heisenberg (1932). "Über den Bau der Atomkerne III". *Zeitschrift für Physik* (in German) **80** (9–10): 587–596. Bibcode:1933ZPhy...80..587H. doi:10.1007/BF01335696.

14.9 External links

- Particle Data Group—Review of Particle Physics (2008).

- Georgia State University—HyperPhysics

- Baryons made thinkable, an interactive visualisation allowing physical properties to be compared

Chapter 15

Lepton

For other uses, see Lepton (disambiguation).

A **lepton** is an elementary, half-integer spin (spin $\frac{1}{2}$) particle that does not undergo strong interactions.[1] Two main classes of leptons exist: charged leptons (also known as the *electron-like* leptons), and neutral leptons (better known as neutrinos). Charged leptons can combine with other particles to form various composite particles such as atoms and positronium, while neutrinos rarely interact with anything, and are consequently rarely observed. The best known of all leptons is the electron.

There are six types of leptons, known as *flavours*, forming three *generations*.[2] The first generation is the *electronic leptons*, comprising the electron (e−) and electron neutrino (ν e); the second is the *muonic leptons*, comprising the muon (μ−) and muon neutrino (ν μ); and the third is the *tauonic leptons*, comprising the tau (τ−) and the tau neutrino (ν τ). Electrons have the least mass of all the charged leptons. The heavier muons and taus will rapidly change into electrons through a process of particle decay: the transformation from a higher mass state to a lower mass state. Thus electrons are stable and the most common charged lepton in the universe, whereas muons and taus can only be produced in high energy collisions (such as those involving cosmic rays and those carried out in particle accelerators).

Leptons have various intrinsic properties, including electric charge, spin, and mass. Unlike quarks however, leptons are not subject to the strong interaction, but they are subject to the other three fundamental interactions: gravitation, electromagnetism (excluding neutrinos, which are electrically neutral), and the weak interaction.

For every lepton flavor there is a corresponding type of antiparticle, known as an antilepton, that differs from the lepton only in that some of its properties have equal magnitude but opposite sign. However, according to certain theories, neutrinos may be their own antiparticle, but it is not currently known whether this is the case or not.

The first charged lepton, the electron, was theorized in the mid-19th century by several scientists[3][4][5] and was discovered in 1897 by J. J. Thomson.[6] The next lepton to be observed was the muon, discovered by Carl D. Anderson in 1936, which was classified as a meson at the time.[7] After investigation, it was realized that the muon did not have the expected properties of a meson, but rather behaved like an electron, only with higher mass. It took until 1947 for the concept of "leptons" as a family of particle to be proposed.[8] The first neutrino, the electron neutrino, was proposed by Wolfgang Pauli in 1930 to explain certain characteristics of beta decay.[8] It was first observed in the Cowan–Reines neutrino experiment conducted by Clyde Cowan and Frederick Reines in 1956.[8][9] The muon neutrino was discovered in 1962 by Leon M. Lederman, Melvin Schwartz and Jack Steinberger,[10] and the tau discovered between 1974 and 1977 by Martin Lewis Perl and his colleagues from the Stanford Linear Accelerator Center and Lawrence Berkeley National Laboratory.[11] The tau neutrino remained elusive until July 2000, when the DONUT collaboration from Fermilab announced its discovery.[12][13]

Leptons are an important part of the Standard Model. Electrons are one of the components of atoms, alongside protons and neutrons. Exotic atoms with muons and taus instead of electrons can also be synthesized, as well as lepton–antilepton particles such as positronium.

15.1   Etymology

The name *lepton* comes from the Greek λεπτός *leptós*, "fine, small, thin" (neuter form: λεπτόν *leptón*);[14][15] the earliest attested form of the word is the Mycenaean Greek 𐀩𐀡𐀵, *re-po-to*, written in Linear B syllabic script.[16] *Lepton* was first used by physicist Léon Rosenfeld in 1948:[17]

> Following a suggestion of Prof. C. Møller, I adopt — as a pendant to "nucleon" — the denomination "lepton" (from λεπτός, small, thin, delicate) to denote a particle of small mass.

The etymology incorrectly implies that all the leptons are of small mass. When Rosenfeld named them, the only known leptons were electrons and muons, which are in fact of small mass — the mass of an electron (0.511 MeV/c^2)[18] and the mass of a muon (with a value of 105.7 MeV/c^2)[19] are fractions of the mass of the "heavy" proton (938.3 MeV/c^2).[20] However, the mass of the tau (discovered in the mid 1970s) (1777 MeV/c^2)[21] is nearly twice that of the proton, and about 3,500 times that of the electron.

15.2   History

See also: Electron § Discovery, Muon § History and Tau (particle) § History
The first lepton identified was the electron, discovered by J.J. Thomson and his team of British physicists in 1897.[22][23]

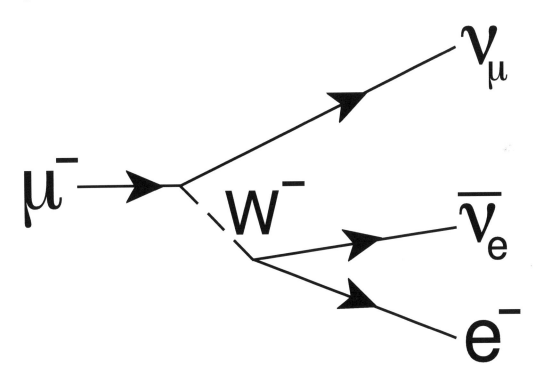

A muon transmutes into a muon neutrino by emitting a W− boson. The W− boson subsequently decays into an electron and an electron antineutrino.

Then in 1930 Wolfgang Pauli postulated the electron neutrino to preserve conservation of energy, conservation of momentum, and conservation of angular momentum in beta decay.[24] Pauli theorized that an undetected particle was carrying away the difference between the energy, momentum, and angular momentum of the initial and observed final particles. The electron neutrino was simply called the neutrino, as it was not yet known that neutrinos came in different flavours (or different "generations").

Nearly 40 years after the discovery of the electron, the muon was discovered by Carl D. Anderson in 1936. Due to its mass, it was initially categorized as a meson rather than a lepton.[25] It later became clear that the muon was much more similar to the electron than to mesons, as muons do not undergo the strong interaction, and thus the muon was reclassified: electrons, muons, and the (electron) neutrino were grouped into a new group of particles – the leptons. In 1962 Leon M. Lederman, Melvin Schwartz and Jack Steinberger showed that more than one type of neutrino exists by first detecting interactions of the muon neutrino, which earned them the 1988 Nobel Prize, although by then the different flavours of neutrino had already been theorized.[26]

The tau was first detected in a series of experiments between 1974 and 1977 by Martin Lewis Perl with his colleagues at the SLAC LBL group.[27] Like the electron and the muon, it too was expected to have an associated neutrino. The first evidence for tau neutrinos came from the observation of "missing" energy and momentum in tau decay, analogous to the "missing" energy and momentum in beta decay leading to the discovery of the electron neutrino. The first detection of tau neutrino interactions was announced in 2000 by the DONUT collaboration at Fermilab, making it the latest particle of the Standard Model to have been directly observed,[28] apart from the Higgs boson, which probably has been discovered in 2012.

Although all present data is consistent with three generations of leptons, some particle physicists are searching for a fourth generation. The current lower limit on the mass of such a fourth charged lepton is 100.8 GeV/c^2,[29] while its associated neutrino would have a mass of at least 45.0 GeV/c^2.[30]

15.3 Properties

15.3.1 Spin and chirality

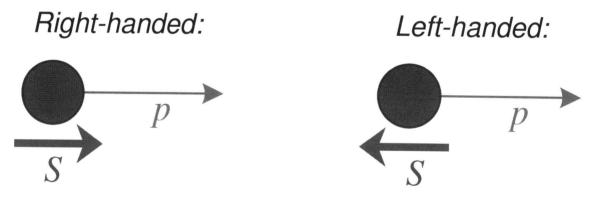

Left-handed and right-handed helicities

Leptons are spin-$\frac{1}{2}$ particles. The spin-statistics theorem thus implies that they are fermions and thus that they are subject to the Pauli exclusion principle; no two leptons of the same species can be in exactly the same state at the same time. Furthermore, it means that a lepton can have only two possible spin states, namely up or down.

A closely related property is chirality, which in turn is closely related to a more easily visualized property called helicity. The helicity of a particle is the direction of its spin relative to its momentum; particles with spin in the same direction as their momentum are called *right-handed* and otherwise they are called *left-handed*. When a particle is mass-less, the direction of its momentum relative to its spin is frame independent, while for massive particles it is possible to 'overtake' the particle by a Lorentz transformation flipping the helicity. Chirality is a technical property (defined through the transformation behaviour under the Poincaré group) that agrees with helicity for (approximately) massless particles and is still well defined for massive particles.

In many quantum field theories—such as quantum electrodynamics and quantum chromodynamics—left and right-handed fermions are identical. However in the Standard Model left-handed and right-handed fermions are treated asymmetrically. Only left-handed fermions participate in the weak interaction, while there are no right-handed neutrinos. This is an example of parity violation. In the literature left-handed fields are often denoted by a capital L subscript (e.g. e–L) and right-handed fields are denoted by a capital R subscript.

15.3.2   Electromagnetic interaction

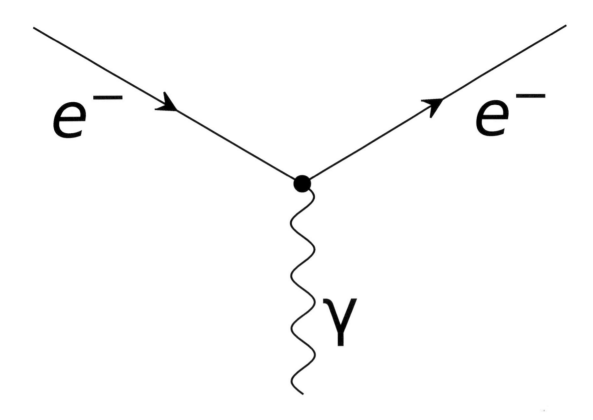

Lepton–photon interaction

One of the most prominent properties of leptons is their electric charge, Q. The electric charge determines the strength of their electromagnetic interactions. It determines the strength of the electric field generated by the particle (see Coulomb's law) and how strongly the particle reacts to an external electric or magnetic field (see Lorentz force). Each generation contains one lepton with $Q = -e$ (conventionally the charge of a particle is expressed in units of the elementary charge) and one lepton with zero electric charge. The lepton with electric charge is commonly simply referred to as a 'charged lepton' while the neutral lepton is called a neutrino. For example the first generation consists of the electron e– with a negative electric charge and the electrically neutral electron neutrino ν
e.

In the language of quantum field theory the electromagnetic interaction of the charged leptons is expressed by the fact that the particles interact with the quantum of the electromagnetic field, the photon. The Feynman diagram of the electron-photon interaction is shown on the right.

Because leptons possess an intrinsic rotation in the form of their spin, charged leptons generate a magnetic field. The size of their magnetic dipole moment μ is given by,

$$\mu = g\frac{Q\hbar}{4m},$$

where m is the mass of the lepton and g is the so-called g-factor for the lepton. First order approximation quantum mechanics predicts that the g-factor is 2 for all leptons. However, higher order quantum effects caused by loops in Feynman diagrams introduce corrections to this value. These corrections, referred to as the anomalous magnetic dipole moment, are very sensitive to the details of a quantum field theory model and thus provide the opportunity for precision tests of the standard model. The theoretical and measured values for the electron anomalous magnetic dipole moment are within agreement within eight significant figures.[31]

15.3.3 Weak interaction

In the Standard Model, the left-handed charged lepton and the left-handed neutrino are arranged in doublet (ν eL, e−L) that transforms in the spinor representation ($T = \frac{1}{2}$) of the weak isospin SU(2) gauge symmetry. This means that these particles are eigenstates of the isospin projection T_3 with eigenvalues $\frac{1}{2}$ and $-\frac{1}{2}$ respectively. In the meantime, the right-handed charged lepton transforms as a weak isospin scalar ($T = 0$) and thus does not participate in the weak interaction, while there is no right-handed neutrino at all.

The Higgs mechanism recombines the gauge fields of the weak isospin SU(2) and the weak hypercharge U(1) symmetries to three massive vector bosons (W+, W−, Z0) mediating the weak interaction, and one massless vector boson, the photon, responsible for the electromagnetic interaction. The electric charge Q can be calculated from the isospin projection T_3 and weak hypercharge YW through the Gell-Mann–Nishijima formula,

$$Q = T_3 + YW/2$$

To recover the observed electric charges for all particles the left-handed weak isospin doublet (ν eL, e−L) must thus have $YW = -1$, while the right-handed isospin scalar e− R must have $YW = -2$. The interaction of the leptons with the massive weak interaction vector bosons is shown in the figure on the left.

15.3.4 Mass

In the Standard Model each lepton starts out with no intrinsic mass. The charged leptons (i.e. the electron, muon, and tau) obtain an effective mass through interaction with the Higgs field, but the neutrinos remain massless. For technical reasons the masslessness of the neutrinos implies that there is no mixing of the different generations of charged leptons as there is for quarks. This is in close agreement with current experimental observations.[32]

However, it is known from experiments – most prominently from observed neutrino oscillations[33] – that neutrinos do in fact have some very small mass, probably less than $2 \ eV/c^2$.[34] This implies the existence of physics beyond the Standard Model. The currently most favoured extension is the so-called seesaw mechanism, which would explain both why the left-handed neutrinos are so light compared to the corresponding charged leptons, and why we have not yet seen any right-handed neutrinos.

15.3.5 Leptonic numbers

Main article: Lepton number

The members of each generation's weak isospin doublet are assigned leptonic numbers that are conserved under the Standard Model.[35] Electrons and electron neutrinos have an *electronic number* of $L_e = 1$, while muons and muon neutrinos have a *muonic number* of $L\mu = 1$, while tau particles and tau neutrinos have a *tauonic number* of $L\tau = 1$. The antileptons have their respective generation's leptonic numbers of −1.

Conservation of the leptonic numbers means that the number of leptons of the same type remains the same, when particles interact. This implies that leptons and antileptons must be created in pairs of a single generation. For example, the following processes are allowed under conservation of leptonic numbers:

e− + e+ → γ + γ,

$$\left(\begin{array}{c} \nu_e \\ e^- \end{array} \right), \left(\begin{array}{c} \nu_\mu \\ \mu^- \end{array} \right), \left(\begin{array}{c} \nu_\tau \\ \tau^- \end{array} \right)$$

Each generation forms a weak isospin doublet.

τ− + τ+ → Z0 + Z0,

but not these:

γ → e− + μ+,

W− → e− + ν

τ,

Z0 → μ− + τ+.

However, neutrino oscillations are known to violate the conservation of the individual leptonic numbers. Such a violation is considered to be smoking gun evidence for physics beyond the Standard Model. A much stronger conservation law is the conservation of the total number of leptons (*L*), conserved even in the case of neutrino oscillations, but even it is still violated by a tiny amount by the chiral anomaly.

15.4 Universality

The coupling of the leptons to gauge bosons are flavour-independent (i.e., the interactions between leptons and gauge bosons are the same for all leptons).[35] This property is called *lepton universality* and has been tested in measurements of the tau and muon lifetimes and of Z boson partial decay widths, particularly at the Stanford Linear Collider (SLC) and Large Electron-Positron Collider (LEP) experiments.[36]:241–243[37]:138

The decay rate (Γ) of muons through the process μ− → e− + ν

e + ν

μ is approximately given by an expression of the form (see muon decay for more details)[35]

$$\Gamma \left(\mu^- \to e^- + \bar{\nu}_e + \nu_\mu \right) = K_1 G_F^2 m_\mu^5,$$

where K_1 is some constant, and *GF* is the Fermi coupling constant. The decay rate of tau particles through the process

τ− → e− + ν

e + ν

τ is given by an expression of the same form[35]

$$\Gamma \left(\tau^- \to e^- + \bar{\nu}_e + \nu_\tau \right) = K_2 G_F^2 m_\tau^5,$$

where K_2 is some constant. Muon–Tauon universality implies that $K_1 = K_2$. On the other hand, electron–muon universality implies[35]

$$\Gamma\left(\tau^{-} \to e^{-} + \bar{\nu}_e + \nu_\tau\right) = \Gamma\left(\tau^{-} \to \mu^{-} + \bar{\nu}_\mu + \nu_\tau\right).$$

This explains why the branching ratios for the electronic mode (17.85%) and muonic (17.36%) mode of tau decay are equal (within error).[21]

Universality also accounts for the ratio of muon and tau lifetimes. The lifetime of a lepton (τ_l) is related to the decay rate by[35]

$$\tau_l = \frac{B\left(l^{-} \to e^{-} + \bar{\nu}_e + \nu_l\right)}{\Gamma\left(l^{-} \to e^{-} + \bar{\nu}_e + \nu_l\right)},$$

where $B(x \to y)$ and $\Gamma(x \to y)$ denotes the branching ratios and the resonance width of the process $x \to y$.

The ratio of tau and muon lifetime is thus given by[35]

$$\frac{\tau_\tau}{\tau_\mu} = \frac{B\left(\tau^{-} \to e^{-} + \bar{\nu}_e + \nu_\tau\right)}{B\left(\mu^{-} \to e^{-} + \bar{\nu}_e + \nu_\mu\right)} \left(\frac{m_\mu}{m_\tau}\right)^5.$$

Using the values of the 2008 *Review of Particle Physics* for the branching ratios of muons[19] and tau[21] yields a lifetime ratio of ~1.29×10^{-7}, comparable to the measured lifetime ratio of ~1.32×10^{-7}. The difference is due to K_1 and K_2 not actually being constants; they depend on the mass of leptons.

15.5 Table of leptons

15.6 See also

- Koide formula

- List of particles

- Preons – hypothetical particles which were once postulated to be subcomponents of quarks and leptons

15.7 Notes

[1] "Lepton (physics)". *Encyclopædia Britannica*. Retrieved 2010-09-29.

[2] R. Nave. "Leptons". *HyperPhysics*. Georgia State University, Department of Physics and Astronomy. Retrieved 2010-09-29.

[3] W.V. Farrar (1969). "Richard Laming and the Coal-Gas Industry, with His Views on the Structure of Matter". *Annals of Science* **25** (3): 243–254. doi:10.1080/00033796900200141.

[4] T. Arabatzis (2006). *Representing Electrons: A Biographical Approach to Theoretical Entities*. University of Chicago Press. pp. 70–74. ISBN 0-226-02421-0.

[5] J.Z. Buchwald, A. Warwick (2001). *Histories of the Electron: The Birth of Microphysics*. MIT Press. pp. 195–203. ISBN 0-262-52424-4.

[6] J.J. Thomson (1897). "Cathode Rays". *Philosophical Magazine* **44** (269): 293. doi:10.1080/14786449708621070.

[7] S.H. Neddermeyer, C.D. Anderson; Anderson (1937). "Note on the Nature of Cosmic-Ray Particles". *Physical Review* **51** (10): 884–886. Bibcode:1937PhRv...51..884N. doi:10.1103/PhysRev.51.884.

[8] "The Reines-Cowan Experiments: Detecting the Poltergeist" (PDF). *Los Alamos Science* **25**: 3. 1997. Retrieved 2010-02-10.

[9] F. Reines, C.L. Cowan, Jr.; Cowan (1956). "The Neutrino". *Nature* **178** (4531): 446. Bibcode:1956Natur.178..446R. doi:10.1038/178446a0.

[10] G. Danby; Gaillard, J-M.; Goulianos, K.; Lederman, L.; Mistry, N.; Schwartz, M.; Steinberger, J.; et al. (1962). "Observation of high-energy neutrino reactions and the existence of two kinds of neutrinos". *Physical Review Letters* **9**: 36. Bibcode:1962PhRvL.. .9...36D.doi:10.1103/PhysRevLett.9.36.

[11] M.L. Perl; Abrams, G.; Boyarski, A.; Breidenbach, M.; Briggs, D.; Bulos, F.; Chinowsky, W.; Dakin, J.; Feldman, G.; Friedberg, C.; Fryberger, D.; Goldhaber, G.; Hanson, G.; Heile, F.; Jean-Marie, B.; Kadyk, J.; Larsen, R.; Litke, A.; Lüke, D.; Lulu, B.; Lüth, V.; Lyon, D.; Morehouse, C.; Paterson, J.; Pierre, F.; Pun, T.; Rapidis, P.; Richter, B.; Sadoulet, B.; et al. (1975). "Evidence for Anomalous Lepton Production in e+e− Annihilation". *Physical Review Letters* **35** (22): 1489. Bibcode:1975PhRvL..35.1489P. doi:10.1103/PhysRevLett.35.1489.

[12] "Physicists Find First Direct Evidence for Tau Neutrino at Fermilab" (Press release). Fermilab. 20 July 2000.

[13] K. Kodama *et al.* (DONUT Collaboration); Kodama; Ushida; Andreopoulos; Saoulidou; Tzanakos; Yager; Baller; Boehnlein; Freeman; Lundberg; Morfin; Rameika; Yun; Song; Yoon; Chung; Berghaus; Kubantsev; Reay; Sidwell; Stanton; Yoshida; Aoki; Hara; Rhee; Ciampa; Erickson; Graham; et al. (2001). "Observation of tau neutrino interactions". *Physics Letters B* **504** (3): 218. arXiv:hep-ex/0012035. Bibcode:2001PhLB..504..218D. doi:10.1016/S0370-2693(01)00307-0.

[14] "lepton". *Online Etymology Dictionary*.

[15] λεπτός. Liddell, Henry George; Scott, Robert; *A Greek–English Lexicon* at the Perseus Project.

[16] Found on the KN L 693 and PY Un 1322 tablets. "The Linear B word re-po-to". Palaeolexicon. Word study tool of ancient languages. Raymoure, K.A. "re-po-to". *Minoan Linear A & Mycenaean Linear B*. Deaditerranean. "KN 693 L (103)". "PY 1322 Un + fr. (Cii)". *DĀMOS: Database of Mycenaean at Oslo*. University of Oslo.

[17] L. Rosenfeld (1948)

[18] C. Amsler *et al.* (2008): Particle listings − e−

[19] C. Amsler *et al.* (2008): Particle listings − µ−

[20] C. Amsler *et al.* (2008): Particle listings − p+

[21] C. Amsler *et al.* (2008): Particle listings − τ−

[22] S. Weinberg (2003)

[23] R. Wilson (1997)

[24] K. Riesselmann (2007)

[25] S.H. Neddermeyer, C.D. Anderson (1937)

[26] I.V. Anicin (2005)

[27] M.L. Perl et al. (1975)

[28] K. Kodama (2001)

[29] C. Amsler *et al.* (2008) Heavy Charged Leptons Searches

[30] C. Amsler *et al.* (2008) Searches for Heavy Neutral Leptons

[31] M.E. Peskin, D.V. Schroeder (1995), p. 197

[32] M.E. Peskin, D.V. Schroeder (1995), p. 27

[33] Y. Fukuda *et al.* (1998)

[34] C.Amsler et al. (2008): Particle listings − Neutrino properties

[35] B.R. Martin, G. Shaw (1992)

[36] J. P. Cumalat (1993). *Physics in Collision 12*. Atlantica Séguier Frontières. ISBN 978-2-86332-129-4.

[37] G Fraser (1 January 1998). *The Particle Century*. CRC Press. ISBN 978-1-4200-5033-2.

[38] J. Peltoniemi, J. Sarkamo (2005)

15.8 References

- C. Amsler *et al.* (Particle Data Group); Amsler; Doser; Antonelli; Asner; Babu; Baer; Band; Barnett; Bergren; Beringer; Bernardi; Bertl; Bichsel; Biebel; Bloch; Blucher; Blusk; Cahn; Carena; Caso; Ceccucci; Chakraborty; Chen; Chivukula; Cowan; Dahl; d'Ambrosio; Damour; et al. (2008). "Review of Particle Physics". *Physics Letters B* **667**: 1. Bibcode:2008PhLB..667....1P. doi:10.1016/j.physletb.2008.07.018.

- I.V. Anicin (2005). "The Neutrino – Its Past, Present and Future". *SFIN (Institute of Physics, Belgrade) year XV, Series A: Conferences, No. A2 (2002) 3–59*: 3172. arXiv:physics/0503172. Bibcode:2005physics...3172A.

- Y.Fukuda; Hayakawa, T.; Ichihara, E.; Inoue, K.; Ishihara, K.; Ishino, H.; Itow, Y.; Kajita, T.; et al. (1998). "Evidence for Oscillation of Atmospheric Neutrinos". *Physical Review Letters* **81** (8): 1562–1567. arXiv:hep-ex/9807003. Bibcode:1998PhRvL..81.1562F. doi:10.1103/PhysRevLett.81.1562.

- K. Kodama; Ushida, N.; Andreopoulos, C.; Saoulidou, N.; Tzanakos, G.; Yager, P.; Baller, B.; Boehnlein, D.; Freeman, W.; Lundberg, B.; Morfin, J.; Rameika, R.; Yun, J.C.; Song, J.S.; Yoon, C.S.; Chung, S.H.; Berghaus, P.; Kubantsev, M.; Reay, N.W.; Sidwell, R.; Stanton, N.; Yoshida, S.; Aoki, S.; Hara, T.; Rhee, J.T.; Ciampa, D.; Erickson, C.; Graham, M.; Heller, K.; et al. (2001). "Observation of tau neutrino interactions". *Physics Letters B* **504** (3): 218. arXiv:hep-ex/0012035. Bibcode:2001PhLB..504..218D. doi:10.1016/S0370-2693(01)00307-0.

- B.R. Martin, G. Shaw (1992). "Chapter 2 – Leptons, quarks and hadrons". *Particle Physics*. John Wiley & Sons. pp. 23–47. ISBN 0-471-92358-3.

- S.H. Neddermeyer, C.D. Anderson; Anderson (1937). "Note on the Nature of Cosmic-Ray Particles". *Physical Review* **51** (10): 884–886. Bibcode:1937PhRv...51..884N. doi:10.1103/PhysRev.51.884.

- J. Peltoniemi, J. Sarkamo (2005). "Laboratory measurements and limits for neutrino properties". *The Ultimate Neutrino Page*. Retrieved 2008-11-07. External link in |work= (help)

- M.L. Perl; Abrams, G.; Boyarski, A.; Breidenbach, M.; Briggs, D.; Bulos, F.; Chinowsky, W.; Dakin, J.; et al. (1975). "Evidence for Anomalous Lepton Production in e+–e− Annihilation". *Physical Review Letters* **35** (22): 1489–1492. Bibcode:1975PhRvL..35.1489P. doi:10.1103/PhysRevLett.35.1489.

- M.E. Peskin, D.V. Schroeder (1995). *Introduction to Quantum Field Theory*. Westview Press. ISBN 0-201-50397-2.

- K. Riesselmann (2007). "Logbook: Neutrino Invention". *Symmetry Magazine* **4** (2).

- L. Rosenfeld (1948). *Nuclear Forces*. Interscience Publishers. p. xvii.

- R. Shankar (1994). "Chapter 2 – Rotational Invariance and Angular Momentum". *Principles of Quantum Mechanics* (2nd ed.). Springer. pp. 305–352. ISBN 978-0-306-44790-7.

- S. Weinberg (2003). *The Discovery of Subatomic Particles*. Cambridge University Press. ISBN 0-521-82351-X.

- R. Wilson (1997). *Astronomy Through the Ages: The Story of the Human Attempt to Understand the Universe*. CRC Press. p. 138. ISBN 0-7484-0748-0.

15.9 External links

- Particle Data Group homepage. The PDG compiles authoritative information on particle properties.

- Leptons, a summary of leptons from *Hyperphysics*.

Chapter 16

Quark

This article is about the particle. For other uses, see Quark (disambiguation).

A **quark** (/ˈkwɔrk/ or /ˈkwɑrk/) is an elementary particle and a fundamental constituent of matter. Quarks combine to form composite particles called hadrons, the most stable of which are protons and neutrons, the components of atomic nuclei.[1] Due to a phenomenon known as *color confinement*, quarks are never directly observed or found in isolation; they can be found only within hadrons, such as baryons (of which protons and neutrons are examples), and mesons.[2][3] For this reason, much of what is known about quarks has been drawn from observations of the hadrons themselves.

Quarks have various intrinsic properties, including electric charge, mass, color charge and spin. Quarks are the only elementary particles in the Standard Model of particle physics to experience all four fundamental interactions, also known as *fundamental forces* (electromagnetism, gravitation, strong interaction, and weak interaction), as well as the only known particles whose electric charges are not integer multiples of the elementary charge.

There are six types of quarks, known as *flavors*: up, down, strange, charm, top, and bottom.[4] Up and down quarks have the lowest masses of all quarks. The heavier quarks rapidly change into up and down quarks through a process of particle decay: the transformation from a higher mass state to a lower mass state. Because of this, up and down quarks are generally stable and the most common in the universe, whereas strange, charm, bottom, and top quarks can only be produced in high energy collisions (such as those involving cosmic rays and in particle accelerators). For every quark flavor there is a corresponding type of antiparticle, known as an *antiquark*, that differs from the quark only in that some of its properties have equal magnitude but opposite sign.

The quark model was independently proposed by physicists Murray Gell-Mann and George Zweig in 1964.[5] Quarks were introduced as parts of an ordering scheme for hadrons, and there was little evidence for their physical existence until deep inelastic scattering experiments at the Stanford Linear Accelerator Center in 1968.[6][7] Accelerator experiments have provided evidence for all six flavors. The top quark was the last to be discovered at Fermilab in 1995.[5]

16.1 Classification

See also: Standard Model

The Standard Model is the theoretical framework describing all the currently known elementary particles. This model contains six flavors of quarks (q), named up (u), down (d), strange (s), charm (c), bottom (b), and top (t).[4] Antiparticles of quarks are called *antiquarks*, and are denoted by a bar over the symbol for the corresponding quark, such as u for an up antiquark. As with antimatter in general, antiquarks have the same mass, mean lifetime, and spin as their respective quarks, but the electric charge and other charges have the opposite sign.[8]

Quarks are spin-$\frac{1}{2}$ particles, implying that they are fermions according to the spin-statistics theorem. They are subject to the Pauli exclusion principle, which states that no two identical fermions can simultaneously occupy the same quantum state. This is in contrast to bosons (particles with integer spin), any number of which can be in the same state.[9] Unlike

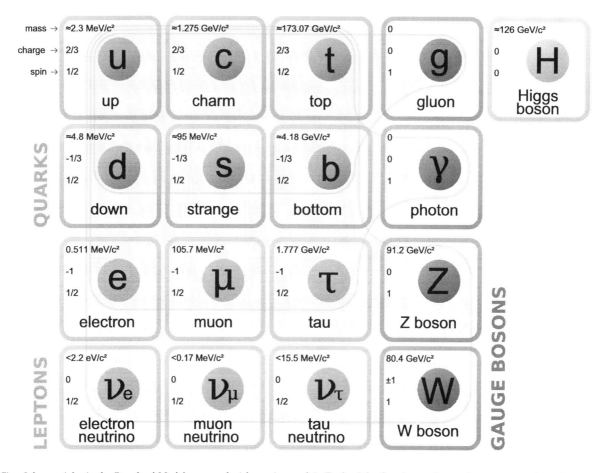

Six of the particles in the Standard Model are quarks (shown in purple). Each of the first three columns forms a generation *of matter.*

leptons, quarks possess color charge, which causes them to engage in the strong interaction. The resulting attraction between different quarks causes the formation of composite particles known as *hadrons* (see "Strong interaction and color charge" below).

The quarks which determine the quantum numbers of hadrons are called *valence quarks*; apart from these, any hadron may contain an indefinite number of virtual (or *sea*) quarks, antiquarks, and gluons which do not influence its quantum numbers.[10] There are two families of hadrons: baryons, with three valence quarks, and mesons, with a valence quark and an antiquark.[11] The most common baryons are the proton and the neutron, the building blocks of the atomic nucleus.[12] A great number of hadrons are known (see list of baryons and list of mesons), most of them differentiated by their quark content and the properties these constituent quarks confer. The existence of "exotic" hadrons with more valence quarks, such as tetraquarks (qqqq) and pentaquarks (qqqqq), has been conjectured[13] but not proven.[nb 1][13][14] However, on 13 July 2015, the LHCb collaboration at CERN reported results consistent with pentaquark states.[15]

Elementary fermions are grouped into three generations, each comprising two leptons and two quarks. The first generation includes up and down quarks, the second strange and charm quarks, and the third bottom and top quarks. All searches for a fourth generation of quarks and other elementary fermions have failed,[16] and there is strong indirect evidence that no more than three generations exist.[nb 2][17] Particles in higher generations generally have greater mass and less stability, causing them to decay into lower-generation particles by means of weak interactions. Only first-generation (up and down) quarks occur commonly in nature. Heavier quarks can only be created in high-energy collisions (such as in those involving cosmic rays), and decay quickly; however, they are thought to have been present during the first fractions of a second after the Big Bang, when the universe was in an extremely hot and dense phase (the quark epoch). Studies of heavier quarks are conducted in artificially created conditions, such as in particle accelerators.[18]

Having electric charge, mass, color charge, and flavor, quarks are the only known elementary particles that engage in all four fundamental interactions of contemporary physics: electromagnetism, gravitation, strong interaction, and weak interaction.[12] Gravitation is too weak to be relevant to individual particle interactions except at extremes of energy (Planck energy) and distance scales (Planck distance). However, since no successful quantum theory of gravity exists, gravitation is not described by the Standard Model.

See the table of properties below for a more complete overview of the six quark flavors' properties.

16.2   History

The quark model was independently proposed by physicists Murray Gell-Mann[19] (pictured) and George Zweig[20][21] in 1964.[5] The proposal came shortly after Gell-Mann's 1961 formulation of a particle classification system known as the *Eightfold Way*—or, in more technical terms, SU(3) flavor symmetry.[22] Physicist Yuval Ne'eman had independently developed a scheme similar to the Eightfold Way in the same year.[23][24]

At the time of the quark theory's inception, the "particle zoo" included, amongst other particles, a multitude of hadrons. Gell-Mann and Zweig posited that they were not elementary particles, but were instead composed of combinations of quarks and antiquarks. Their model involved three flavors of quarks, up, down, and strange, to which they ascribed properties such as spin and electric charge.[19][20][21] The initial reaction of the physics community to the proposal was mixed. There was particular contention about whether the quark was a physical entity or a mere abstraction used to explain concepts that were not fully understood at the time.[25]

In less than a year, extensions to the Gell-Mann–Zweig model were proposed. Sheldon Lee Glashow and James Bjorken predicted the existence of a fourth flavor of quark, which they called *charm*. The addition was proposed because it allowed for a better description of the weak interaction (the mechanism that allows quarks to decay), equalized the number of known quarks with the number of known leptons, and implied a mass formula that correctly reproduced the masses of the known mesons.[26]

In 1968, deep inelastic scattering experiments at the Stanford Linear Accelerator Center (SLAC) showed that the proton contained much smaller, point-like objects and was therefore not an elementary particle.[6][7][27] Physicists were reluctant to firmly identify these objects with quarks at the time, instead calling them "partons"—a term coined by Richard Feynman.[28][29][30] The objects that were observed at SLAC would later be identified as up and down quarks as the other flavors were discovered.[31] Nevertheless, "parton" remains in use as a collective term for the constituents of hadrons (quarks, antiquarks, and gluons).

The strange quark's existence was indirectly validated by SLAC's scattering experiments: not only was it a necessary component of Gell-Mann and Zweig's three-quark model, but it provided an explanation for the kaon (K) and pion (π) hadrons discovered in cosmic rays in 1947.[32]

In a 1970 paper, Glashow, John Iliopoulos and Luciano Maiani presented the so-called GIM mechanism to explain the experimental non-observation of flavor-changing neutral currents. This theoretical model required the existence of the as-yet undiscovered charm quark.[33][34] The number of supposed quark flavors grew to the current six in 1973, when Makoto Kobayashi and Toshihide Maskawa noted that the experimental observation of CP violation[nb 3][35] could be explained if there were another pair of quarks.

Charm quarks were produced almost simultaneously by two teams in November 1974 (see November Revolution)—one at SLAC under Burton Richter, and one at Brookhaven National Laboratory under Samuel Ting. The charm quarks were observed bound with charm antiquarks in mesons. The two parties had assigned the discovered meson two different symbols, J and ψ; thus, it became formally known as the J/ψ meson. The discovery finally convinced the physics community of the quark model's validity.[30]

In the following years a number of suggestions appeared for extending the quark model to six quarks. Of these, the 1975 paper by Haim Harari[36] was the first to coin the terms *top* and *bottom* for the additional quarks.[37]

In 1977, the bottom quark was observed by a team at Fermilab led by Leon Lederman.[38][39] This was a strong indicator of the top quark's existence: without the top quark, the bottom quark would have been without a partner. However, it was not until 1995 that the top quark was finally observed, also by the CDF[40] and DØ[41] teams at Fermilab.[5] It had a mass much larger than had been previously expected,[42] almost as large as that of a gold atom.[43]

Murray Gell-Mann at TED in 2007. Gell-Mann and George Zweig proposed the quark model in 1964.

16.3 Etymology

For some time, Gell-Mann was undecided on an actual spelling for the term he intended to coin, until he found the word *quark* in James Joyce's book *Finnegans Wake*:

> Three quarks for Muster Mark!
> Sure he has not got much of a bark

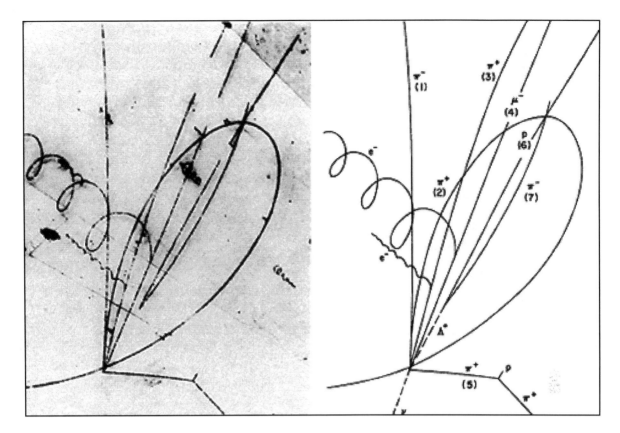

Photograph of the event that led to the discovery of the Σ++
c baryon, at the Brookhaven National Laboratory in 1974

> And sure any he has it's all beside the mark.
> — James Joyce, *Finnegans Wake*[44]

Gell-Mann went into further detail regarding the name of the quark in his book *The Quark and the Jaguar*:[45]

> In 1963, when I assigned the name "quark" to the fundamental constituents of the nucleon, I had the sound first, without the spelling, which could have been "kwork". Then, in one of my occasional perusals of *Finnegans Wake*, by James Joyce, I came across the word "quark" in the phrase "Three quarks for Muster Mark". Since "quark" (meaning, for one thing, the cry of the gull) was clearly intended to rhyme with "Mark", as well as "bark" and other such words, I had to find an excuse to pronounce it as "kwork". But the book represents the dream of a publican named Humphrey Chimpden Earwicker. Words in the text are typically drawn from several sources at once, like the "portmanteau" words in *Through the Looking-Glass*. From time to time, phrases occur in the book that are partially determined by calls for drinks at the bar. I argued, therefore, that perhaps one of the multiple sources of the cry "Three quarks for Muster Mark" might be "Three quarts for Mister Mark", in which case the pronunciation "kwork" would not be totally unjustified. In any case, the number three fitted perfectly the way quarks occur in nature.

Zweig preferred the name *ace* for the particle he had theorized, but Gell-Mann's terminology came to prominence once the quark model had been commonly accepted.[46]

The quark flavors were given their names for several reasons. The up and down quarks are named after the up and down components of isospin, which they carry.[47] Strange quarks were given their name because they were discovered to be components of the strange particles discovered in cosmic rays years before the quark model was proposed; these particles were deemed "strange" because they had unusually long lifetimes.[48] Glashow, who coproposed charm quark with Bjorken, is quoted as saying, "We called our construct the 'charmed quark', for we were fascinated and pleased by the symmetry it brought to the subnuclear world."[49] The names "bottom" and "top", coined by Harari, were chosen because they are "logical partners for up and down quarks".[36][37][48] In the past, bottom and top quarks were sometimes referred to as "beauty" and "truth" respectively, but these names have somewhat fallen out of use.[50] While "truth" never did catch on, accelerator complexes devoted to massive production of bottom quarks are sometimes called "beauty factories".[51]

16.4 Properties

16.4.1 Electric charge

See also: Electric charge

Quarks have fractional electric charge values – either $\frac{1}{3}$ or $\frac{2}{3}$ times the elementary charge (e), depending on flavor. Up, charm, and top quarks (collectively referred to as *up-type quarks*) have a charge of $+\frac{2}{3}$ e, while down, strange, and bottom quarks (*down-type quarks*) have $-\frac{1}{3}$ e. Antiquarks have the opposite charge to their corresponding quarks; up-type antiquarks have charges of $-\frac{2}{3}$ e and down-type antiquarks have charges of $+\frac{1}{3}$ e. Since the electric charge of a hadron is the sum of the charges of the constituent quarks, all hadrons have integer charges: the combination of three quarks (baryons), three antiquarks (antibaryons), or a quark and an antiquark (mesons) always results in integer charges.[52] For example, the hadron constituents of atomic nuclei, neutrons and protons, have charges of 0 e and +1 e respectively; the neutron is composed of two down quarks and one up quark, and the proton of two up quarks and one down quark.[12]

16.4.2 Spin

See also: Spin (physics)

Spin is an intrinsic property of elementary particles, and its direction is an important degree of freedom. It is sometimes visualized as the rotation of an object around its own axis (hence the name "spin"), though this notion is somewhat misguided at subatomic scales because elementary particles are believed to be point-like.[53]

Spin can be represented by a vector whose length is measured in units of the reduced Planck constant $\hbar$ (pronounced "h bar"). For quarks, a measurement of the spin vector component along any axis can only yield the values $+\hbar/2$ or $-\hbar/2$; for this reason quarks are classified as spin-$\frac{1}{2}$ particles.[54] The component of spin along a given axis – by convention the z axis – is often denoted by an up arrow ↑ for the value $+\frac{1}{2}$ and down arrow ↓ for the value $-\frac{1}{2}$, placed after the symbol for flavor. For example, an up quark with a spin of $+\frac{1}{2}$ along the z axis is denoted by u↑.[55]

16.4.3 Weak interaction

Main article: Weak interaction
 A quark of one flavor can transform into a quark of another flavor only through the weak interaction, one of the four fundamental interactions in particle physics. By absorbing or emitting a W boson, any up-type quark (up, charm, and top quarks) can change into any down-type quark (down, strange, and bottom quarks) and vice versa. This flavor transformation mechanism causes the radioactive process of beta decay, in which a neutron (n) "splits" into a proton (p), an electron (e−) and an electron antineutrino (ν
e) (see picture). This occurs when one of the down quarks in the neutron (udd) decays into an up quark by emitting a virtual W− boson, transforming the neutron into a proton (uud). The W− boson then decays into an electron and an electron antineutrino.[56]

Both beta decay and the inverse process of *inverse beta decay* are routinely used in medical applications such as positron emission tomography (PET) and in experiments involving neutrino detection.

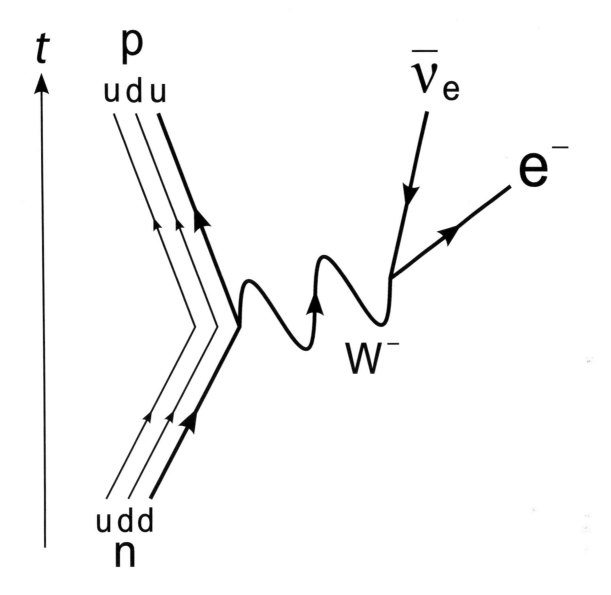

Feynman diagram of beta decay with time flowing upwards. The CKM matrix (discussed below) encodes the probability of this and other quark decays.

While the process of flavor transformation is the same for all quarks, each quark has a preference to transform into the quark of its own generation. The relative tendencies of all flavor transformations are described by a mathematical table, called the Cabibbo–Kobayashi–Maskawa matrix (CKM matrix). Enforcing unitarity, the approximate magnitudes of the entries of the CKM matrix are:[57]

$$\begin{bmatrix} |V_{ud}| & |V_{us}| & |V_{ub}| \\ |V_{cd}| & |V_{cs}| & |V_{cb}| \\ |V_{td}| & |V_{ts}| & |V_{tb}| \end{bmatrix} \approx \begin{bmatrix} 0.974 & 0.225 & 0.003 \\ 0.225 & 0.973 & 0.041 \\ 0.009 & 0.040 & 0.999 \end{bmatrix},$$

where Vij represents the tendency of a quark of flavor i to change into a quark of flavor j (or vice versa).[nb 4]

There exists an equivalent weak interaction matrix for leptons (right side of the W boson on the above beta decay diagram), called the Pontecorvo–Maki–Nakagawa–Sakata matrix (PMNS matrix).[58] Together, the CKM and PMNS matrices

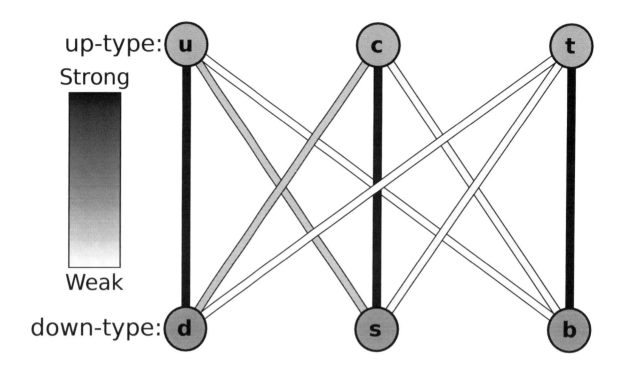

up-type:

Strong

Weak

down-type:

The strengths of the weak interactions between the six quarks. The "intensities" of the lines are determined by the elements of the CKM matrix.

describe all flavor transformations, but the links between the two are not yet clear.[59]

16.4.4 Strong interaction and color charge

See also: Color charge and Strong interaction

According to quantum chromodynamics (QCD), quarks possess a property called *color charge*. There are three types of color charge, arbitrarily labeled *blue*, *green*, and *red*.[nb 5] Each of them is complemented by an anticolor – *antiblue*, *antigreen*, and *antired*. Every quark carries a color, while every antiquark carries an anticolor.[60]

The system of attraction and repulsion between quarks charged with different combinations of the three colors is called strong interaction, which is mediated by force carrying particles known as *gluons*; this is discussed at length below. The theory that describes strong interactions is called quantum chromodynamics (QCD). A quark, which will have a single color value, can form a bound system with an antiquark carrying the corresponding anticolor. The result of two attracting quarks will be color neutrality: a quark with color charge ξ plus an antiquark with color charge $-\xi$ will result in a color charge of 0 (or "white" color) and the formation of a meson. This is analogous to the additive color model in basic optics. Similarly, the combination of three quarks, each with different color charges, or three antiquarks, each with anticolor charges, will result in the same "white" color charge and the formation of a baryon or antibaryon.[61]

In modern particle physics, gauge symmetries – a kind of symmetry group – relate interactions between particles (see gauge theories). Color SU(3) (commonly abbreviated to SU(3)$_c$) is the gauge symmetry that relates the color charge in quarks and is the defining symmetry for quantum chromodynamics.[62] Just as the laws of physics are independent of which directions in space are designated x, y, and z, and remain unchanged if the coordinate axes are rotated to a new orientation, the physics of quantum chromodynamics is independent of which directions in three-dimensional color space are identified as blue, red, and green. SU(3)$_c$ color transformations correspond to "rotations" in color space (which, mathematically speaking, is a complex space). Every quark flavor f, each with subtypes fB, fG, fR corresponding to the quark colors,[63] forms a triplet: a three-component quantum field which transforms under the fundamental representation of $SU(3)_c$.[64] The requirement that $SU(3)_c$ should be local – that is, that its transformations be allowed to vary with space and time – determines the properties of the strong interaction, in particular the existence of eight gluon types to act as its force carriers.[62][65]

16.4.5 Mass

See also: Invariant mass

Two terms are used in referring to a quark's mass: *current quark mass* refers to the mass of a quark by itself, while *constituent quark mass* refers to the current quark mass plus the mass of the gluon particle field surrounding the quark.[66] These masses typically have very different values. Most of a hadron's mass comes from the gluons that bind the constituent quarks together, rather than from the quarks themselves. While gluons are inherently massless, they possess energy – more specifically, quantum chromodynamics binding energy (QCBE) – and it is this that contributes so greatly to the overall mass of the hadron (see mass in special relativity). For example, a proton has a mass of approximately 938 MeV/c^2, of which the rest mass of its three valence quarks only contributes about 11 MeV/c^2; much of the remainder can be attributed to the gluons' QCBE.[67][68]

The Standard Model posits that elementary particles derive their masses from the Higgs mechanism, which is related to the Higgs boson. Physicists hope that further research into the reasons for the top quark's large mass of ~173 GeV/c^2, almost the mass of a gold atom,[67][69] might reveal more about the origin of the mass of quarks and other elementary particles.[70]

16.4.6 Table of properties

See also: Flavor (particle physics)

The following table summarizes the key properties of the six quarks. Flavor quantum numbers (isospin (I_3), charm (C), strangeness (S, not to be confused with spin), topness (T), and bottomness (B')) are assigned to certain quark flavors, and denote qualities of quark-based systems and hadrons. The baryon number (B) is $+\frac{1}{3}$ for all quarks, as baryons are made of three quarks. For antiquarks, the electric charge (Q) and all flavor quantum numbers (B, I_3, C, S, T, and B') are of opposite sign. Mass and total angular momentum (J; equal to spin for point particles) do not change sign for the antiquarks.

J = total angular momentum, B = baryon number, Q = electric charge, I_3 = isospin, C = charm, S = strangeness, T = topness, B' = bottomness.

* Notation such as 173210±510 ± 710 denotes two types of measurement uncertainty. In the case of the top quark, the first uncertainty is statistical in nature, and the second is systematic.

16.5 Interacting quarks

See also: Color confinement and Gluon

As described by quantum chromodynamics, the strong interaction between quarks is mediated by gluons, massless vector gauge bosons. Each gluon carries one color charge and one anticolor charge. In the standard framework of particle interactions (part of a more general formulation known as perturbation theory), gluons are constantly exchanged between quarks through a virtual emission and absorption process. When a gluon is transferred between quarks, a color change occurs in both; for example, if a red quark emits a red–antigreen gluon, it becomes green, and if a green quark absorbs a red–antigreen gluon, it becomes red. Therefore, while each quark's color constantly changes, their strong interaction is preserved.[71][72][73]

Since gluons carry color charge, they themselves are able to emit and absorb other gluons. This causes *asymptotic freedom*: as quarks come closer to each other, the chromodynamic binding force between them weakens.[74] Conversely, as the distance between quarks increases, the binding force strengthens. The color field becomes stressed, much as an elastic band is stressed when stretched, and more gluons of appropriate color are spontaneously created to strengthen the field. Above a certain energy threshold, pairs of quarks and antiquarks are created. These pairs bind with the quarks being separated, causing new hadrons to form. This phenomenon is known as *color confinement*: quarks never appear in isolation.[75][76] This process of hadronization occurs before quarks, formed in a high energy collision, are able to interact in any other way. The only exception is the top quark, which may decay before it hadronizes.[77]

16.5.1 Sea quarks

Hadrons, along with the *valence quarks* (q
v) that contribute to their quantum numbers, contain virtual quark–antiquark (qq) pairs known as *sea quarks* (q
s). Sea quarks form when a gluon of the hadron's color field splits; this process also works in reverse in that the annihilation of two sea quarks produces a gluon. The result is a constant flux of gluon splits and creations colloquially known as "the sea".[78] Sea quarks are much less stable than their valence counterparts, and they typically annihilate each other within the interior of the hadron. Despite this, sea quarks can hadronize into baryonic or mesonic particles under certain circumstances.[79]

16.5.2 Other phases of quark matter

Main article: QCD matter
 Under sufficiently extreme conditions, quarks may become deconfined and exist as free particles. In the course of asymptotic freedom, the strong interaction becomes weaker at higher temperatures. Eventually, color confinement would be lost and an extremely hot plasma of freely moving quarks and gluons would be formed. This theoretical phase of matter is called quark–gluon plasma.[82] The exact conditions needed to give rise to this state are unknown and have been the subject of a great deal of speculation and experimentation. A recent estimate puts the needed temperature at $(1.90\pm0.02)\times10^{12}$ kelvin.[83] While a state of entirely free quarks and gluons has never been achieved (despite numerous attempts by CERN in the 1980s and 1990s),[84] recent experiments at the Relativistic Heavy Ion Collider have yielded evidence for liquid-like quark matter exhibiting "nearly perfect" fluid motion.[85]

The quark–gluon plasma would be characterized by a great increase in the number of heavier quark pairs in relation to the number of up and down quark pairs. It is believed that in the period prior to 10^{-6} seconds after the Big Bang (the quark epoch), the universe was filled with quark–gluon plasma, as the temperature was too high for hadrons to be stable.[86]

Given sufficiently high baryon densities and relatively low temperatures – possibly comparable to those found in neutron stars – quark matter is expected to degenerate into a Fermi liquid of weakly interacting quarks. This liquid would be characterized by a condensation of colored quark Cooper pairs, thereby breaking the local $SU(3)_c$ symmetry. Because quark Cooper pairs harbor color charge, such a phase of quark matter would be color superconductive; that is, color charge would be able to pass through it with no resistance.[87]

16.6   See also

- Color–flavor locking

- Neutron magnetic moment

- Leptons

- Preons – Hypothetical particles which were once postulated to be subcomponents of quarks and leptons

- Quarkonium – Mesons made of a quark and antiquark of the same flavor

- Quark star – A hypothetical degenerate neutron star with extreme density

- Quark–lepton complementarity – Possible fundamental relation between quarks and leptons

16.7   Notes

[1] Several research groups claimed to have proven the existence of tetraquarks and pentaquarks in the early 2000s. While the status of tetraquarks is still under debate, all known pentaquark candidates have previously been established as non-existent.

[2] The main evidence is based on the resonance width of the Z0 boson, which constrains the 4th generation neutrino to have a mass greater than ~45 GeV/c^2. This would be highly contrasting with the other three generations' neutrinos, whose masses cannot exceed 2 MeV/c^2.

[3] CP violation is a phenomenon which causes weak interactions to behave differently when left and right are swapped (P symmetry) and particles are replaced with their corresponding antiparticles (C symmetry).

[4] The actual probability of decay of one quark to another is a complicated function of (amongst other variables) the decaying quark's mass, the masses of the decay products, and the corresponding element of the CKM matrix. This probability is directly proportional (but not equal) to the magnitude squared ($|Vij|^2$) of the corresponding CKM entry.

[5] Despite its name, color charge is not related to the color spectrum of visible light.

16.8   References

[1] "Quark (subatomic particle)". *Encyclopædia Britannica*. Retrieved 2008-06-29.

[2] R. Nave. "Confinement of Quarks". *HyperPhysics*. Georgia State University, Department of Physics and Astronomy. Retrieved 2008-06-29.

[3] R. Nave. "Bag Model of Quark Confinement". *HyperPhysics*. Georgia State University, Department of Physics and Astronomy. Retrieved 2008-06-29.

[4] R. Nave. "Quarks". *HyperPhysics*. Georgia State University, Department of Physics and Astronomy. Retrieved 2008-06-29.

[5] B. Carithers, P. Grannis (1995). "Discovery of the Top Quark" (PDF). *Beam Line* (SLAC) **25** (3): 4–16. Retrieved 2008-09-23.

[6] E.D. Bloom; et al. (1969). "High-Energy Inelastic *e–p* Scattering at 6° and 10°". *Physical Review Letters* **23** (16): 930–934. Bibcode:1969PhRvL..23..930B. doi:10.1103/PhysRevLett.23.930.

[7] M. Breidenbach; et al. (1969). "Observed Behavior of Highly Inelastic Electron–Proton Scattering". *Physical Review Letters* **23** (16): 935–939. Bibcode:1969PhRvL..23..935B. doi:10.1103/PhysRevLett.23.935.

[8] S.S.M. Wong (1998). *Introductory Nuclear Physics* (2nd ed.). Wiley Interscience. p. 30. ISBN 0-471-23973-9.

[9] K.A. Peacock (2008). *The Quantum Revolution*. Greenwood Publishing Group. p. 125. ISBN 0-313-33448-X.

[10] B. Povh, C. Scholz, K. Rith, F. Zetsche (2008). *Particles and Nuclei*. Springer. p. 98. ISBN 3-540-79367-4.

[11] Section 6.1. in P.C.W. Davies (1979). *The Forces of Nature*. Cambridge University Press. ISBN 0-521-22523-X.

[12] M. Munowitz (2005). *Knowing*. Oxford University Press. p. 35. ISBN 0-19-516737-6.

[13] W.-M. Yao (Particle Data Group); et al. (2006). "Review of Particle Physics: Pentaquark Update" (PDF). *Journal of Physics G* **33** (1): 1–1232. arXiv:astro-ph/0601168. Bibcode:2006JPhG...33....1Y. doi:10.1088/0954-3899/33/1/001.

[14] C. Amsler (Particle Data Group); et al. (2008). "Review of Particle Physics: Pentaquarks" (PDF). *Physics Letters B* **667** (1): 1–1340. Bibcode:2008PhLB..667....1P. doi:10.1016/j.physletb.2008.07.018.
C. Amsler (Particle Data Group); et al. (2008). "Review of Particle Physics: New Charmonium-Like States" (PDF). *Physics Letters B* **667** (1): 1–1340. Bibcode:2008PhLB..667....1P. doi:10.1016/j.physletb.2008.07.018.
E.V. Shuryak (2004). *The QCD Vacuum, Hadrons and Superdense Matter*. World Scientific. p. 59. ISBN 981-238-574-6.

[15] R. Aaij (LHCb collaboration); et al. (2015). "Observation of J/ψp resonances consistent with pentaquark states in Λ0
      b→J/ψK−
      p decays". *Physical Review Letters* **115** (7). doi:10.1103/PhysRevLett.115.072001.

[16] C. Amsler (Particle Data Group); et al. (2008). "Review of Particle Physics: b′ (4th Generation) Quarks, Searches for" (PDF).
      Physics Letters B **667** (1): 1–1340. Bibcode:2008PhLB..667....1P. doi:10.1016/j.physletb.2008.07.018.
      C. Amsler (Particle Data Group); et al. (2008). "Review of Particle Physics: t′ (4th Generation) Quarks, Searches for" (PDF).
      Physics Letters B **667** (1): 1–1340. Bibcode:2008PhLB..667....1P. doi:10.1016/j.physletb.2008.07.018.

[17] D. Decamp; Deschizeaux, B.; Lees, J.-P.; Minard, M.-N.; Crespo, J.M.; Delfino, M.; Fernandez, E.; Martinez, M.; et al. (1989).
      "Determination of the number of light neutrino species". *Physics Letters B* **231** (4): 519. Bibcode:1989PhLB..231..519D.
      doi:10.1016/0370-2693(89)90704-1.
      A. Fisher (1991). "Searching for the Beginning of Time: Cosmic Connection". *Popular Science* **238** (4): 70.
      J.D. Barrow (1997) [1994]. "The Singularity and Other Problems". *The Origin of the Universe* (Reprint ed.). Basic Books.
      ISBN 978-0-465-05314-8.

[18] D.H. Perkins (2003). *Particle Astrophysics*. Oxford University Press. p. 4. ISBN 0-19-850952-9.

[19] M. Gell-Mann (1964). "A Schematic Model of Baryons and Mesons". *Physics Letters* **8** (3): 214–215. Bibcode:1964PhL......8..2
      doi:10.1016/S0031-9163(64)92001-3.

[20] G. Zweig (1964). "An SU(3) Model for Strong Interaction Symmetry and its Breaking" (PDF). *CERN Report No.8182/TH.401*.

[21] G. Zweig (1964). "An SU(3) Model for Strong Interaction Symmetry and its Breaking: II". *CERN Report No.8419/TH.412*.

[22] M. Gell-Mann (2000) [1964]. "The Eightfold Way: A theory of strong interaction symmetry". In M. Gell-Mann, Y. Ne'eman.
      The Eightfold Way. Westview Press. p. 11. ISBN 0-7382-0299-1.
      Original: M. Gell-Mann (1961). "The Eightfold Way: A theory of strong interaction symmetry". *Synchrotron Laboratory
      Report CTSL-20* (California Institute of Technology).

[23] Y. Ne'eman (2000) [1964]. "Derivation of strong interactions from gauge invariance". In M. Gell-Mann, Y. Ne'eman. *The
      Eightfold Way*. Westview Press. ISBN 0-7382-0299-1.
      Original Y. Ne'eman (1961). "Derivation of strong interactions from gauge invariance". *Nuclear Physics* **26** (2): 222. Bibcode:
      1961NucPh..26..222N.doi:10.1016/0029-5582(61)90134-1.

[24] R.C. Olby, G.N. Cantor (1996). *Companion to the History of Modern Science*. Taylor & Francis. p. 673. ISBN 0-415-14578-3.

[25] A. Pickering (1984). *Constructing Quarks*. University of Chicago Press. pp. 114–125. ISBN 0-226-66799-5.

[26] B.J. Bjorken, S.L. Glashow; Glashow (1964). "Elementary Particles and SU(4)". *Physics Letters* **11** (3): 255–257. Bibcode:1964
      doi:10.1016/0031-9163(64)90433-0.

[27] J.I. Friedman. "The Road to the Nobel Prize". Hue University. Retrieved 2008-09-29.

[28] R.P. Feynman (1969). "Very High-Energy Collisions of Hadrons". *Physical Review Letters* **23** (24): 1415–1417. Bibcode:1969
      doi:10.1103/PhysRevLett.23.1415.

[29] S. Kretzer; et al. (2004). "CTEQ6 Parton Distributions with Heavy Quark Mass Effects". *Physical Review D* **69** (11): 114005.
      arXiv:hep-ph/0307022. Bibcode:2004PhRvD..69k4005K. doi:10.1103/PhysRevD.69.114005.

[30] D.J. Griffiths (1987). *Introduction to Elementary Particles*. John Wiley & Sons. p. 42. ISBN 0-471-60386-4.

[31] M.E. Peskin, D.V. Schroeder (1995). *An introduction to quantum field theory*. Addison–Wesley. p. 556. ISBN 0-201-50397-2.

[32] V.V. Ezhela (1996). *Particle physics*. Springer. p. 2. ISBN 1-56396-642-5.

[33] S.L. Glashow, J. Iliopoulos, L. Maiani; Iliopoulos; Maiani (1970). "Weak Interactions with Lepton–Hadron Symmetry".
      Physical Review D **2** (7): 1285–1292. Bibcode:1970PhRvD...2.1285G. doi:10.1103/PhysRevD.2.1285.

[34] D.J. Griffiths (1987). *Introduction to Elementary Particles*. John Wiley & Sons. p. 44. ISBN 0-471-60386-4.

[35] M. Kobayashi, T. Maskawa; Maskawa (1973). "CP-Violation in the Renormalizable Theory of Weak Interaction". *Progress of
      Theoretical Physics* **49** (2): 652–657. Bibcode:1973PThPh..49..652K. doi:10.1143/PTP.49.652.

[36] H. Harari (1975). "A new quark model for hadrons". *Physics Letters B* **57B** (3): 265. Bibcode:1975PhLB...57..265H. doi:10.1016/0370-2693(75)90072-6.

[37] K.W. Staley (2004). *The Evidence for the Top Quark*. Cambridge University Press. pp. 31–33. ISBN 978-0-521-82710-2.

[38] S.W. Herb; et al. (1977). "Observation of a Dimuon Resonance at 9.5 GeV in 400-GeV Proton-Nucleus Collisions". *Physical Review Letters* **39** (5): 252. Bibcode:1977PhRvL..39..252H. doi:10.1103/PhysRevLett.39.252.

[39] M. Bartusiak (1994). *A Positron named Priscilla*. National Academies Press. p. 245. ISBN 0-309-04893-1.

[40] F. Abe (CDF Collaboration); et al. (1995). "Observation of Top Quark Production in pp Collisions with the Collider Detector at Fermilab". *Physical Review Letters* **74** (14): 2626–2631. Bibcode:1995PhRvL..74.2626A. doi:10.1103/PhysRevLett.74.2626. PMID 10057978.

[41] S. Abachi (DØ Collaboration); et al. (1995). "Search for High Mass Top Quark Production in pp Collisions at √s = 1.8 TeV". *Physical Review Letters* **74** (13): 2422–2426. Bibcode:1995PhRvL..74.2422A. doi:10.1103/PhysRevLett.74.2422.

[42] K.W. Staley (2004). *The Evidence for the Top Quark*. Cambridge University Press. p. 144. ISBN 0-521-82710-8.

[43] "New Precision Measurement of Top Quark Mass". Brookhaven National Laboratory News. 2004. Retrieved 2013-11-03.

[44] J. Joyce (1982) [1939]. *Finnegans Wake*. Penguin Books. p. 383. ISBN 0-14-006286-6.

[45] M. Gell-Mann (1995). *The Quark and the Jaguar: Adventures in the Simple and the Complex*. Henry Holt and Co. p. 180. ISBN 978-0-8050-7253-2.

[46] J. Gleick (1992). *Genius: Richard Feynman and modern physics*. Little Brown and Company. p. 390. ISBN 0-316-90316-7.

[47] J.J. Sakurai (1994). S.F Tuan, ed. *Modern Quantum Mechanics* (Revised ed.). Addison–Wesley. p. 376. ISBN 0-201-53929-2.

[48] D.H. Perkins (2000). *Introduction to high energy physics*. Cambridge University Press. p. 8. ISBN 0-521-62196-8.

[49] M. Riordan (1987). *The Hunting of the Quark: A True Story of Modern Physics*. Simon & Schuster. p. 210. ISBN 978-0-671-50466-3.

[50] F. Close (2006). *The New Cosmic Onion*. CRC Press. p. 133. ISBN 1-58488-798-2.

[51] J.T. Volk; et al. (1987). "Letter of Intent for a Tevatron Beauty Factory" (PDF). Fermilab Proposal #783.

[52] G. Fraser (2006). *The New Physics for the Twenty-First Century*. Cambridge University Press. p. 91. ISBN 0-521-81600-9.

[53] "The Standard Model of Particle Physics". BBC. 2002. Retrieved 2009-04-19.

[54] F. Close (2006). *The New Cosmic Onion*. CRC Press. pp. 80–90. ISBN 1-58488-798-2.

[55] D. Lincoln (2004). *Understanding the Universe*. World Scientific. p. 116. ISBN 981-238-705-6.

[56] "Weak Interactions". *Virtual Visitor Center*. Stanford Linear Accelerator Center. 2008. Retrieved 2008-09-28.

[57] K. Nakamura; et al. (2010). "Review of Particles Physics: The CKM Quark-Mixing Matrix" (PDF). *J. Phys. G* **37** (75021): 150.

[58] Z. Maki, M. Nakagawa, S. Sakata (1962). "Remarks on the Unified Model of Elementary Particles". *Progress of Theoretical Physics* **28** (5): 870. Bibcode:1962PThPh..28..870M. doi:10.1143/PTP.28.870.

[59] B.C. Chauhan, M. Picariello, J. Pulido, E. Torrente-Lujan (2007). "Quark–lepton complementarity, neutrino and standard model data predict θPMNS
13 = 9°+1°
−2°". *European Physical Journal* **C50** (3): 573–578. arXiv:hep-ph/0605032. Bibcode:2007EPJC...50..573C. doi:10.1140/epjc/s10052-007-0212-z.

[60] R. Nave. "The Color Force". *HyperPhysics*. Georgia State University, Department of Physics and Astronomy. Retrieved 2009-04-26.

[61] B.A. Schumm (2004). *Deep Down Things*. Johns Hopkins University Press. pp. 131–132. ISBN 0-8018-7971-X. OCLC 55229065.

[62] Part III of M.E. Peskin, D.V. Schroeder (1995). *An Introduction to Quantum Field Theory.* Addison–Wesley. ISBN 0-201-50397-2.

[63] V. Icke (1995). *The force of symmetry.* Cambridge University Press. p. 216. ISBN 0-521-45591-X.

[64] M.Y. Han (2004). *A story of light.* World Scientific. p. 78. ISBN 981-256-034-3.

[65] C. Sutton. "Quantum chromodynamics (physics)". *Encyclopædia Britannica Online.* Retrieved 2009-05-12.

[66] A. Watson (2004). *The Quantum Quark.* Cambridge University Press. pp. 285–286. ISBN 0-521-82907-0.

[67] K.A. Olive *et al.* (Particle Data Group), Chin. Phys. **C38**, 090001 (2014) (URL: http://pdg.lbl.gov)

[68] W. Weise, A.M. Green (1984). *Quarks and Nuclei.* World Scientific. pp. 65–66. ISBN 9971-966-61-1.

[69] D. McMahon (2008). *Quantum Field Theory Demystified.* McGraw–Hill. p. 17. ISBN 0-07-154382-1.

[70] S.G. Roth (2007). *Precision electroweak physics at electron–positron colliders.* Springer. p. VI. ISBN 3-540-35164-7.

[71] R.P. Feynman (1985). *QED: The Strange Theory of Light and Matter* (1st ed.). Princeton University Press. pp. 136–137. ISBN 0-691-08388-6.

[72] M. Veltman (2003). *Facts and Mysteries in Elementary Particle Physics.* World Scientific. pp. 45–47. ISBN 981-238-149-X.

[73] F. Wilczek, B. Devine (2006). *Fantastic Realities.* World Scientific. p. 85. ISBN 981-256-649-X.

[74] F. Wilczek, B. Devine (2006). *Fantastic Realities.* World Scientific. pp. 400ff. ISBN 981-256-649-X.

[75] M. Veltman (2003). *Facts and Mysteries in Elementary Particle Physics.* World Scientific. pp. 295–297. ISBN 981-238-149-X.

[76] T. Yulsman (2002). *Origin.* CRC Press. p. 55. ISBN 0-7503-0765-X.

[77] F. Garberson (2008). "Top Quark Mass and Cross Section Results from the Tevatron". arXiv:0808.0273 [hep-ex].

[78] J. Steinberger (2005). *Learning about Particles.* Springer. p. 130. ISBN 3-540-21329-5.

[79] C.-Y. Wong (1994). *Introduction to High-energy Heavy-ion Collisions.* World Scientific. p. 149. ISBN 981-02-0263-6.

[80] S.B. Rüester, V. Werth, M. Buballa, I.A. Shovkovy, D.H. Rischke; Werth; Buballa; Shovkovy; Rischke (2005). "The phase diagram of neutral quark matter: Self-consistent treatment of quark masses". *Physical Review D* **72** (3): 034003. arXiv:hep-ph/0503184. Bibcode:2005PhRvD..72c4004R. doi:10.1103/PhysRevD.72.034004.

[81] M.G. Alford, K. Rajagopal, T. Schaefer, A. Schmitt; Schmitt; Rajagopal; Schäfer (2008). "Color superconductivity in dense quark matter".*Reviews of Modern Physics***80**(4): 1455–1515.arXiv:0709.4635.Bibcode:2008RvMP...80.1455A.doi:10.1103.

[82]S. Mrowczynski (1998). "Quark–Gluon Plasma".*Acta Physica Polonica B***29**: 3711.arXiv:nucl-th/9905005.Bibcode:1998Ac

[83] Z. Fodor, S.D. Katz; Katz (2004). "Critical point of QCD at finite T and μ, lattice results for physical quark masses". *Journal of High Energy Physics* **2004** (4): 50. arXiv:hep-lat/0402006. Bibcode:2004JHEP...04..050F. doi:10.1088/1126-6708/2004/04/050.

[84] U. Heinz, M. Jacob (2000). "Evidence for a New State of Matter: An Assessment of the Results from the CERN Lead Beam Programme". arXiv:nucl-th/0002042.

[85] "RHIC Scientists Serve Up "Perfect" Liquid". Brookhaven National Laboratory News. 2005. Retrieved 2009-05-22.

[86] T. Yulsman (2002). *Origins: The Quest for Our Cosmic Roots.* CRC Press. p. 75. ISBN 0-7503-0765-X.

[87] A. Sedrakian, J.W. Clark, M.G. Alford (2007). *Pairing in fermionic systems.* World Scientific. pp. 2–3. ISBN 981-256-907-3.

16.9   Further reading

- A. Ali, G. Kramer; Kramer (2011). "JETS and QCD: A historical review of the discovery of the quark and gluon jets and its impact on QCD". *European Physical Journal H* **36** (2): 245. arXiv:1012.2288. Bibcode:2011EPJH...36 ..245A.doi:10.1140/epjh/e2011-10047-1.

- D.J. Griffiths (2008). *Introduction to Elementary Particles* (2nd ed.). Wiley–VCH. ISBN 3-527-40601-8.

- I.S. Hughes (1985). *Elementary particles* (2nd ed.). Cambridge University Press. ISBN 0-521-26092-2.

- R. Oerter (2005). *The Theory of Almost Everything: The Standard Model, the Unsung Triumph of Modern Physics.* Pi Press. ISBN 0-13-236678-9.

- A. Pickering (1984). *Constructing Quarks: A Sociological History of Particle Physics.* The University of Chicago Press. ISBN 0-226-66799-5.

- B. Povh (1995). *Particles and Nuclei: An Introduction to the Physical Concepts.* Springer–Verlag. ISBN 0-387-59439-6.

- M. Riordan (1987). *The Hunting of the Quark: A true story of modern physics.* Simon & Schuster. ISBN 0-671-64884-5.

- B.A. Schumm (2004). *Deep Down Things: The Breathtaking Beauty of Particle Physics.* Johns Hopkins University Press. ISBN 0-8018-7971-X.

16.10   External links

- 1969 Physics Nobel Prize lecture by Murray Gell-Mann

- 1976 Physics Nobel Prize lecture by Burton Richter

- 1976 Physics Nobel Prize lecture by Samuel C.C. Ting

- 2008 Physics Nobel Prize lecture by Makoto Kobayashi

- 2008 Physics Nobel Prize lecture by Toshihide Maskawa

- The Top Quark And The Higgs Particle by T.A. Heppenheimer – A description of CERN's experiment to count the families of quarks.

- Bowley, Roger; Copeland, Ed. "Quarks". *Sixty Symbols*. Brady Haran for the University of Nottingham.

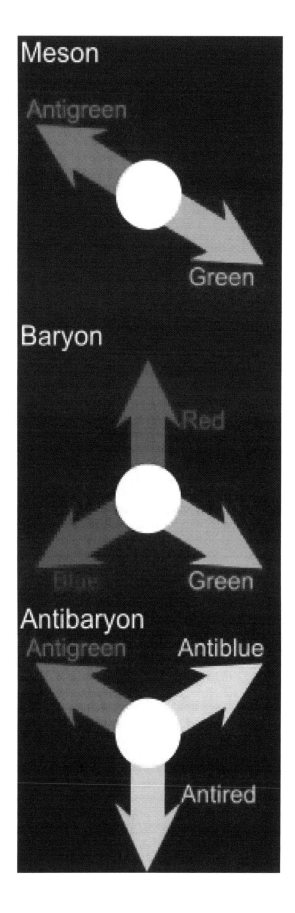

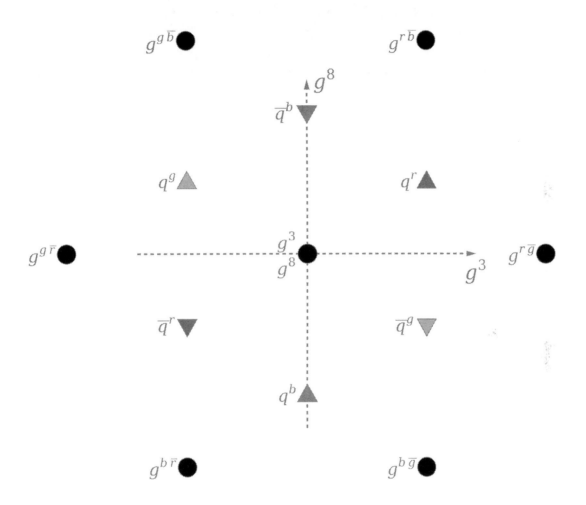

The pattern of strong charges for the three colors of quark, three antiquarks, and eight gluons (with two of zero charge overlapping).

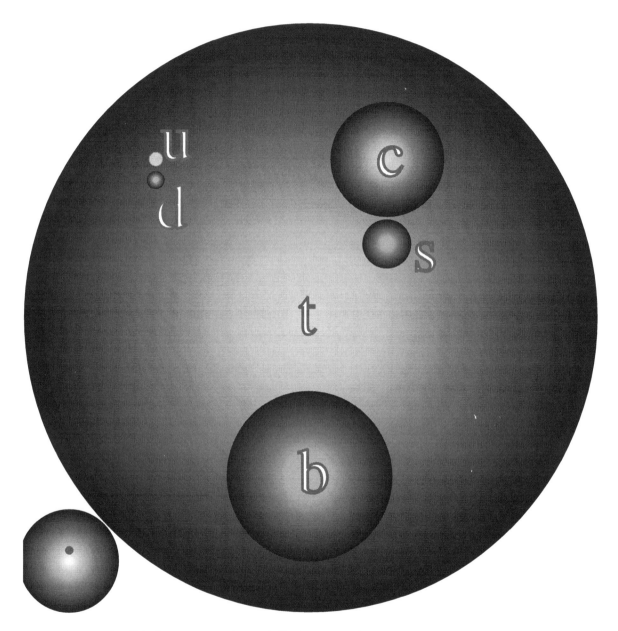

Current quark masses for all six flavors in comparison, as balls of proportional volumes. Proton and electron (red) are shown in bottom left corner for scale

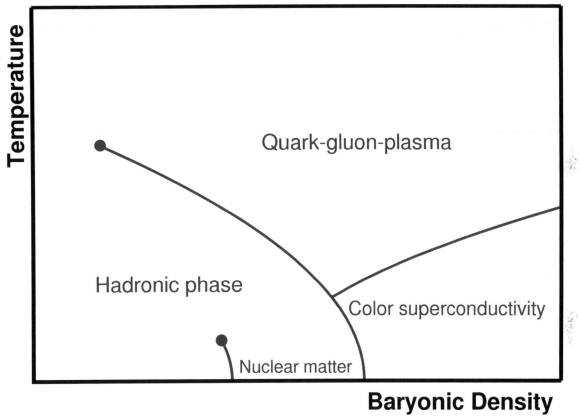

A qualitative rendering of the phase diagram of quark matter. The precise details of the diagram are the subject of ongoing research.[80][81]

Chapter 17

Up quark

The **up quark** or **u quark** (symbol: *u*) is the lightest of all quarks, a type of elementary particle, and a major constituent of matter. It, along with the down quark, forms the neutrons (one up quark, two down quarks) and protons (two up quarks, one down quark) of atomic nuclei. It is part of the first generation of matter, has an electric charge of $+\frac{2}{3}$ *e* and a bare mass of 1.8–3.0 MeV/c^2. Like all quarks, the up quark is an elementary fermion with spin-$\frac{1}{2}$, and experiences all four fundamental interactions: gravitation, electromagnetism, weak interactions, and strong interactions. The antiparticle of the up quark is the **up antiquark** (sometimes called *antiup quark* or simply *antiup*), which differs from it only in that some of its properties have equal magnitude but opposite sign.

Its existence (along with that of the down and strange quarks) was postulated in 1964 by Murray Gell-Mann and George Zweig to explain the *Eightfold Way* classification scheme of hadrons. The up quark was first observed by experiments at the Stanford Linear Accelerator Center in 1968.

17.1 History

In the beginnings of particle physics (first half of the 20th century), hadrons such as protons, neutrons and pions were thought to be elementary particles. However, as new hadrons were discovered, the 'particle zoo' grew from a few particles in the early 1930s and 1940s to several dozens of them in the 1950s. The relationships between each of them were unclear until 1961, when Murray Gell-Mann[2] and Yuval Ne'eman[3] (independently of each other) proposed a hadron classification scheme called the *Eightfold Way*, or in more technical terms, SU(3) flavor symmetry.

This classification scheme organized the hadrons into isospin multiplets, but the physical basis behind it was still unclear. In 1964, Gell-Mann[4] and George Zweig[5][6] (independently of each other) proposed the quark model, then consisting only of up, down, and strange quarks.[7] However, while the quark model explained the Eightfold Way, no direct evidence of the existence of quarks was found until 1968 at the Stanford Linear Accelerator Center.[8][9] Deep inelastic scattering experiments indicated that protons had substructure, and that protons made of three more-fundamental particles explained the data (thus confirming the quark model).[10]

At first people were reluctant to describe the three bodies as quarks, instead preferring Richard Feynman's parton description,[11][12][13] but over time the quark theory became accepted (see *November Revolution*).[14]

17.2 Mass

Despite being extremely common, the bare mass of the up quark is not well determined, but probably lies between 1.8 and 3.0 MeV/c^2.[1] Lattice QCD calculations give a more precise value: 2.01±0.14 MeV/c^2.[15]

When found in mesons (particles made of one quark and one antiquark) or baryons (particles made of three quarks), the 'effective mass' (or 'dressed' mass) of quarks becomes greater because of the binding energy caused by the gluon field between each quark (see mass–energy equivalence).The bare mass of up quarks is so light, it cannot be straightforwardly calculated because relativistic effects have to be taken into account.

17.3   See also

- Down quark

- Isospin

- Quark model

- Quantum Mechanics

17.4   References

[1] J. Beringer (Particle Data Group); et al. (2012). "PDGLive Particle Summary 'Quarks (u, d, s, c, b, t, b', t', Free)'" (PDF). Particle Data Group. Retrieved 2013-02-21.

[2] M. Gell-Mann (2000) [1964]. "The Eightfold Way: A theory of strong interaction symmetry". In M. Gell-Mann, Y. Ne'eman. *The Eightfold Way*. Westview Press. p. 11. ISBN 0-7382-0299-1.
Original: M. Gell-Mann (1961). "The Eightfold Way: A theory of strong interaction symmetry". *Synchrotron Laboratory Report CTSL-20* (California Institute of Technology)

[3] Y. Ne'eman (2000) [1964]. "Derivation of strong interactions from gauge invariance". In M. Gell-Mann, Y. Ne'eman. *The Eightfold Way*. Westview Press. ISBN 0-7382-0299-1.
Original Y. Ne'eman (1961). "Derivation of strong interactions from gauge invariance". *Nuclear Physics* **26** (2): 222. Bibcode: 1961NucPh..26..222N.doi:10.1016/0029-5582(61)90134-1.

[4] M. Gell-Mann (1964). "A Schematic Model of Baryons and Mesons". *Physics Letters* **8** (3): 214–215. Bibcode:1964PhL......8..2 14G.doi:10.1016/S0031-9163(64)92001-3.

[5] G. Zweig (1964). "An SU(3) Model for Strong Interaction Symmetry and its Breaking". *CERN Report No.8181/Th 8419.*

[6] G. Zweig (1964). "An SU(3) Model for Strong Interaction Symmetry and its Breaking: II". *CERN Report No.8419/Th 8412.*

[7] B. Carithers, P. Grannis (1995). "Discovery of the Top Quark" (PDF). *Beam Line* (SLAC) **25** (3): 4–16. Retrieved 2008-09-23.

[8] E. D. Bloom; Coward, D.; Destaebler, H.; Drees, J.; Miller, G.; Mo, L.; Taylor, R.; Breidenbach, M.; et al. (1969). "High-Energy Inelastic *e–p* Scattering at 6° and 10°". *Physical Review Letters* **23** (16): 930–934. Bibcode:1969PhRvL..23..930B. doi:10.1103/PhysRevLett.23.930.

[9] M. Breidenbach; Friedman, J.; Kendall, H.; Bloom, E.; Coward, D.; Destaebler, H.; Drees, J.; Mo, L.; Taylor, R.; et al. (1969). "Observed Behavior of Highly Inelastic Electron–Proton Scattering". *Physical Review Letters* **23** (16): 935–939. Bibcode:1969PhRvL..23..935B. doi:10.1103/PhysRevLett.23.935.

[10] J. I. Friedman. "The Road to the Nobel Prize". Hue University. Retrieved 2008-09-29.

[11] R. P. Feynman (1969). "Very High-Energy Collisions of Hadrons". *Physical Review Letters* **23** (24): 1415–1417. Bibcode:1969 doi:10.1103/PhysRevLett.23.1415.

[12] S. Kretzer; Lai, H.; Olness, Fredrick; Tung, W.; et al. (2004). "CTEQ6 Parton Distributions with Heavy Quark Mass Effects". *Physical Review D* **69** (11): 114005.arXiv:hep-ph/0307022.Bibcode:2004PhRvD..69k4005K.doi:10.1103/PhysRevD.69.114

[13] D. J. Griffiths (1987). *Introduction to Elementary Particles*. John Wiley & Sons. p. 42. ISBN 0-471-60386-4.

[14] M. E. Peskin, D. V. Schroeder (1995). *An introduction to quantum field theory*. Addison–Wesley. p. 556. ISBN 0-201-50397-2.

[15] Cho, Adrian (April 2010). "Mass of the Common Quark Finally Nailed Down". Science Magazine.

17.5 Further reading

- A. Ali, G. Kramer; Kramer (2011). "JETS and QCD: A historical review of the discovery of the quark and gluon jets and its impact on QCD". *European Physical Journal H* **36** (2): 245. arXiv:1012.2288. Bibcode:2011EPJH...36 ..245A.doi:10.1140/epjh/e2011-10047-1.

- R. Nave. "Quarks". *HyperPhysics*. Georgia State University, Department of Physics and Astronomy. Retrieved 2008-06-29.

- A. Pickering (1984). *Constructing Quarks.* University of Chicago Press. pp. 114–125. ISBN 0-226-66799-5.

Chapter 18

Down quark

The **down quark** or **d quark** (symbol: d) is the second-lightest of all quarks, a type of elementary particle, and a major constituent of matter. Together with the up quark, it forms the neutrons (one up quark, two down quarks) and protons (two up quarks, one down quark) of atomic nuclei. It is part of the first generation of matter, has an electric charge of $-\frac{1}{3}$ e and a bare mass of 4.8+0.5 −0.3 MeV/c^2.[1] Like all quarks, the down quark is an elementary fermion with spin-$\frac{1}{2}$, and experiences all four fundamental interactions: gravitation, electromagnetism, weak interactions, and strong interactions. The antiparticle of the down quark is the **down antiquark** (sometimes called *antidown quark* or simply *antidown*), which differs from it only in that some of its properties have equal magnitude but opposite sign.

Its existence (along with that of the up and strange quarks) was postulated in 1964 by Murray Gell-Mann and George Zweig to explain the *Eightfold Way* classification scheme of hadrons. The down quark was first observed by experiments at the Stanford Linear Accelerator Center in 1968.

18.1 History

In the beginnings of particle physics (first half of the 20th century), hadrons such as protons, neutrons, and pions were thought to be elementary particles. However, as new hadrons were discovered, the 'particle zoo' grew from a few particles in the early 1930s and 1940s to several dozens of them in the 1950s. The relationships between each of them was unclear until 1961, when Murray Gell-Mann[2] and Yuval Ne'eman[3] (independently of each other) proposed a hadron classification scheme called the *Eightfold Way*, or in more technical terms, SU(3) flavor symmetry.

This classification scheme organized the hadrons into isospin multiplets, but the physical basis behind it was still unclear. In 1964, Gell-Mann[4] and George Zweig[5][6] (independently of each other) proposed the quark model, then consisting only of up, down, and strange quarks.[7] However, while the quark model explained the Eightfold Way, no direct evidence of the existence of quarks was found until 1968 at the Stanford Linear Accelerator Center.[8][9] Deep inelastic scattering experiments indicated that protons had substructure, and that protons made of three more-fundamental particles explained the data (thus confirming the quark model).[10]

At first people were reluctant to identify the three-bodies as quarks, instead preferring Richard Feynman's parton description, [11][12][13]but over time the quark theory became accepted(see*November Revolution*).[14]

18.2 Mass

Despite being extremely common, the bare mass of the down quark is not well determined, but probably lies between 4.5 and 5.3$10^0$ MeV/c^2.[1] Lattice QCD calculations give a more precise value: 4.79±0.16 MeV/c^2.[15]

When found in mesons (particles made of one quark and one antiquark) or baryons (particles made of three quarks), the 'effective mass' (or 'dressed' mass) of quarks becomes greater because of the binding energy caused by the gluon field between quarks (see mass–energy equivalence). For example, the effective mass of down quarks in a proton is around 330 MeV/c^2. Because the bare mass of down quarks is so small, it cannot be straightforwardly calculated because relativistic effects have to be taken into account.

18.3 See also

- Up quark

- Isospin

- Quark model

18.4 References

[1] J. Beringer (Particle Data Group); et al. (2013). "PDGLive Particle Summary 'Quarks (u, d, s, c, b, t, b′, t′, Free)'" (PDF). Particle Data Group. Retrieved 2013-07-23.

[2] M. Gell-Mann (2000) [1964]. "The Eightfold Way: A theory of strong interaction symmetry". In M. Gell-Mann, Y. Ne'eman. *The Eightfold Way*. Westview Press. p. 11. ISBN 0-7382-0299-1.
Original: M. Gell-Mann (1961). "The Eightfold Way: A theory of strong interaction symmetry". *Synchrotron Laboratory Report CTSL-20* (California Institute of Technology).

[3] Y. Ne'eman (2000) [1964]. "Derivation of strong interactions from gauge invariance". In M. Gell-Mann, Y. Ne'eman. *The Eightfold Way*. Westview Press. ISBN 0-7382-0299-1.
Original Y. Ne'eman (1961). "Derivation of strong interactions from gauge invariance". *Nuclear Physics* **26** (2): 222. Bibcode: 1961NucPh..26..222N.doi:10.1016/0029-5582(61)90134-1.

[4] M. Gell-Mann (1964). "A Schematic Model of Baryons and Mesons".*Physics Letters***8**(3): 214–215.Bibcode:1964PhL......8.. doi:10.1016/S0031-9163(64)92001-3.

[5] G. Zweig (1964). "An SU(3) Model for Strong Interaction Symmetry and its Breaking". *CERN Report No.8181/Th 8419*.

[6] G. Zweig (1964). "An SU(3) Model for Strong Interaction Symmetry and its Breaking: II". *CERN Report No.8419/Th 8412*.

[7] B. Carithers, P. Grannis (1995). "Discovery of the Top Quark" (PDF). *Beam Line* (SLAC) **25** (3): 4–16. Retrieved 2008-09-23.

[8] E. D. Bloom; Coward, D.; Destaebler, H.; Drees, J.; Miller, G.; Mo, L.; Taylor, R.; Breidenbach, M.; et al. (1969). "High-Energy Inelastic *e–p* Scattering at 6° and 10°". *Physical Review Letters* **23** (16): 930–934. Bibcode:1969PhRvL..23..930B. doi:10.1103/PhysRevLett.23.930.

[9] M. Breidenbach; Friedman, J.; Kendall, H.; Bloom, E.; Coward, D.; Destaebler, H.; Drees, J.; Mo, L.; Taylor, R.; et al. (1969). "Observed Behavior of Highly Inelastic Electron–Proton Scattering". *Physical Review Letters* **23** (16): 935–939. Bibcode:1969PhRvL..23..935B. doi:10.1103/PhysRevLett.23.935.

[10] J. I. Friedman. "The Road to the Nobel Prize". Hue University. Retrieved 2008-09-29.

[11] R. P. Feynman (1969). "Very High-Energy Collisions of Hadrons".*Physical Review Letters***23**(24): 1415–1417.Bibcode:1969 doi:10.1103/PhysRevLett.23.1415.

[12] S. Kretzer; Lai, H.; Olness, Fredrick; Tung, W.; et al. (2004). "CTEQ6 Parton Distributions with Heavy Quark Mass Effects". *Physical Review D***69**(11): 114005.arXiv:hep-ph/0307022.Bibcode:2004PhRvD..69k4005K.doi:10.1103/PhysRevD.69.

[13] D. J. Griffiths (1987). *Introduction to Elementary Particles*. John Wiley & Sons. p. 42. ISBN 0-471-60386-4.

[14] M. E. Peskin, D. V. Schroeder (1995). *An introduction to quantum field theory*. Addison–Wesley. p. 556. ISBN 0-201-50397-2.

[15] Cho, Adrian (April 2010). "Mass of the Common Quark Finally Nailed Down". Science Magazine.

18.5   Further reading

- A. Ali, G. Kramer; Kramer (2011). "JETS and QCD: A historical review of the discovery of the quark and gluon jets and its impact on QCD". *European Physical Journal H* **36** (2): 245. arXiv:1012.2288. Bibcode:2011EPJH...36 ..245A.doi:10.1140/epjh/e2011-10047-1.

- R. Nave. "Quarks". *HyperPhysics*. Georgia State University, Department of Physics and Astronomy. Retrieved 2008-06-29.

- A. Pickering (1984). *Constructing Quarks*. University of Chicago Press. pp. 114–125. ISBN 0-226-66799-5.

Chapter 19

Strange quark

The **strange quark** or **s quark** (from its symbol, *s*) is the third-lightest of all quarks, a type of elementary particle. Strange quarks are found in subatomic particles called hadrons. Example of hadrons containing strange quarks include kaons (K), strange D mesons (D
s), Sigma baryons (Σ), and other strange particles.

Along with the charm quark, it is part of the second generation of matter, and has an electric charge of $-\frac{1}{3}\,e$ and a bare mass of $95{+}5\ {-}5\,\mathrm{MeV}/c^2$.

[1] Like all quarks, the strange quark is an elementary fermion with spin-$\frac{1}{2}$, and experiences all four fundamental interactions: gravitation, electromagnetism, weak interactions, and strong interactions. The antiparticle of the strange quark is the **strange antiquark** (sometimes called *antistrange quark* or simply *antistrange*), which differs from it only in that some of its properties have equal magnitude but opposite sign.

The first strange particle (a particle containing a strange quark) was discovered in 1947 (kaons), but the existence of the strange quark itself (and that of the up and down quarks) was only postulated in 1964 by Murray Gell-Mann and George Zweig to explain the *Eightfold Way* classification scheme of hadrons. The first evidence for the existence of quarks came in 1968, in deep inelastic scattering experiments at the Stanford Linear Accelerator Center. These experiments confirmed the existence of up and down quarks, and by extension, strange quarks, as they were required to explain the Eightfold Way.

19.1   History

In the beginnings of particle physics (first half of the 20th century), hadrons such as protons, neutron and pions were thought to be elementary particles. However, new hadrons were discovered, the 'particle zoo' grew from a few particles in the early 1930s and 1940s to several dozens of them in the 1950s. However some particles were much longer lived than others; most particles decayed through the strong interaction and had lifetimes of around 10^{-23} seconds. But when they decayed through the weak interactions, they had lifetimes of around 10^{-10} seconds to decay. While studying these decays Murray Gell-Mann (in 1953)[2][3] and Kazuhiko Nishijima (in 1955)[4] developed the concept of *strangeness* (which Nishijima called *eta-charge*, after the eta meson (η)) which explained the 'strangeness' of the longer-lived particles. The Gell-Mann–Nishijima formula is the result of these efforts to understand strange decays.

However, the relationships between each particles and the physical basis behind the strangeness property was still unclear. In 1961, Gell-Mann[5] and Yuval Ne'eman[6] (independently of each other) proposed a hadron classification scheme called the *Eightfold Way*, or in more technical terms, SU(3) flavor symmetry. This ordered hadrons into isospin multiplets. The physical basis behind both isospin and strangeness was only explained in 1964, when Gell-Mann[7] and George Zweig[8][9] (independently of each other) proposed the quark model, then consisting only of up, down, and strange quarks.[10] Up and down quarks were the carriers of isospin, while the strange quark carried strangeness. While the quark model explained the Eightfold Way, no direct evidence of the existence of quarks was found until 1968 at the Stanford Linear Accelerator Center.[11][12] Deep inelastic scattering experiments indicated that protons had substructure, and that protons made of three more-fundamental particles explained the data (thus confirming the quark model).[13]

At first people were reluctant to identify the three-bodies as quarks, instead preferring Richard Feynman's parton description, [14][15][16]but over time the quark theory became accepted(see*November Revolution*).[17]

19.2   See also

- Quark model

- Strange matter

- Strangeness production

- Strangelet

- Strange star

19.3   References

[1] J. Beringer (Particle Data Group); et al. (2012). "PDGLive Particle Summary 'Quarks (u, d, s, c, b, t, b′, t′, Free)'" (PDF). Particle Data Group. Retrieved 2012-11-30.

[2] M. Gell-Mann (1953). "Isotopic Spin and New Unstable Particles". *Physical Review* **92** (3): 833. Bibcode:1953PhRv...92..833G. doi:10.1103/PhysRev.92.833.

[3] G. Johnson (2000). *Strange Beauty: Murray Gell-Mann and the Revolution in Twentieth-Century Physics*. Random House. p. 119. ISBN 0-679-43764-9. By the end of the summer... [Gell-Mann] completed his first paper, "Isotopic Spin and Curious Particles" and send it of to *Physical Review*. The editors hated the title, so he amended it to "Strange Particles". They wouldn't go for that either—never mind that almost everybody used the term—suggesting instead "Isotopic Spin and New Unstable Particles".

[4] K. Nishijima, Kazuhiko (1955). "Charge Independence Theory of V Particles". *Progress of Theoretical Physics* **13** (3): 285. Bibcode:1955PThPh..13..285N. doi:10.1143/PTP.13.285.

[5] M. Gell-Mann (2000) [1964]. "The Eightfold Way: A theory of strong interaction symmetry". In M. Gell-Mann, Y. Ne'eman. *The Eightfold Way*. Westview Press. p. 11. ISBN 0-7382-0299-1.
     Original: M. Gell-Mann (1961). "The Eightfold Way: A theory of strong interaction symmetry". *Synchrotron Laboratory Report CTSL-20* (California Institute of Technology)

[6] Y. Ne'eman (2000) [1964]. "Derivation of strong interactions from gauge invariance". In M. Gell-Mann, Y. Ne'eman. *The Eightfold Way*. Westview Press. ISBN 0-7382-0299-1.
     Original Y. Ne'eman (1961). "Derivation of strong interactions from gauge invariance". *Nuclear Physics* **26** (2): 222. Bibcode: doi:10.1016/0029-5582(61)90134-1.

[7] M. Gell-Mann (1964). "A Schematic Model of Baryons and Mesons".*Physics Letters***8**(3): 214–215.Bibcode:1964PhL......8.. doi:10.1016/S0031-9163(64)92001-3.

[8] G. Zweig (1964). "An SU(3) Model for Strong Interaction Symmetry and its Breaking". *CERN Report No.8181/Th 8419*.

[9] G. Zweig (1964). "An SU(3) Model for Strong Interaction Symmetry and its Breaking: II". *CERN Report No.8419/Th 8412*.

[10] B. Carithers, P. Grannis (1995). "Discovery of the Top Quark" (PDF). *Beam Line* (SLAC) **25** (3): 4–16. Retrieved 2008-09-23.

[11] E. D. Bloom; Coward, D.; Destaebler, H.; Drees, J.; Miller, G.; Mo, L.; Taylor, R.; Breidenbach, M.; et al. (1969). "High-Energy Inelastic *e–p* Scattering at 6° and 10°". *Physical Review Letters* **23** (16): 930–934. Bibcode:1969PhRvL..23..930B. doi:10.1103/PhysRevLett.23.930.

[12] M. Breidenbach; Friedman, J.; Kendall, H.; Bloom, E.; Coward, D.; Destaebler, H.; Drees, J.; Mo, L.; Taylor, R.; et al. (1969). "Observed Behavior of Highly Inelastic Electron–Proton Scattering". *Physical Review Letters* **23** (16): 935–939. Bibcode:1969PhRvL..23..935B. doi:10.1103/PhysRevLett.23.935.

[13] J. I. Friedman. "The Road to the Nobel Prize". Hue University. Retrieved 2008-09-29.

[14] R. P. Feynman (1969). "Very High-Energy Collisions of Hadrons". *Physical Review Letters* **23**(24): 1415–1417. Bibcode:1969 doi:10.1103/PhysRevLett.23.1415.

[15] S. Kretzer; Lai, H.; Olness, Fredrick; Tung, W.; et al. (2004). "CTEQ6 Parton Distributions with Heavy Quark Mass Effects". *Physical Review D* **69**(11):114005.arXiv:hep-th/0307022.Bibcode:2004PhRvD..69k4005K.doi:10.1103/PhysRevD.69.

[16] D. J. Griffiths (1987). *Introduction to Elementary Particles.* John Wiley & Sons. p. 42. ISBN 0-471-60386-4.

[17] M. E. Peskin, D. V. Schroeder (1995). *An introduction to quantum field theory.* Addison–Wesley. p. 556. ISBN 0-201-50397-2.

19.4 Further reading

- R. Nave. "Quarks". *HyperPhysics.* Georgia State University, Department of Physics and Astronomy. Retrieved 2008-06-29.

- A. Pickering (1984). *Constructing Quarks.* University of Chicago Press. pp. 114–125. ISBN 0-226-66799-5.

Chapter 20

Charm quark

The **charm quark** or **c quark** (from its symbol, c) is the third most massive of all quarks, a type of elementary particle. Charm quarks are found in hadrons, which are subatomic particles made of quarks. Example of hadrons containing charm quarks include the J/ψ meson (J/ψ), D mesons (D), charmed Sigma baryons (Σ c), and other charmed particles.

It, along with the strange quark is part of the second generation of matter, and has an electric charge of $+\frac{2}{3}\, e$ and a bare mass of 1.29+0.05 −0.11 GeV/c^2.[1] Like all quarks, the charm quark is an elementary fermion with spin-$\frac{1}{2}$, and experiences all four fundamental interactions: gravitation, electromagnetism, weak interactions, and strong interactions. The antiparticle of the charm quark is the **charm antiquark** (sometimes called *anticharm quark* or simply *anticharm*), which differs from it only in that some of its properties have equal magnitude but opposite sign.

The existence of a fourth quark had been speculated by a number of authors around 1964 (for instance by James Bjorken and Sheldon Glashow[4]), but its prediction is usually credited to Sheldon Glashow, John Iliopoulos and Luciano Maiani in 1970 (see GIM mechanism).[5] The first charmed particle (a particle containing a charm quark) to be discovered was the J/ψ meson. It was discovered by a team at the Stanford Linear Accelerator Center (SLAC), led by Burton Richter,[6] and one at the Brookhaven National Laboratory (BNL), led by Samuel Ting.[7]

The 1974 discovery of the J/ψ (and thus the charm quark) ushered in a series of breakthroughs which are collectively known as the *November Revolution*.

20.1 Hadrons containing charm quarks

Main articles: List of baryons and list of mesons

Some of the hadrons containing charm quarks include:

- D mesons contain a charm quark (or its antiparticle) and an up or down quark.

- D s mesons contain a charm quark and a strange quark.

- There are many charmonium states, for example the J/ψ particle. These consist of a charm quark and its antiparticle.

- Charmed baryons have been observed, and are named in analogy with strange baryons (e.g. Λ+ c).

20.2 See also

- Quark model

20.3 Notes

[1] K. Nakamura (Particle Data Group); et al. (2011). "PDGLive Particle Summary 'Quarks (u, d, s, c, b, t, b′, t′, Free)'" (PDF). Particle Data Group. Retrieved 2011-08-08.

[2] R. Nave. "Transformation of Quark Flavors by the Weak Interaction". Retrieved 2010-12-06. The c quark has about 5% probability of decaying into a d quark instead of an s quark.

[3] K. Nakamura *et al.* (Particle Data Group); et al. (2010). "Review of Particles Physics: The CKM Quark-Mixing Matrix" (PDF). *Journal of Physics G* **37** (75021): 150.

[4] B.J. Bjorken, S.L. Glashow; Glashow (1964). "Elementary particles and SU(4)". *Physics Letters* **11**(3): 255–257. Bibcode:1964 doi:10.1016/0031-9163(64)90433-0.

[5] S.L. Glashow, J. Iliopoulos, L. Maiani; Iliopoulos; Maiani (1970). "Weak Interactions with Lepton–Hadron Symmetry". *Physical Review D* **2** (7): 1285–1292. Bibcode:1970PhRvD...2.1285G. doi:10.1103/PhysRevD.2.1285.

[6] J.-E. Augustin; et al. (1974). "Discovery of a Narrow Resonance in e^+e^- Annihilation". *Physical Review Letters* **33** (23): 1406. Bibcode:1974PhRvL..33.1406A. doi:10.1103/PhysRevLett.33.1406.

[7] J.J. Aubert; et al. (1974). "Experimental Observation of a Heavy Particle J". *Physical Review Letters* **33**(23): 1404. Bibcode: doi:10.1103/PhysRevLett.33.1404.

20.4 Further reading

- R. Nave. "Quarks". *HyperPhysics*. Georgia State University, Department of Physics and Astronomy. Retrieved 2008-06-29.

- A. Pickering (1984). *Constructing Quarks*. University of Chicago Press. pp. 114–125. ISBN 0-226-66799-5.

Chapter 21

Bottom quark

The **bottom quark** or **b quark**, also known as the **beauty quark**, is a third-generation quark with a charge of $-\frac{1}{3}\,e$. Although all quarks are described in a similar way by quantum chromodynamics, the bottom quark's large bare mass (around 4.2 GeV/c^2,[3] a bit more than four times the mass of a proton), combined with low values of the CKM matrix elements V_{ub} and V_{cb}, gives it a distinctive signature that makes it relatively easy to identify experimentally (using a technique called B-tagging). Because three generations of quark are required for CP violation (see CKM matrix), mesons containing the bottom quark are the easiest particles to use to investigate the phenomenon; such experiments are being performed at the BaBar, Belle and LHCb experiments.

The bottom quark is also notable because it is a product in almost all top quark decays, and is a frequent decay product for the Higgs boson. The bottom quark was theorized in 1973 by physicists Makoto Kobayashi and Toshihide Maskawa to explain CP violation.[1] The name "bottom" was introduced in 1975 by Haim Harari.[4][5] The bottom quark was discovered in 1977 by the Fermilab E288 experiment team led by Leon M. Lederman, when collisions produced bottomonium.[2][6][7] Kobayashi and Maskawa won the 2008 Nobel Prize in Physics for their explanation of CP-violation.[8][9] On its discovery, there were efforts to name the bottom quark "beauty", but "bottom" became the predominant usage.

The bottom quark can decay into either an up quark or charm quark via the weak interaction. Both these decays are suppressed by the CKM matrix, making lifetimes of most bottom particles ($\sim 10^{-12}$ s) somewhat higher than those of charmed particles ($\sim 10^{-13}$ s), but lower than those of strange particles (from $\sim 10^{-10}$ to $\sim 10^{-8}$ s).

21.1   Hadrons containing bottom quarks

Main articles: list of baryons and list of mesons

Some of the hadrons containing bottom quarks include:

- B mesons contain a bottom quark (or its antiparticle) and an up or down quark.

- B
  c and B
  s mesons contain a bottom quark along with a charm quark or strange quark respectively.

- There are many bottomonium states, for example the Υ meson and $\chi_b(3P)$, the first particle discovered in LHC. These consist of a bottom quark and its antiparticle.

- Bottom baryons have been observed, and are named in analogy with strange baryons (e.g. $\Lambda 0$
  b).

174

21.2 See also

- Quark model

21.3 References

[1] M. Kobayashi; T. Maskawa (1973). "CP-Violation in the Renormalizable Theory of Weak Interaction". *Progress of Theoretical Physics* **49** (2): 652–657. Bibcode:1973PThPh..49..652K. doi:10.1143/PTP.49.652.

[2] "Discoveries at Fermilab – Discovery of the Bottom Quark" (Press release). Fermilab. 7 August 1977. Retrieved 2009-07-24.

[3] J. Beringer (Particle Data Group); et al. (2012). "PDGLive Particle Summary 'Quarks (u, d, s, c, b, t, b′, t′, Free)'" (PDF). Particle Data Group. Retrieved 2012-12-18.

[4] H. Harari (1975). "A new quark model for hadrons". *Physics Letters B***57**(3): 265.Bibcode:1975PhLB...57..265H.doi:10.1016 2693(75)90072-6.

[5] K.W. Staley (2004). *The Evidence for the Top Quark*. Cambridge University Press. pp. 31–33. ISBN 978-0-521-82710-2.

[6] L.M. Lederman (2005). "Logbook: Bottom Quark". *Symmetry Magazine* **2** (8).

[7] S.W. Herb; Hom, D.; Lederman, L.; Sens, J.; Snyder, H.; Yoh, J.; Appel, J.; Brown, B.; Brown, C.; Innes, W.; Ueno, K.; Yamanouchi, T.; Ito, A.; Jöstlein, H.; Kaplan, D.; Kephart, R.; et al. (1977). "Observation of a Dimuon Resonance at 9.5 GeV in 400-GeV Proton-Nucleus Collisions". *Physical Review Letters* **39** (5): 252. Bibcode:1977PhRvL..39..252H. doi:10.1103/PhysRevLett.39.252.

[8] 2008 Physics Nobel Prize lecture by Makoto Kobayashi

[9] 2008 Physics Nobel Prize lecture by Toshihide Maskawa

21.4 Further reading

- L. Lederman (1978). "The Upsilon Particle". *Scientific American* **239** (4): 72. doi:10.1038/scientificamerican1078-72.

- R. Nave. "Quarks". *HyperPhysics*. Georgia State University, Department of Physics and Astronomy. Retrieved 2008-06-29.

- A. Pickering (1984). *Constructing Quarks*. University of Chicago Press. pp. 114–125. ISBN 0-226-66799-5.

- J. Yoh (1997). "The Discovery of the b Quark at Fermilab in 1977: The Experiment Coordinator's Story" (PDF). *Proceedings of Twenty Beautiful Years of Bottom Physics*. Fermilab. Retrieved 2009-07-24.

21.5 External links

- History of the discovery of the bottom quark / Upsilon meson

Chapter 22

Top quark

The **top quark**, also known as the **t quark** (symbol: t) or **truth quark**, is the most massive of all observed elementary particles. Like all quarks, the top quark is an elementary fermion with spin-$\frac{1}{2}$, and experiences all four fundamental interactions: gravitation, electromagnetism, weak interactions, and strong interactions. It has an electric charge of $+\frac{2}{3}$ e,[2] It has a massive mass of 173.34 ± 0.27 (stat) ± 0.71 (syst)10^0 GeV/c^2,[1] which is about the same mass as an atom of tungsten. The antiparticle of the top quark is the **top antiquark** (symbol: t, sometimes called *antitop quark* or simply *antitop*), which differs from it only in that some of its properties have equal magnitude but opposite sign.

The top quark interacts primarily by the strong interaction, but can only decay through the weak force. It decays to a W boson and either a bottom quark (most frequently), a strange quark, or, on the rarest of occasions, a down quark. The Standard Model predicts its mean lifetime to be roughly 5×10^{-25} s.[3] This is about a twentieth of the timescale for strong interactions, and therefore it does not form hadrons, giving physicists a unique opportunity to study a "bare" quark (all other quarks hadronize, meaning that they combine with other quarks to form hadrons, and can only be observed as such). Because it is so massive, the properties of the top quark allow predictions to be made of the mass of the Higgs boson under certain extensions of the Standard Model (see Mass and coupling to the Higgs boson below). As such, it is extensively studied as a means to discriminate between competing theories.

Its existence (and that of the bottom quark) was postulated in 1973 by Makoto Kobayashi and Toshihide Maskawa to explain the observed CP violations in kaon decay,[4] and was discovered in 1995 by the CDF[5] and DØ[6] experiments at Fermilab. Kobayashi and Maskawa won the 2008 Nobel Prize in Physics for the prediction of the top and bottom quark, which together form the third generation of quarks.[7]

22.1   History

In 1973, Makoto Kobayashi and Toshihide Maskawa predicted the existence of a third generation of quarks to explain observed CP violations in kaon decay.[4] The names top and bottom were introduced by Haim Harari in 1975,[8][9] to match the names of the first generation of quarks (up and down) reflecting the fact that the two were the 'up' and 'down' component of a weak isospin doublet.[10] The top quark was sometimes called *truth quark* in the past, but over time *top quark* became the predominant use.[11]

The proposal of Kobayashi and Maskawa heavily relied on the GIM mechanism put forward by Sheldon Lee Glashow, John Iliopoulos and Luciano Maiani,[12] which predicted the existence of the then still unobserved charm quark. When in November 1974 teams at Brookhaven National Laboratory (BNL) and the Stanford Linear Accelerator Center (SLAC) simultaneously announced the discovery of the J/ψ meson, it was soon after identified as a bound state of the missing charm quark with its antiquark. This discovery allowed the GIM mechanism to become part of the Standard Model.[13] With the acceptance of the GIM mechanism, Kobayashi and Maskawa's prediction also gained in credibility. Their case was further strengthened by the discovery of the tau by Martin Lewis Perl's team at SLAC between 1974 and 1978.[14] This announced a third generation of leptons, breaking the new symmetry between leptons and quarks introduced by the GIM mechanism. Restoration of the symmetry implied the existence of a fifth and sixth quark.

It was in fact not long until a fifth quark, the bottom, was discovered by the E288 experiment team, led by Leon Lederman at Fermilab in 1977.[15][16][17] This strongly suggested that there must also be a sixth quark, the top, to complete the pair. It was known that this quark would be heavier than the bottom, requiring more energy to create in particle collisions, but the general expectation was that the sixth quark would soon be found. However, it took another 18 years before the existence of the top was confirmed.[18]

Early searches for the top quark at SLAC and DESY (in Hamburg) came up empty-handed. When, in the early eighties, the Super Proton Synchrotron (SPS) at CERN discovered the W boson and the Z boson, it was again felt that the discovery of the top was imminent. As the SPS gained competition from the Tevatron at Fermilab there was still no sign of the missing particle, and it was announced by the group at CERN that the top mass must be at least 41 GeV/c^2. After a race between CERN and Fermilab to discover the top, the accelerator at CERN reached its limits without creating a single top, pushing the lower bound on its mass up to 77 GeV/c^2.[18]

The Tevatron was (until the start of LHC operation at CERN in 2009) the only hadron collider powerful enough to produce top quarks. In order to be able to confirm a future discovery, a second detector, the DØ detector, was added to the complex (in addition to the Collider Detector at Fermilab (CDF) already present). In October 1992, the two groups found their first hint of the top, with a single creation event that appeared to contain the top. In the following years, more evidence was collected and on April 22, 1994, the CDF group submitted their paper presenting tentative evidence for the existence of a top quark with a mass of about 175 GeV/c^2. In the meantime, DØ had found no more evidence than the suggestive event in 1992. A year later, on March 2, 1995, after having gathered more evidence and a reanalysis of the DØ data (who had been searching for a much lighter top), the two groups jointly reported the discovery of the top with a certainty of 99.9998% at a mass of 176±18 GeV/c^2.[5][6][18]

In the years leading up to the top quark discovery, it was realized that certain precision measurements of the electroweak vector boson masses and couplings are very sensitive to the value of the top quark mass. These effects become much larger for higher values of the top mass and therefore could indirectly see the top quark even if it could not be directly detected in any experiment at the time. The largest effect from the top quark mass was on the T parameter and by 1994 the precision of these indirect measurements had led to a prediction of the top quark mass to be between 145 GeV/c^2 and 185 GeV/c^2.

The Discovery of the Top Quark Finding the sixth quark involved the world's most energetic collisions and a cast of thousands

by Tony M. Liss and Paul L. Tipton[19]It is the development of techniques such as the (VIOLENT COLLISION between a proton and an antiproton (center) creates a top quark (red) and an antitop (blue). These decay to other particles, typically producing a number of jets and possibly an electron or positron) that ultimately allowed such precision calculations that led to Gerardus 't Hooft and Martinus Veltman winning the Nobel Prize in physics in 1999.[20][21]

22.2 Properties

- At the final Tevatron energy of 1.96 TeV, top–antitop pairs were produced with a cross section of about 7 picobarns (pb).[22] The Standard Model prediction (at next-to-leading order with m_t = 175 GeV/c^2) is 6.7–7.5 pb.

- The W bosons from top quark decays carry polarization from the parent particle, hence pose themselves as a unique probe to top polarization.

- In the Standard Model, the top quark is predicted to have a spin quantum number of $1/2$ and electric charge $+2/3$. A first measurement of the top quark charge has been published, resulting in approximately 90% confidence limit that the top quark charge is indeed $+2/3$.[23]

22.3 Production

Because top quarks are very massive, large amounts of energy are needed to create one. The only way to achieve such high energies is through high energy collisions. These occur naturally in the Earth's upper atmosphere as cosmic rays collide with particles in the air, or can be created in a particle accelerator. In 2011, after the Tevatron ceased operations, the Large Hadron Collider at CERN became the only accelerator that generates a beam of sufficient energy to produce top quarks, with a center-of-mass energy of 7 TeV.

There are multiple processes that can lead to the production of a top quark. The most common is production of a top–antitop pair via strong interactions. In a collision, a highly energetic gluon is created, which subsequently decays into a top and antitop. This process was responsible for the majority of the top events at Tevatron and was the process observed when the top was first discovered in 1995.[24] It is also possible to produce pairs of top–antitop through the decay of an intermediate photon or Z-boson. However, these processes are predicted to be much rarer and have a virtually identical experimental signature in a hadron collider like Tevatron.

A distinctly different process is the production of single tops via weak interaction. This can happen in two ways (called channels): either an intermediate W-boson decays into a top and antibottom quark ("s-channel") or a bottom quark (probably created in a pair through the decay of a gluon) transforms to top quark by exchanging a W-boson with an up or down quark ("t-channel"). The first evidence for these processes was published by the DØ collaboration in December 2006,[25] and in March 2009 the CDF[26] and DØ[24] collaborations released twin papers with the definitive observation of these processes. The main significance of measuring these production processes is that their frequency is directly proportional to the $|V_{tb}|^2$ component of the CKM matrix.

22.4   Decay

Because of its enormous mass, the top quark is extremely short-lived with a predicted lifetime of only 5×10^{-25} s.[3] As a result top quarks do not have time to form hadrons before they decay, as other quarks do which provides physicists with the unique opportunity to study the behavior of a "bare" quark. The only known way the top quark can decay is through the weak interaction producing a W-boson and a down-type quark (down, strange, or bottom).

In particular, it is possible to directly determine the branching ratio $\Gamma(W^+b) / \Gamma(W^+q\ (q = b,s,d))$. The best current determination of this ratio is 0.91 ± 0.04.[27] Since this ratio is equal to $|V_{tb}|^2$ according to the Standard Model, this gives another way of determining the CKM element $|V_{tb}|$, or in combination with the determination of $|V_{tb}|$ from single top production provides tests for the assumption that the CKM matrix is unitary.[28]

The Standard Model also allows more exotic decays, but only at one loop level, meaning that they are extremely suppressed. In particular, it is possible for a top quark to decay into another up-type quark (an up or a charm) by emitting a photon or a Z-boson.[29] Searches for these exotic decay modes have provided no evidence for their existence in accordance with expectations from the Standard Model. The branching ratios for these decays have been determined to be less than 5.9 in 1,000 for photonic decay and less than 2.1 in 1,000 for Z-boson decay at 95% confidence.[27]

22.5   Mass and coupling to the Higgs boson

The Standard Model describes fermion masses through the Higgs mechanism. The Higgs boson has a Yukawa coupling to the left- and right-handed top quarks. After electroweak symmetry breaking (when the Higgs acquires a vacuum expectation value), the left- and right-handed components mix, becoming a mass term.

$$\mathcal{L} = y_t h q u^c \rightarrow \frac{y_t v}{\sqrt{2}}(1 + h^0/v)u u^c$$

The top quark Yukawa coupling has a value of

$$y_t = \sqrt{2}m_t/v \simeq 1$$

where $v = 246$ GeV is the value of the Higgs vacuum expectation value.

22.5.1 Yukawa couplings

See also: Beta function (physics)

In the Standard Model, all of the quark and lepton Yukawa couplings are small compared to the top quark Yukawa coupling. Understanding this hierarchy in the fermion masses is an open problem in theoretical physics. Yukawa couplings are not constants and their values change depending on the energy scale (distance scale) at which they are measured. The dynamics of Yukawa couplings are determined by the renormalization group equation.

One of the prevailing views in particle physics is that the size of the top quark Yukawa coupling is determined by the renormalization group, leading to the "quasi-infrared fixed point."

The Yukawa couplings of the up, down, charm, strange and bottom quarks, are hypothesized to have small values at the extremely high energy scale of grand unification, 10^{15} GeV. They increase in value at lower energy scales, at which the quark masses are generated by the Higgs. The slight growth is due to corrections from the QCD coupling. The corrections from the Yukawa couplings are negligible for the lower mass quarks.

If, however, a quark Yukawa coupling has a large value at very high energies, its Yukawa corrections will evolve and cancel against the QCD corrections. This is known as a (quasi-) infrared fixed point. No matter what the initial starting value of the coupling is, if it is sufficiently large it will reach this fixed point value. The corresponding quark mass is then predicted.

The top quark Yukawa coupling lies very near the infrared fixed point of the Standard Model. The renormalization group equation is:

$$\mu \frac{\partial}{\partial \mu} y_t \approx \frac{y_t}{16\pi^2} \left(\frac{9}{2} y_t^2 - 8g_3^2 - \frac{9}{4} g_2^2 - \frac{17}{20} g_1^2 \right),$$

where g_3 is the color gauge coupling, g_2 is the weak isospin gauge coupling, and g_1 is the weak hypercharge gauge coupling. This equation describes how the Yukawa coupling changes with energy scale μ. Solutions to this equation for large initial values y_t cause the right-hand side of the equation to quickly approach zero, locking y_t to the QCD coupling g_3. The value of the fixed point is fairly precisely determined in the Standard Model, leading to a top quark mass of 230 GeV. However, if there is more than one Higgs doublet, the mass value will be reduced by Higgs mixing angle effects in an unpredicted way.

In the minimal supersymmetric extension of the Standard Model (MSSM), there are two Higgs doublets and the renormalization group equation for the top quark Yukawa coupling is slightly modified:

$$\mu \frac{\partial}{\partial \mu} y_t \approx \frac{y_t}{16\pi^2} \left(6y_t^2 + y_b^2 - \frac{16}{3} g_3^2 - 3g_2^2 - \frac{13}{15} g_1^2 \right),$$

where y_b is the bottom quark Yukawa coupling. This leads to a fixed point where the top mass is smaller, 170–200 GeV. The uncertainty in this prediction arises because the bottom quark Yukawa coupling can be amplified in the MSSM. Some theorists believe this is supporting evidence for the MSSM.

The quasi-infrared fixed point has subsequently formed the basis of top quark condensation theories of electroweak symmetry breaking in which the Higgs boson is composite at *extremely* short distance scales, composed of a pair of top and antitop quarks.

22.6 See also

- CDF experiment
- Topness
- Top quark condensate

- Topcolor

- Quark model

22.7   References

[1] The ATLAS, CDF, CMS, D0 Collaborations (2014). "First combination of Tevatron and LHC measurements of the top-quark mass". Retrieved 2014-03-19.

[2] S. Willenbrock (2003). "The Standard Model and the Top Quark". In H.B Prosper and B. Danilov (eds.). *Techniques and Concepts of High-Energy Physics XII*. NATO Science Series **123**. Kluwer Academic. pp. 1–41. arXiv:hep-ph/0211067v3. ISBN 1-4020-1590-9.

[3] A. Quadt (2006). "Top quark physics at hadron colliders".*European Physical Journal C***48**(3): 835–1000.Bibcode:2006EPJC doi:10.1140/epjc/s2006-02631-6.

[4] M. Kobayashi, T. Maskawa (1973). "*CP*-Violation in the Renormalizable Theory of Weak Interaction". *Progress of Theoretical Physics* **49** (2): 652. Bibcode:1973PThPh..49..652K. doi:10.1143/PTP.49.652.

[5] F. Abe *et al.* (CDF Collaboration) (1995). "Observation of Top Quark Production in pp Collisions with the Collider Detector at Fermilab". *Physical Review Letters* **74** (14): 2626–2631. Bibcode:1995PhRvL..74.2626A. doi:10.1103/PhysRevLett.74.2626. PMID 10057978.

[6] S. Abachi *et al.* (DØ Collaboration) (1995). "Search for High Mass Top Quark Production in pp Collisions at √s = 1.8 TeV". *Physical Review Letters* **74** (13): 2422–2426. Bibcode:1995PhRvL..74.2422A. doi:10.1103/PhysRevLett.74.2422.

[7] "2008 Nobel Prize in Physics". The Nobel Foundation. 2008. Retrieved 2009-09-11.

[8] H. Harari (1975). "A new quark model for hadrons".*Physics Letters B***57**(3): 265.Bibcode:1975PhLB...57..265H.doi:10.10 2693(75)90072-6.

[9] K.W. Staley (2004). *The Evidence for the Top Quark*. Cambridge University Press. pp. 31–33. ISBN 978-0-521-82710-2.

[10] D.H. Perkins (2000). *Introduction to high energy physics*. Cambridge University Press. p. 8. ISBN 0-521-62196-8.

[11] F. Close (2006). *The New Cosmic Onion*. CRC Press. p. 133. ISBN 1-58488-798-2.

[12] S.L. Glashow, J. Iliopoulous, L. Maiani (1970). "Weak Interactions with Lepton–Hadron Symmetry". *Physical Review D* **2** (7): 1285–1292. Bibcode:1970PhRvD...2.1285G. doi:10.1103/PhysRevD.2.1285.

[13] A. Pickering (1999). *Constructing Quarks: A Sociological History of Particle Physics*. University of Chicago Press. pp. 253–254. ISBN 978-0-226-66799-7.

[14] M.L. Perl; et al. (1975). "Evidence for Anomalous Lepton Production in e+e− Annihilation". *Physical Review Letters* **35** (22): 1489. Bibcode:1975PhRvL..35.1489P. doi:10.1103/PhysRevLett.35.1489.

[15] "Discoveries at Fermilab – Discovery of the Bottom Quark" (Press release). Fermilab. 7 August 1977. Retrieved 2009-07-24.

[16] L.M. Lederman (2005). "Logbook: Bottom Quark". *Symmetry Magazine* **2** (8).

[17] S.W. Herb; et al. (1977). "Observation of a Dimuon Resonance at 9.5 GeV in 400-GeV Proton-Nucleus Collisions". *Physical Review Letters* **39** (5): 252. Bibcode:1977PhRvL..39..252H. doi:10.1103/PhysRevLett.39.252.

[18] T.M. Liss, P.L. Tipton (1997). "The Discovery of the Top Quark" (PDF). *Scientific American*: 54–59.

[19] The Discovery of the Top Quark Finding the sixth quark involved the world's most energetic collisions and a cast of thousandsby Tony M. Liss and Paul L. Tipton http://lphe.epfl.ch/~{}mtran/seminaires/Cours_Master_Bordeaux/Articles/SciAmTop.pdf

[20] "The Nobel Prize in Physics 1999". The Nobel Foundation. Retrieved 2009-09-10.

[21] "The Nobel Prize in Physics 1999, Press Release" (Press release). The Nobel Foundation. 12 October 1999. Retrieved 2009-09-10.

[22] D. Chakraborty (DØ and CDF collaborations) (2002). *Top quark and W/Z results from the Tevatron* (PDF). Rencontres de Moriond. p. 26.

[23] V.M. Abazov *et al.* (DØ Collaboration) (2007). "Experimental discrimination between charge 2*e*/3 top quark and charge 4*e*/3 exotic quark production scenarios". *Physical Review Letters* **98** (4): 041801. arXiv:hep-ex/0608044. Bibcode:2007PhRvL..98d1801A.doi:10.1103/PhysRevLett.98.041801.PMID17358756.

[24] V.M. Abazov *et al.* (DØ Collaboration) (2009). "Observation of Single Top Quark Production". *Physical Review Letters* **103** (9). arXiv:0903.0850. Bibcode:2009PhRvL.103i2001A. doi:10.1103/PhysRevLett.103.092001.

[25] V.M. Abazov *et al.* (DØ Collaboration) (2007). "Evidence for production of single top quarks and first direct measurement of |V$_{tb}$|". *Physical Review Letters* **98** (18): 181802. arXiv:hep-ex/0612052. Bibcode:2007PhRvL..98r1802A. doi:10.1103/PhysRevLett.98.181802.PMID17501561.

[26] T. Aaltonen *et al.* (CDF Collaboration) (2009). "First Observation of Electroweak Single Top Quark Production". *Physical Review Letters* **103** (9). arXiv:0903.0885. Bibcode:2009PhRvL.103i2002A. doi:10.1103/PhysRevLett.103.092002.

[27] J. Beringer *et al.* (Particle Data Group) (2012). "PDGLive Particle Summary 'Quarks (u, d, s, c, b, t, b', t', Free)'" (PDF). Particle Data Group. Retrieved 2013-07-23.

[28] V.M. Abazov *et al.* (DØ Collaboration) (2008). "Simultaneous measurement of the ratio B(t→Wb)/B(t→Wq) and the top-quark pair production cross section with the DØ detector at √s = 1.96 TeV". *Physical Review Letters* **100** (19): 192003. arXiv:0801.1326. Bibcode:2008PhRvL.100s2003A. doi:10.1103/PhysRevLett.100.192003.

[29] S. Chekanov *et al.* (ZEUS Collaboration) (2003). "Search for single-top production in ep collisions at HERA". *Physics Letters B* **559** (3–4): 153. arXiv:hep-ex/0302010. Bibcode:2003PhLB..559..153Z. doi:10.1016/S0370-2693(03)00333-2.

22.8 Further reading

- Frank Fiedler; for the D0; CDF Collaborations (June 2005). "Top Quark Production and Properties at the Tevatron". arXiv:hep-ex/0506005 [hep-ex].

- R. Nave. "Quarks". *HyperPhysics*. Georgia State University, Department of Physics and Astronomy. Retrieved 2008-06-29.

- A. Pickering (1984). *Constructing Quarks*. University of Chicago Press. pp. 114–125. ISBN 0-226-66799-5.

22.9 External links

- Top quark on arxiv.org

- Tevatron Electroweak Working Group

- Top quark information on Fermilab website

- Logbook pages from CDF and DZero collaborations' top quark discovery

- Scientific American article on the discovery of the top quark

- Public Homepage of Top Quark Analysis Results from DØ Collaboration at Fermilab

- Public Homepage of Top Quark Analysis Results from CDF Collaboration at Fermilab

- Harvard Magazine article about the 1994 top quark discovery

- 1999 Nobel Prize in Physics

Chapter 23

Electron

For other uses, see Electron (disambiguation).

The **electron** is a subatomic particle, symbol e− or β−, with a negative elementary electric charge.[8] Electrons belong to the first generation of the lepton particle family,[9] and are generally thought to be elementary particles because they have no known components or substructure.[1] The electron has a mass that is approximately 1/1836 that of the proton.[10] Quantum mechanical properties of the electron include an intrinsic angular momentum (spin) of a half-integer value in units of $\hbar$, which means that it is a fermion. Being fermions, no two electrons can occupy the same quantum state, in accordance with the Pauli exclusion principle.[9] Like all matter, electrons have properties of both particles and waves, and so can collide with other particles and can be diffracted like light. The wave properties of electrons are easier to observe with experiments than those of other particles like neutrons and protons because electrons have a lower mass and hence a higher De Broglie wavelength for typical energies.

Many physical phenomena involve electrons in an essential role, such as electricity, magnetism, and thermal conductivity, and they also participate in gravitational, electromagnetic and weak interactions.[11] An electron generates an electric field surrounding it. An electron moving relative to an observer generates a magnetic field. External magnetic fields deflect an electron. Electrons radiate or absorb energy in the form of photons when accelerated. Laboratory instruments are capable of containing and observing individual electrons as well as electron plasma using electromagnetic fields, whereas dedicated telescopes can detect electron plasma in outer space. Electrons are involved in many applications such as electronics, welding, cathode ray tubes, electron microscopes, radiation therapy, lasers, gaseous ionization detectors and particle accelerators.

Interactions involving electrons and other subatomic particles are of interest in fields such as chemistry and nuclear physics. The Coulomb force interaction between positive protons inside atomic nuclei and negative electrons composes atoms. Ionization or changes in the proportions of particles changes the binding energy of the system. The exchange or sharing of the electrons between two or more atoms is the main cause of chemical bonding.[12] British natural philosopher Richard Laming first hypothesized the concept of an indivisible quantity of electric charge to explain the chemical properties of atoms in 1838;[3] Irish physicist George Johnstone Stoney named this charge 'electron' in 1891, and J. J. Thomson and his team of British physicists identified it as a particle in 1897.[5][13][14] Electrons can also participate in nuclear reactions, such as nucleosynthesis in stars, where they are known as beta particles. Electrons may be created through beta decay of radioactive isotopes and in high-energy collisions, for instance when cosmic rays enter the atmosphere. The antiparticle of the electron is called the positron; it is identical to the electron except that it carries electrical and other charges of the opposite sign. When an electron collides with a positron, both particles may be totally annihilated, producing gamma ray photons.

23.1 History

See also: History of electromagnetism

The ancient Greeks noticed that amber attracted small objects when rubbed with fur. Along with lightning, this phenomenon is one of humanity's earliest recorded experiences with electricity. [15] In his 1600 treatise *De Magnete*, the English scientist William Gilbert coined the New Latin term *electricus*, to refer to this property of attracting small objects after being rubbed. [16] Both *electric* and *electricity* are derived from the Latin *ēlectrum* (also the root of the alloy of the same name), which came from the Greek word for amber, ἤλεκτρον (*ēlektron*).

In the early 1700s, Francis Hauksbee and French chemist Charles François de Fay independently discovered what they believed were two kinds of frictional electricity—one generated from rubbing glass, the other from rubbing resin. From this, Du Fay theorized that electricity consists of two electrical fluids, *vitreous* and *resinous*, that are separated by friction, and that neutralize each other when combined.[17] A decade later Benjamin Franklin proposed that electricity was not from different types of electrical fluid, but the same electrical fluid under different pressures. He gave them the modern charge nomenclature of positive and negative respectively.[18] Franklin thought of the charge carrier as being positive, but he did not correctly identify which situation was a surplus of the charge carrier, and which situation was a deficit.[19]

Between 1838 and 1851, British natural philosopher Richard Laming developed the idea that an atom is composed of a core of matter surrounded by subatomic particles that had unit electric charges.[2] Beginning in 1846, German physicist William Weber theorized that electricity was composed of positively and negatively charged fluids, and their interaction was governed by the inverse square law. After studying the phenomenon of electrolysis in 1874, Irish physicist George Johnstone Stoney suggested that there existed a "single definite quantity of electricity", the charge of a monovalent ion. He was able to estimate the value of this elementary charge e by means of Faraday's laws of electrolysis.[20] However, Stoney believed these charges were permanently attached to atoms and could not be removed. In 1881, German physicist Hermann von Helmholtz argued that both positive and negative charges were divided into elementary parts, each of which "behaves like atoms of electricity".[3]

Stoney initially coined the term *electrolion* in 1881. Ten years later, he switched to *electron* to describe these elementary charges, writing in 1894: "... an estimate was made of the actual amount of this most remarkable fundamental unit of electricity, for which I have since ventured to suggest the name *electron*". A 1906 proposal to change to *electrion* failed because Hendrik Lorentz preferred to keep *electron*.[21][22] The word *electron* is a combination of the words *electric* and *ion*.[23] The suffix *-on* which is now used to designate other subatomic particles, such as a proton or neutron, is in turn derived from electron.[24][25]

23.1.1 Discovery

The German physicist Johann Wilhelm Hittorf studied electrical conductivity in rarefied gases: in 1869, he discovered a glow emitted from the cathode that increased in size with decrease in gas pressure. In 1876, the German physicist Eugen Goldstein showed that the rays from this glow cast a shadow, and he dubbed the rays cathode rays.[27] During the 1870s, the English chemist and physicist Sir William Crookes developed the first cathode ray tube to have a high vacuum inside.[28] He then showed that the luminescence rays appearing within the tube carried energy and moved from the cathode to the anode. Furthermore, by applying a magnetic field, he was able to deflect the rays, thereby demonstrating that the beam behaved as though it were negatively charged.[29][30] In 1879, he proposed that these properties could be explained by what he termed 'radiant matter'. He suggested that this was a fourth state of matter, consisting of negatively charged molecules that were being projected with high velocity from the cathode.[31]

The German-born British physicist Arthur Schuster expanded upon Crookes' experiments by placing metal plates parallel to the cathode rays and applying an electric potential between the plates. The field deflected the rays toward the positively charged plate, providing further evidence that the rays carried negative charge. By measuring the amount of deflection for a given level of current, in 1890 Schuster was able to estimate the charge-to-mass ratio of the ray components. However, this produced a value that was more than a thousand times greater than what was expected, so little credence was given to his calculations at the time.[29][32]

In 1892 Hendrik Lorentz suggested that the mass of these particles (electrons) could be a consequence of their electric

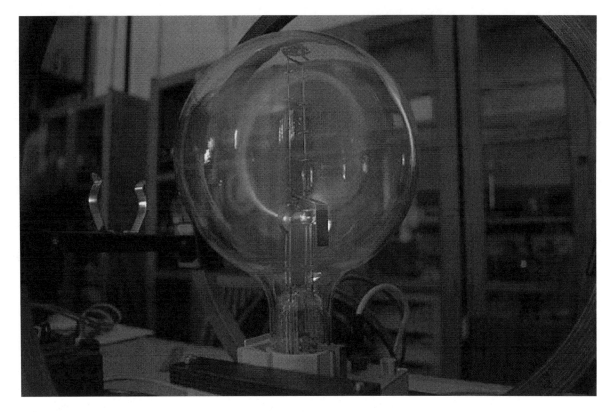

A beam of electrons deflected in a circle by a magnetic field[26]

charge.[33]

In 1896, the British physicist J. J. Thomson, with his colleagues John S. Townsend and H. A. Wilson,[13] performed experiments indicating that cathode rays really were unique particles, rather than waves, atoms or molecules as was believed earlier.[5] Thomson made good estimates of both the charge e and the mass m, finding that cathode ray particles, which he called "corpuscles," had perhaps one thousandth of the mass of the least massive ion known: hydrogen.[5][14] He showed that their charge to mass ratio, e/m, was independent of cathode material. He further showed that the negatively charged particles produced by radioactive materials, by heated materials and by illuminated materials were universal.[5][34] The name electron was again proposed for these particles by the Irish physicist George F. Fitzgerald, and the name has since gained universal acceptance.[29]

While studying naturally fluorescing minerals in 1896, the French physicist Henri Becquerel discovered that they emitted radiation without any exposure to an external energy source. These radioactive materials became the subject of much interest by scientists, including the New Zealand physicist Ernest Rutherford who discovered they emitted particles. He designated these particles alpha and beta, on the basis of their ability to penetrate matter.[35] In 1900, Becquerel showed that the beta rays emitted by radium could be deflected by an electric field, and that their mass-to-charge ratio was the same as for cathode rays.[36] This evidence strengthened the view that electrons existed as components of atoms.[37][38]

The electron's charge was more carefully measured by the American physicists Robert Millikan and Harvey Fletcher in their oil-drop experiment of 1909, the results of which were published in 1911. This experiment used an electric field to prevent a charged droplet of oil from falling as a result of gravity. This device could measure the electric charge from as few as 1–150 ions with an error margin of less than 0.3%. Comparable experiments had been done earlier by Thomson's team,[5] using clouds of charged water droplets generated by electrolysis,[13] and in 1911 by Abram Ioffe, who independently obtained the same result as Millikan using charged microparticles of metals, then published his results in 1913.[39] However, oil drops were more stable than water drops because of their slower evaporation rate, and thus more suited to precise experimentation over longer periods of time.[40]

Around the beginning of the twentieth century, it was found that under certain conditions a fast-moving charged particle caused a condensation of supersaturated water vapor along its path. In 1911, Charles Wilson used this principle to devise his cloud chamber so he could photograph the tracks of charged particles, such as fast-moving electrons.[41]

23.1.2 Atomic theory

See also: The proton–electron model of the nucleus

By 1914, experiments by physicists Ernest Rutherford, Henry Moseley, James Franck and Gustav Hertz had largely established the structure of an atom as a dense nucleus of positive charge surrounded by lower-mass electrons.[42] In 1913, Danish physicist Niels Bohr postulated that electrons resided in quantized energy states, with the energy determined by the angular momentum of the electron's orbits about the nucleus. The electrons could move between these states, or orbits, by the emission or absorption of photons at specific frequencies. By means of these quantized orbits, he accurately explained the spectral lines of the hydrogen atom.[43] However, Bohr's model failed to account for the relative intensities of the spectral lines and it was unsuccessful in explaining the spectra of more complex atoms.[42]

Chemical bonds between atoms were explained by Gilbert Newton Lewis, who in 1916 proposed that a covalent bond between two atoms is maintained by a pair of electrons shared between them.[44] Later, in 1927, Walter Heitler and Fritz London gave the full explanation of the electron-pair formation and chemical bonding in terms of quantum mechanics.[45] In 1919, the American chemist Irving Langmuir elaborated on the Lewis' static model of the atom and suggested that all electrons were distributed in successive "concentric (nearly) spherical shells, all of equal thickness".[46] The shells were, in turn, divided by him in a number of cells each containing one pair of electrons. With this model Langmuir was able to qualitatively explain the chemical properties of all elements in the periodic table,[45] which were known to largely repeat themselves according to the periodic law.[47]

In 1924, Austrian physicist Wolfgang Pauli observed that the shell-like structure of the atom could be explained by a set of four parameters that defined every quantum energy state, as long as each state was inhabited by no more than a single electron. (This prohibition against more than one electron occupying the same quantum energy state became known as the Pauli exclusion principle.)[48] The physical mechanism to explain the fourth parameter, which had two distinct possible values, was provided by the Dutch physicists Samuel Goudsmit and George Uhlenbeck. In 1925, Goudsmit and Uhlenbeck suggested that an electron, in addition to the angular momentum of its orbit, possesses an intrinsic angular momentum and magnetic dipole moment.[42][49] The intrinsic angular momentum became known as spin, and explained the previously mysterious splitting of spectral lines observed with a high-resolution spectrograph; this phenomenon is known as fine structure splitting.[50]

23.1.3 Quantum mechanics

See also: History of quantum mechanics

In his 1924 dissertation *Recherches sur la théorie des quanta* (Research on Quantum Theory), French physicist Louis de Broglie hypothesized that all matter possesses a de Broglie wave similar to light.[51] That is, under the appropriate conditions, electrons and other matter would show properties of either particles or waves. The corpuscular properties of a particle are demonstrated when it is shown to have a localized position in space along its trajectory at any given moment.[52] Wave-like nature is observed, for example, when a beam of light is passed through parallel slits and creates interference patterns. In 1927, the interference effect was found in a beam of electrons by English physicist George Paget Thomson with a thin metal film and by American physicists Clinton Davisson and Lester Germer using a crystal of nickel.[53]

De Broglie's prediction of a wave nature for electrons led Erwin Schrödinger to postulate a wave equation for electrons moving under the influence of the nucleus in the atom. In 1926, this equation, the Schrödinger equation, successfully described how electron waves propagated.[54] Rather than yielding a solution that determined the location of an electron over time, this wave equation also could be used to predict the probability of finding an electron near a position, especially a position near where the electron was bound in space, for which the electron wave equations did not change in time. This approach led to a second formulation of quantum mechanics (the first being by Heisenberg in 1925), and solutions of Schrödinger's equation, like Heisenberg's, provided derivations of the energy states of an electron in a hydrogen atom that were equivalent to those that had been derived first by Bohr in 1913, and that were known to reproduce the hydrogen spectrum.[55] Once spin and the interaction between multiple electrons were considered, quantum mechanics later made it possible to predict the configuration of electrons in atoms with higher atomic numbers than hydrogen.[56]

In 1928, building on Wolfgang Pauli's work, Paul Dirac produced a model of the electron – the Dirac equation, consistent with relativity theory, by applying relativistic and symmetry considerations to the hamiltonian formulation of the quantum mechanics of the electro-magnetic field.[57] To resolve some problems within his relativistic equation, in 1930 Dirac developed a model of the vacuum as an infinite sea of particles having negative energy, which was dubbed the Dirac sea. This led him to predict the existence of a positron, the antimatter counterpart of the electron.[58] This particle was discovered in 1932 by Carl Anderson, who proposed calling standard electrons *negatrons*, and using *electron* as a generic term to describe both the positively and negatively charged variants.

In 1947 Willis Lamb, working in collaboration with graduate student Robert Retherford, found that certain quantum states of hydrogen atom, which should have the same energy, were shifted in relation to each other, the difference being the Lamb shift. About the same time, Polykarp Kusch, working with Henry M. Foley, discovered the magnetic moment of the electron is slightly larger than predicted by Dirac's theory. This small difference was later called anomalous magnetic dipole moment of the electron. This difference was later explained by the theory of quantum electrodynamics, developed by Sin-Itiro Tomonaga, Julian Schwinger and Richard Feynman in the late 1940s.[59]

23.1.4   Particle accelerators

With the development of the particle accelerator during the first half of the twentieth century, physicists began to delve deeper into the properties of subatomic particles.[60] The first successful attempt to accelerate electrons using electromagnetic induction was made in 1942 by Donald Kerst. His initial betatron reached energies of 2.3 MeV, while subsequent betatrons achieved 300 MeV. In 1947, synchrotron radiation was discovered with a 70 MeV electron synchrotron at General Electric. This radiation was caused by the acceleration of electrons, moving near the speed of light, through a magnetic field.[61]

With a beam energy of 1.5 GeV, the first high-energy particle collider was ADONE, which began operations in 1968.[62] This device accelerated electrons and positrons in opposite directions, effectively doubling the energy of their collision when compared to striking a static target with an electron.[63] The Large Electron–Positron Collider (LEP) at CERN, which was operational from 1989 to 2000, achieved collision energies of 209 GeV and made important measurements for the Standard Model of particle physics.[64][65]

23.1.5   Confinement of individual electrons

Individual electrons can now be easily confined in ultra small ($L = 20$ nm, $W = 20$ nm) CMOS transistors operated at cryogenic temperature over a range of −269 °C (4 K) to about −258 °C (15 K).[66] The electron wavefunction spreads in a semiconductor lattice and negligibly interacts with the valence band electrons, so it can be treated in the single particle formalism, by replacing its mass with the effective mass tensor.

23.2   Characteristics

23.2.1   Classification

In the Standard Model of particle physics, electrons belong to the group of subatomic particles called leptons, which are believed to be fundamental or elementary particles. Electrons have the lowest mass of any charged lepton (or electrically charged particle of any type) and belong to the first-generation of fundamental particles.[67] The second and third generation contain charged leptons, the muon and the tau, which are identical to the electron in charge, spin and interactions, but are more massive. Leptons differ from the other basic constituent of matter, the quarks, by their lack of strong interaction. All members of the lepton group are fermions, because they all have half-odd integer spin; the electron has spin 1/2.[68]

23.2.2 Fundamental properties

The invariant mass of an electron is approximately 9.109×10^{-31} kilograms,[69] or 5.489×10^{-4} atomic mass units. On the basis of Einstein's principle of mass–energy equivalence, this mass corresponds to a rest energy of 0.511 MeV. The ratio between the mass of a proton and that of an electron is about 1836.[10][70] Astronomical measurements show that the proton-to-electron mass ratio has held the same value for at least half the age of the universe, as is predicted by the Standard Model.[71]

Electrons have an electric charge of -1.602×10^{-19} coulomb,[69] which is used as a standard unit of charge for sub-atomic particles, and is also called the elementary charge. This elementary charge has a relative standard uncertainty of 2.2×10^{-8}.[69] Within the limits of experimental accuracy, the electron charge is identical to the charge of a proton, but with the opposite sign.[72] As the symbol e is used for the elementary charge, the electron is commonly symbolized by e−, where the minus sign indicates the negative charge. The positron is symbolized by e+ because it has the same properties as the electron but with a positive rather than negative charge.[68][69]

The electron has an intrinsic angular momentum or spin of 1/2.[69] This property is usually stated by referring to the electron as a spin-1/2 particle.[68] For such particles the spin magnitude is $\sqrt{3}/2\ \hbar$.[note 3] while the result of the measurement of a projection of the spin on any axis can only be $\pm\hbar/2$. In addition to spin, the electron has an intrinsic magnetic moment along its spin axis.[69] It is approximately equal to one Bohr magneton,[73][note 4] which is a physical constant equal to $9.27400915(23) \times 10^{-24}$ joules per tesla.[69] The orientation of the spin with respect to the momentum of the electron defines the property of elementary particles known as helicity.[74]

The electron has no known substructure.[1][75] and it is assumed to be a point particle with a point charge and no spatial extent.[9] In classical physics, the angular momentum and magnetic moment of an object depend upon its physical dimensions. Hence, the concept of a dimensionless electron possessing these properties contrasts to experimental observations in Penning traps which point to finite non-zero radius of the electron. A possible explanation of this paradoxical situation is given below in the "Virtual particles" subsection by taking into consideration the Foldy-Wouthuysen transformation.

The issue of the radius of the electron is a challenging problem of the modern theoretical physics. The admission of the hypothesis of a finite radius of the electron is incompatible to the premises of the theory of relativity. On the other hand, a point-like electron (zero radius) generates serious mathematical difficulties due to the self-energy of the electron tending to infinity.[76] These aspects have been analyzed in detail by Dmitri Ivanenko and Arseny Sokolov.

Observation of a single electron in a Penning trap shows the upper limit of the particle's radius is 10^{-22} meters.[77] There *is* a physical constant called the "classical electron radius", with the much larger value of 2.8179×10^{-15} m, greater than the radius of the proton. However, the terminology comes from a simplistic calculation that ignores the effects of quantum mechanics; in reality, the so-called classical electron radius has little to do with the true fundamental structure of the electron.[78][note 5]

There are elementary particles that spontaneously decay into less massive particles. An example is the muon, which decays into an electron, a neutrino and an antineutrino, with a mean lifetime of 2.2×10^{-6} seconds. However, the electron is thought to be stable on theoretical grounds: the electron is the least massive particle with non-zero electric charge, so its decay would violate charge conservation.[79] The experimental lower bound for the electron's mean lifetime is 6.6×10^{28} years, at a 90% confidence level.[7][80][81]

23.2.3 Quantum properties

As with all particles, electrons can act as waves. This is called the wave–particle duality and can be demonstrated using the double-slit experiment.

The wave-like nature of the electron allows it to pass through two parallel slits simultaneously, rather than just one slit as would be the case for a classical particle. In quantum mechanics, the wave-like property of one particle can be described mathematically as a complex-valued function, the wave function, commonly denoted by the Greek letter psi (ψ). When the absolute value of this function is squared, it gives the probability that a particle will be observed near a location—a probability density.[82]:162–218

Electrons are identical particles because they cannot be distinguished from each other by their intrinsic physical properties. In quantum mechanics, this means that a pair of interacting electrons must be able to swap positions without an observable change to the state of the system. The wave function of fermions, including electrons, is antisymmetric, meaning that it changes sign when two electrons are swapped; that is, $\psi(r_1, r_2) = -\psi(r_2, r_1)$, where the variables r_1 and r_2 correspond to the first and second electrons, respectively. Since the absolute value is not changed by a sign swap, this corresponds to equal probabilities. Bosons, such as the photon, have symmetric wave functions instead.[82]:162–218

In the case of antisymmetry, solutions of the wave equation for interacting electrons result in a zero probability that each pair will occupy the same location or state. This is responsible for the Pauli exclusion principle, which precludes any two electrons from occupying the same quantum state. This principle explains many of the properties of electrons. For example, it causes groups of bound electrons to occupy different orbitals in an atom, rather than all overlapping each other in the same orbit.[82]:162–218

23.2.4   Virtual particles

Main article: Virtual particle

In a simplified picture, every photon spends some time as a combination of a virtual electron plus its antiparticle, the virtual positron, which rapidly annihilate each other shortly thereafter.[83] The combination of the energy variation needed to create these particles, and the time during which they exist, fall under the threshold of detectability expressed by the Heisenberg uncertainty relation, $\Delta E \cdot \Delta t \geq \hbar$. In effect, the energy needed to create these virtual particles, ΔE, can be "borrowed" from the vacuum for a period of time, Δt, so that their product is no more than the reduced Planck constant, $\hbar \approx 6.6\times10^{-16}$ eV·s. Thus, for a virtual electron, Δt is at most 1.3×10^{-21} s.[84]

While an electron–positron virtual pair is in existence, the coulomb force from the ambient electric field surrounding an electron causes a created positron to be attracted to the original electron, while a created electron experiences a repulsion. This causes what is called vacuum polarization. In effect, the vacuum behaves like a medium having a dielectric permittivity more than unity. Thus the effective charge of an electron is actually smaller than its true value, and the charge decreases with increasing distance from the electron.[85][86] This polarization was confirmed experimentally in 1997 using the Japanese TRISTAN particle accelerator.[87] Virtual particles cause a comparable shielding effect for the mass of the electron.[88]

The interaction with virtual particles also explains the small (about 0.1%) deviation of the intrinsic magnetic moment of the electron from the Bohr magneton (the anomalous magnetic moment).[73][89] The extraordinarily precise agreement of this predicted difference with the experimentally determined value is viewed as one of the great achievements of quantum electrodynamics.[90]

The apparent paradox (mentioned above in the properties subsection) of a point particle electron having intrinsic angular momentum and magnetic moment can be explained by the formation of virtual photons in the electric field generated by the electron. These photons cause the electron to shift about in a jittery fashion (known as zitterbewegung),[91] which results in a net circular motion with precession. This motion produces both the spin and the magnetic moment of the electron.[91][92] In atoms, this creation of virtual photons explains the Lamb shift observed in spectral lines.[85]

23.2.5   Interaction

An electron generates an electric field that exerts an attractive force on a particle with a positive charge, such as the proton, and a repulsive force on a particle with a negative charge. The strength of this force is determined by Coulomb's inverse square law.[93] When an electron is in motion, it generates a magnetic field.[82]:140 The Ampère-Maxwell law relates the magnetic field to the mass motion of electrons (the current) with respect to an observer. This property of induction supplies the magnetic field that drives an electric motor.[94] The electromagnetic field of an arbitrary moving charged particle is expressed by the Liénard–Wiechert potentials, which are valid even when the particle's speed is close to that of light (relativistic).

When an electron is moving through a magnetic field, it is subject to the Lorentz force that acts perpendicularly to the plane defined by the magnetic field and the electron velocity. This centripetal force causes the electron to follow a helical trajectory through the field at a radius called the gyroradius. The acceleration from this curving motion induces the electron to radiate energy in the form of synchrotron radiation.[82]:160[95][note 6] The energy emission in turn causes a recoil of the electron, known as the Abraham–Lorentz–Dirac Force, which creates a friction that slows the electron. This force is caused by a back-reaction of the electron's own field upon itself.[96]

Photons mediate electromagnetic interactions between particles in quantum electrodynamics. An isolated electron at a constant velocity cannot emit or absorb a real photon; doing so would violate conservation of energy and momentum. Instead, virtual photons can transfer momentum between two charged particles. This exchange of virtual photons, for example, generates the Coulomb force.[97] Energy emission can occur when a moving electron is deflected by a charged particle, such as a proton. The acceleration of the electron results in the emission of Bremsstrahlung radiation.[98]

An inelastic collision between a photon (light) and a solitary (free) electron is called Compton scattering. This collision results in a transfer of momentum and energy between the particles, which modifies the wavelength of the photon by an amount called the Compton shift.[note 7] The maximum magnitude of this wavelength shift is $h/m_e c$, which is known as the Compton wavelength.[99] For an electron, it has a value of 2.43×10^{-12} m.[69] When the wavelength of the light is long (for instance, the wavelength of the visible light is 0.4–0.7 μm) the wavelength shift becomes negligible. Such interaction between the light and free electrons is called Thomson scattering or Linear Thomson scattering.[100]

The relative strength of the electromagnetic interaction between two charged particles, such as an electron and a proton, is given by the fine-structure constant. This value is a dimensionless quantity formed by the ratio of two energies: the electrostatic energy of attraction (or repulsion) at a separation of one Compton wavelength, and the rest energy of the charge. It is given by $\alpha \approx 7.297353 \times 10^{-3}$, which is approximately equal to 1/137.[69]

When electrons and positrons collide, they annihilate each other, giving rise to two or more gamma ray photons. If the electron and positron have negligible momentum, a positronium atom can form before annihilation results in two or three gamma ray photons totalling 1.022 MeV.[101][102] On the other hand, high-energy photons may transform into an electron and a positron by a process called pair production, but only in the presence of a nearby charged particle, such as a nucleus.[103][104]

In the theory of electroweak interaction, the left-handed component of electron's wavefunction forms a weak isospin doublet with the electron neutrino. This means that during weak interactions, electron neutrinos behave like electrons. Either member of this doublet can undergo a charged current interaction by emitting or absorbing a W and be converted into the other member. Charge is conserved during this reaction because the W boson also carries a charge, canceling out any net change during the transmutation. Charged current interactions are responsible for the phenomenon of beta decay in a radioactive atom. Both the electron and electron neutrino can undergo a neutral current interaction via a Z0 exchange, and this is responsible for neutrino-electron elastic scattering.[105]

23.2.6 Atoms and molecules

Main article: Atom

An electron can be *bound* to the nucleus of an atom by the attractive Coulomb force. A system of one or more electrons bound to a nucleus is called an atom. If the number of electrons is different from the nucleus' electrical charge, such an atom is called an ion. The wave-like behavior of a bound electron is described by a function called an atomic orbital. Each orbital has its own set of quantum numbers such as energy, angular momentum and projection of angular momentum, and only a discrete set of these orbitals exist around the nucleus. According to the Pauli exclusion principle each orbital can be occupied by up to two electrons, which must differ in their spin quantum number.

Electrons can transfer between different orbitals by the emission or absorption of photons with an energy that matches the difference in potential.[106] Other methods of orbital transfer include collisions with particles, such as electrons, and the Auger effect.[107] To escape the atom, the energy of the electron must be increased above its binding energy to the atom. This occurs, for example, with the photoelectric effect, where an incident photon exceeding the atom's ionization energy is absorbed by the electron.[108]

The orbital angular momentum of electrons is quantized. Because the electron is charged, it produces an orbital magnetic moment that is proportional to the angular momentum. The net magnetic moment of an atom is equal to the vector sum of orbital and spin magnetic moments of all electrons and the nucleus. The magnetic moment of the nucleus is negligible compared with that of the electrons. The magnetic moments of the electrons that occupy the same orbital (so called, paired electrons) cancel each other out.[109]

The chemical bond between atoms occurs as a result of electromagnetic interactions, as described by the laws of quantum mechanics.[110] The strongest bonds are formed by the sharing or transfer of electrons between atoms, allowing the formation of molecules.[12] Within a molecule, electrons move under the influence of several nuclei, and occupy molecular orbitals; much as they can occupy atomic orbitals in isolated atoms.[111] A fundamental factor in these molecular structures is the existence of electron pairs. These are electrons with opposed spins, allowing them to occupy the same molecular orbital without violating the Pauli exclusion principle (much like in atoms). Different molecular orbitals have different spatial distribution of the electron density. For instance, in bonded pairs (i.e. in the pairs that actually bind atoms together) electrons can be found with the maximal probability in a relatively small volume between the nuclei. On the contrary, in non-bonded pairs electrons are distributed in a large volume around nuclei.[112]

23.2.7   Conductivity

If a body has more or fewer electrons than are required to balance the positive charge of the nuclei, then that object has a net electric charge. When there is an excess of electrons, the object is said to be negatively charged. When there are fewer electrons than the number of protons in nuclei, the object is said to be positively charged. When the number of electrons and the number of protons are equal, their charges cancel each other and the object is said to be electrically neutral. A macroscopic body can develop an electric charge through rubbing, by the triboelectric effect.[116]

Independent electrons moving in vacuum are termed *free* electrons. Electrons in metals also behave as if they were free. In reality the particles that are commonly termed electrons in metals and other solids are quasi-electrons—quasiparticles, which have the same electrical charge, spin and magnetic moment as real electrons but may have a different mass.[117] When free electrons—both in vacuum and metals—move, they produce a net flow of charge called an electric current, which generates a magnetic field. Likewise a current can be created by a changing magnetic field. These interactions are described mathematically by Maxwell's equations.[118]

At a given temperature, each material has an electrical conductivity that determines the value of electric current when an electric potential is applied. Examples of good conductors include metals such as copper and gold, whereas glass and Teflon are poor conductors. In any dielectric material, the electrons remain bound to their respective atoms and the material behaves as an insulator. Most semiconductors have a variable level of conductivity that lies between the extremes of conduction and insulation.[119] On the other hand, metals have an electronic band structure containing partially filled electronic bands. The presence of such bands allows electrons in metals to behave as if they were free or delocalized electrons. These electrons are not associated with specific atoms, so when an electric field is applied, they are free to move like a gas (called Fermi gas)[120] through the material much like free electrons.

Because of collisions between electrons and atoms, the drift velocity of electrons in a conductor is on the order of millimeters per second. However, the speed at which a change of current at one point in the material causes changes in currents in other parts of the material, the velocity of propagation, is typically about 75% of light speed.[121] This occurs because electrical signals propagate as a wave, with the velocity dependent on the dielectric constant of the material.[122]

Metals make relatively good conductors of heat, primarily because the delocalized electrons are free to transport thermal energy between atoms. However, unlike electrical conductivity, the thermal conductivity of a metal is nearly independent of temperature. This is expressed mathematically by the Wiedemann–Franz law,[120] which states that the ratio of thermal conductivity to the electrical conductivity is proportional to the temperature. The thermal disorder in the metallic lattice increases the electrical resistivity of the material, producing a temperature dependence for electric current.[123]

When cooled below a point called the critical temperature, materials can undergo a phase transition in which they lose all resistivity to electric current, in a process known as superconductivity. In BCS theory, this behavior is modeled by pairs of electrons entering a quantum state known as a Bose–Einstein condensate. These Cooper pairs have their motion coupled to nearby matter via lattice vibrations called phonons, thereby avoiding the collisions with atoms that normally create electrical resistance.[124] (Cooper pairs have a radius of roughly 100 nm, so they can overlap each other.)[125] However, the mechanism by which higher temperature superconductors operate remains uncertain.

Electrons inside conducting solids, which are quasi-particles themselves, when tightly confined at temperatures close to absolute zero, behave as though they had split into three other quasiparticles: spinons, orbitons and holons.[126][127] The former carries spin and magnetic moment, the next carries its orbital location while the latter electrical charge.

23.2.8   Motion and energy

According to Einstein's theory of special relativity, as an electron's speed approaches the speed of light, from an observer's point of view its relativistic mass increases, thereby making it more and more difficult to accelerate it from within the observer's frame of reference. The speed of an electron can approach, but never reach, the speed of light in a vacuum, c. However, when relativistic electrons—that is, electrons moving at a speed close to c—are injected into a dielectric medium such as water, where the local speed of light is significantly less than c, the electrons temporarily travel faster than light in the medium. As they interact with the medium, they generate a faint light called Cherenkov radiation.[128]

The effects of special relativity are based on a quantity known as the Lorentz factor, defined as $\gamma = 1/\sqrt{1 - v^2/c^2}$ where v is the speed of the particle. The kinetic energy K_e of an electron moving with velocity v is:

$$K_e = (\gamma - 1)m_e c^2,$$

where m_e is the mass of electron. For example, the Stanford linear accelerator can accelerate an electron to roughly 51 GeV.[129] Since an electron behaves as a wave, at a given velocity it has a characteristic de Broglie wavelength. This is given by $\lambda_e = h/p$ where h is the Planck constant and p is the momentum.[51] For the 51 GeV electron above, the wavelength is about 2.4×10^{-17} m, small enough to explore structures well below the size of an atomic nucleus.[130]

23.3   Formation

The Big Bang theory is the most widely accepted scientific theory to explain the early stages in the evolution of the Universe.[131] For the first millisecond of the Big Bang, the temperatures were over 10 billion Kelvin and photons had mean energies over a million electronvolts. These photons were sufficiently energetic that they could react with each other to form pairs of electrons and positrons. Likewise, positron-electron pairs annihilated each other and emitted energetic photons:

$$\gamma + \gamma \leftrightarrow e+ + e-$$

An equilibrium between electrons, positrons and photons was maintained during this phase of the evolution of the Universe. After 15 seconds had passed, however, the temperature of the universe dropped below the threshold where electron-positron formation could occur. Most of the surviving electrons and positrons annihilated each other, releasing gamma radiation that briefly reheated the universe.[132]

For reasons that remain uncertain, during the process of leptogenesis there was an excess in the number of electrons over positrons.[133] Hence, about one electron in every billion survived the annihilation process. This excess matched the excess of protons over antiprotons, in a condition known as baryon asymmetry, resulting in a net charge of zero for the universe.[134][135] The surviving protons and neutrons began to participate in reactions with each other—in the process known as nucleosynthesis, forming isotopes of hydrogen and helium, with trace amounts of lithium. This process peaked after about five minutes.[136] Any leftover neutrons underwent negative beta decay with a half-life of about a thousand seconds, releasing a proton and electron in the process,

$$n \rightarrow p + e- + \nu_e$$

For about the next 300000–400000 years, the excess electrons remained too energetic to bind with atomic nuclei.[137] What followed is a period known as recombination, when neutral atoms were formed and the expanding universe became transparent to radiation.[138]

Roughly one million years after the big bang, the first generation of stars began to form.[138] Within a star, stellar nucleosynthesis results in the production of positrons from the fusion of atomic nuclei. These antimatter particles immediately annihilate with electrons, releasing gamma rays. The net result is a steady reduction in the number of electrons, and a matching increase in the number of neutrons. However, the process of stellar evolution can result in the synthesis of radioactive isotopes. Selected isotopes can subsequently undergo negative beta decay, emitting an electron and antineutrino from the nucleus.[139] An example is the cobalt-60 (^{60}Co) isotope, which decays to form nickel-60 (60Ni).[140]

At the end of its lifetime, a star with more than about 20 solar masses can undergo gravitational collapse to form a black hole.[141] According to classical physics, these massive stellar objects exert a gravitational attraction that is strong enough to prevent anything, even electromagnetic radiation, from escaping past the Schwarzschild radius. However, quantum mechanical effects are believed to potentially allow the emission of Hawking radiation at this distance. Electrons (and positrons) are thought to be created at the event horizon of these stellar remnants.

When pairs of virtual particles (such as an electron and positron) are created in the vicinity of the event horizon, the random spatial distribution of these particles may permit one of them to appear on the exterior; this process is called quantum tunnelling. The gravitational potential of the black hole can then supply the energy that transforms this virtual particle into a real particle, allowing it to radiate away into space.[142] In exchange, the other member of the pair is given negative energy, which results in a net loss of mass-energy by the black hole. The rate of Hawking radiation increases with decreasing mass, eventually causing the black hole to evaporate away until, finally, it explodes.[143]

Cosmic rays are particles traveling through space with high energies. Energy events as high as 3.0×10^{20} eV have been recorded.[144] When these particles collide with nucleons in the Earth's atmosphere, a shower of particles is generated, including pions.[145] More than half of the cosmic radiation observed from the Earth's surface consists of muons. The particle called a muon is a lepton produced in the upper atmosphere by the decay of a pion.

$$\pi- \rightarrow \mu- + \nu$$
$$\mu$$

A muon, in turn, can decay to form an electron or positron.[146]

$$\mu- \rightarrow e- + \nu$$
$$e + \nu$$
$$\mu$$

23.4 Observation

Remote observation of electrons requires detection of their radiated energy. For example, in high-energy environments such as the corona of a star, free electrons form a plasma that radiates energy due to Bremsstrahlung radiation. Electron gas can undergo plasma oscillation, which is waves caused by synchronized variations in electron density, and these produce energy emissions that can be detected by using radio telescopes.[148]

The frequency of a photon is proportional to its energy. As a bound electron transitions between different energy levels of an atom, it absorbs or emits photons at characteristic frequencies. For instance, when atoms are irradiated by a source with a broad spectrum, distinct absorption lines appear in the spectrum of transmitted radiation. Each element or molecule displays a characteristic set of spectral lines, such as the hydrogen spectral series. Spectroscopic measurements of the strength and width of these lines allow the composition and physical properties of a substance to be determined.[149][150]

In laboratory conditions, the interactions of individual electrons can be observed by means of particle detectors, which allow measurement of specific properties such as energy, spin and charge.[108] The development of the Paul trap and Penning trap allows charged particles to be contained within a small region for long durations. This enables precise measurements of the particle properties. For example, in one instance a Penning trap was used to contain a single electron for a period of 10 months.[151] The magnetic moment of the electron was measured to a precision of eleven digits, which, in 1980, was a greater accuracy than for any other physical constant.[152]

The first video images of an electron's energy distribution were captured by a team at Lund University in Sweden, February 2008. The scientists used extremely short flashes of light, called attosecond pulses, which allowed an electron's motion to be observed for the first time.[153][154]

The distribution of the electrons in solid materials can be visualized by angle-resolved photoemission spectroscopy (ARPES). This technique employs the photoelectric effect to measure the reciprocal space—a mathematical representation of periodic structures that is used to infer the original structure. ARPES can be used to determine the direction, speed and scattering of electrons within the material.[155]

23.5 Plasma applications

23.5.1 Particle beams

Electron beams are used in welding.[157] They allow energy densities up to 10^7 W·cm^{-2} across a narrow focus diameter of 0.1–1.3 mm and usually require no filler material. This welding technique must be performed in a vacuum to prevent the electrons from interacting with the gas before reaching their target, and it can be used to join conductive materials that would otherwise be considered unsuitable for welding.[158][159]

Electron-beam lithography (EBL) is a method of etching semiconductors at resolutions smaller than a micrometer.[160] This technique is limited by high costs, slow performance, the need to operate the beam in the vacuum and the tendency of the electrons to scatter in solids. The last problem limits the resolution to about 10 nm. For this reason, EBL is primarily used for the production of small numbers of specialized integrated circuits.[161]

Electron beam processing is used to irradiate materials in order to change their physical properties or sterilize medical and food products.[162] Electron beams fluidise or quasi-melt glasses without significant increase of temperature on intensive irradiation: e.g. intensive electron radiation causes a many orders of magnitude decrease of viscosity and stepwise decrease of its activation energy.[163]

Linear particle accelerators generate electron beams for treatment of superficial tumors in radiation therapy. Electron therapy can treat such skin lesions as basal-cell carcinomas because an electron beam only penetrates to a limited depth before being absorbed, typically up to 5 cm for electron energies in the range 5–20 MeV. An electron beam can be used to supplement the treatment of areas that have been irradiated by X-rays.[164][165]

Particle accelerators use electric fields to propel electrons and their antiparticles to high energies. These particles emit synchrotron radiation as they pass through magnetic fields. The dependency of the intensity of this radiation upon spin polarizes the electron beam—a process known as the Sokolov–Ternov effect.[note 8] Polarized electron beams can be useful for various experiments. Synchrotron radiation can also cool the electron beams to reduce the momentum spread of the particles. Electron and positron beams are collided upon the particles' accelerating to the required energies; particle detectors observe the resulting energy emissions, which particle physics studies .[166]

23.5.2 Imaging

Low-energy electron diffraction (LEED) is a method of bombarding a crystalline material with a collimated beam of electrons and then observing the resulting diffraction patterns to determine the structure of the material. The required energy of the electrons is typically in the range 20–200 eV.[167] The reflection high-energy electron diffraction (RHEED) technique uses the reflection of a beam of electrons fired at various low angles to characterize the surface of crystalline materials. The beam energy is typically in the range 8–20 keV and the angle of incidence is 1–4°.[168][169]

The electron microscope directs a focused beam of electrons at a specimen. Some electrons change their properties, such as movement direction, angle, and relative phase and energy as the beam interacts with the material. Microscopists can record these changes in the electron beam to produce atomically resolved images of the material.[170] In blue light, conventional optical microscopes have a diffraction-limited resolution of about 200 nm.[171] By comparison, electron microscopes are limited by the de Broglie wavelength of the electron. This wavelength, for example, is equal to 0.0037 nm for electrons accelerated across a 100,000-volt potential.[172] The Transmission Electron Aberration-Corrected Microscope is capable of sub-0.05 nm resolution, which is more than enough to resolve individual atoms.[173] This capability makes the electron microscope a useful laboratory instrument for high resolution imaging. However, electron microscopes are expensive instruments that are costly to maintain.

Two main types of electron microscopes exist: transmission and scanning. Transmission electron microscopes function like overhead projectors, with a beam of electrons passing through a slice of material then being projected by lenses on a photographic slide or a charge-coupled device. Scanning electron microscopes rasteri a finely focused electron beam, as in a TV set, across the studied sample to produce the image. Magnifications range from 100× to 1,000,000× or higher for both microscope types. The scanning tunneling microscope uses quantum tunneling of electrons from a sharp metal tip into the studied material and can produce atomically resolved images of its surface.[174][175][176]

23.5.3   Other applications

In the free-electron laser (FEL), a relativistic electron beam passes through a pair of undulators that contain arrays of dipole magnets whose fields point in alternating directions. The electrons emit synchrotron radiation that coherently interacts with the same electrons to strongly amplify the radiation field at the resonance frequency. FEL can emit a coherent high-brilliance electromagnetic radiation with a wide range of frequencies, from microwaves to soft X-rays. These devices may find manufacturing, communication and various medical applications, such as soft tissue surgery.[177]

Electrons are important in cathode ray tubes, which have been extensively used as display devices in laboratory instruments, computer monitors and television sets.[178] In a photomultiplier tube, every photon striking the photocathode initiates an avalanche of electrons that produces a detectable current pulse.[179] Vacuum tubes use the flow of electrons to manipulate electrical signals, and they played a critical role in the development of electronics technology. However, they have been largely supplanted by solid-state devices such as the transistor.[180]

23.6   See also

- Anyon
- Electride
- Electron bubble
- Exoelectron emission
- g-factor
- Periodic systems of small molecules
- Spintronics
- Stern–Gerlach experiment
- Townsend discharge
- Zeeman effect
- List of particles
- Lepton

23.7   Notes

[1] The fractional version's denominator is the inverse of the decimal value (along with its relative standard uncertainty of 4.2×10^{-13} u).

[2] The electron's charge is the negative of elementary charge, which has a positive value for the proton.

[3] This magnitude is obtained from the spin quantum number as

$$S = \sqrt{s(s+1)} \cdot \frac{h}{2\pi}$$
$$= \frac{\sqrt{3}}{2}\hbar$$

for quantum number *s* = 1/2.
See: Gupta, M.C. (2001). *Atomic and Molecular Spectroscopy*. New Age Publishers. p. 81. ISBN 81-224-1300-5.

[4] Bohr magneton:

$$\mu_{\mathrm{B}} = \frac{e\hbar}{2m_{\mathrm{e}}}.$$

[5] The classical electron radius is derived as follows. Assume that the electron's charge is spread uniformly throughout a spherical volume. Since one part of the sphere would repel the other parts, the sphere contains electrostatic potential energy. This energy is assumed to equal the electron's rest energy, defined by special relativity ($E = mc^2$).
From electrostatics theory, the potential energy of a sphere with radius *r* and charge *e* is given by:

$$E_{\mathrm{p}} = \frac{e^2}{8\pi\varepsilon_0 r},$$

where ε_0 is the vacuum permittivity. For an electron with rest mass m_0, the rest energy is equal to:

$$E_{\mathrm{p}} = m_0 c^2,$$

where *c* is the speed of light in a vacuum. Setting them equal and solving for *r* gives the classical electron radius.
See: Haken, H.; Wolf, H.C.; Brewer, W.D. (2005). *The Physics of Atoms and Quanta: Introduction to Experiments and Theory*. Springer. p. 70. ISBN 3-540-67274-5.

[6] Radiation from non-relativistic electrons is sometimes termed cyclotron radiation.

[7] The change in wavelength, $\Delta\lambda$, depends on the angle of the recoil, θ, as follows,

$$\Delta\lambda = \frac{h}{m_{\mathrm{e}}c}(1 - \cos\theta),$$

where *c* is the speed of light in a vacuum and m_{e} is the electron mass. See Zombeck (2007: 393, 396).

[8] The polarization of an electron beam means that the spins of all electrons point into one direction. In other words, the projections of the spins of all electrons onto their momentum vector have the same sign.

23.8   References

[1] Eichten, E.J.; Peskin, M.E.; Peskin, M. (1983). "New Tests for Quark and Lepton Substructure". *Physical Review Letters* **50** (11): 811–814. Bibcode:1983PhRvL..50..811E. doi:10.1103/PhysRevLett.50.811.

[2] Farrar, W.V. (1969). "Richard Laming and the Coal-Gas Industry, with His Views on the Structure of Matter". *Annals of Science* **25** (3): 243–254. doi:10.1080/00033796900200141.

[3] Arabatzis, T. (2006). *Representing Electrons: A Biographical Approach to Theoretical Entities*. University of Chicago Press. pp. 70–74. ISBN 0-226-02421-0.

[4] Buchwald, J.Z.; Warwick, A. (2001). *Histories of the Electron: The Birth of Microphysics*. MIT Press. pp. 195–203. ISBN 0-262-52424-4.

[5] Thomson, J.J. (1897). "Cathode Rays". *Philosophical Magazine* **44** (269): 293. doi:10.1080/14786449708621070.

[6] P.J. Mohr, B.N. Taylor, and D.B. Newell (2011), "The 2010 CODATA Recommended Values of the Fundamental Physical Constants" (Web Version 6.0). This database was developed by J. Baker, M. Douma, and S. Kotochigova. Available: http://physics.nist.gov/constants [Thursday, 02-Jun-2011 21:00:12 EDT]. National Institute of Standards and Technology, Gaithersburg, MD 20899.

[7] Agostini M. et al. (Borexino Coll.) (2015). "Test of Electric Charge Conservation with Borexino". *Physical Review Letters* **115** (23): 231802. arXiv:1509.01223. doi:10.1103/PhysRevLett.115.231802.

[8] "JERRY COFF". Retrieved 10 September 2010.

[9] Curtis, L.J. (2003). *Atomic Structure and Lifetimes: A Conceptual Approach*. Cambridge University Press. p. 74. ISBN 0-521-53635-9.

[10] "CODATA value: proton-electron mass ratio". *2006 CODATA recommended values.* National Institute of Standards and Technology. Retrieved 2009-07-18.

[11] Anastopoulos, C. (2008). *Particle Or Wave: The Evolution of the Concept of Matter in Modern Physics.* Princeton University Press. pp. 236–237. ISBN 0-691-13512-6.

[12] Pauling, L.C. (1960). *The Nature of the Chemical Bond and the Structure of Molecules and Crystals: an introduction to modern structural chemistry* (3rd ed.). Cornell University Press. pp. 4–10. ISBN 0-8014-0333-2.

[13] Dahl (1997:122–185).

[14] Wilson, R. (1997). *Astronomy Through the Ages: The Story of the Human Attempt to Understand the Universe.* CRC Press. p. 138. ISBN 0-7484-0748-0.

[15] Shipley, J.T. (1945). *Dictionary of Word Origins.* The Philosophical Library. p. 133. ISBN 0-88029-751-4.

[16] Baigrie, B. (2006). *Electricity and Magnetism: A Historical Perspective.* Greenwood Press. pp. 7–8. ISBN 0-313-33358-0.

[17] Keithley, J.F. (1999). *The Story of Electrical and Magnetic Measurements: From 500 B.C. to the 1940s.* IEEE Press. pp. 15, 20. ISBN 0-7803-1193-0.

[18] "Benjamin Franklin (1706–1790)". *Eric Weisstein's World of Biography.* Wolfram Research. Retrieved 2010-12-16.

[19] Myers, R.L. (2006). *The Basics of Physics.* Greenwood Publishing Group. p. 242. ISBN 0-313-32857-9.

[20] Barrow, J.D. (1983). "Natural Units Before Planck". *Quarterly Journal of the Royal Astronomical Society* **24**: 24–26. Bibcode:19

[21] Sōgo Okamura (1994). *History of Electron Tubes.* IOS Press. p. 11. ISBN 978-90-5199-145-1. Retrieved 29 May 2015. In 1881, Stoney named this electromagnetic 'electrolion'. It came to be called 'electron' from 1891. [...] In 1906, the suggestion to call cathode ray particles 'electrions' was brought up but through the opinion of Lorentz of Holland 'electrons' came to be widely used.

[22] Stoney, G.J. (1894). "Of the "Electron," or Atom of Electricity". *Philosophical Magazine* **38**(5): 418–420. doi:10.1080/14786

[23] "electron, n.2". OED Online. March 2013. Oxford University Press. Accessed 12 April 2013

[24] Soukhanov, A.H. ed. (1986). *Word Mysteries & Histories.* Houghton Mifflin Company. p. 73. ISBN 0-395-40265-4.

[25] Guralnik, D.B. ed. (1970). *Webster's New World Dictionary.* Prentice Hall. p. 450.

[26] Born, M.; Blin-Stoyle, R.J.; Radcliffe, J.M. (1989). *Atomic Physics.* Courier Dover. p. 26. ISBN 0-486-65984-4.

[27] Dahl (1997:55–58).

[28] DeKosky, R.K. (1983). "William Crookes and the quest for absolute vacuum in the 1870s". *Annals of Science* **40** (1): 1–18. doi:10.1080/00033798300200101.

[29] Leicester, H.M. (1971). *The Historical Background of Chemistry.* Courier Dover. pp. 221–222. ISBN 0-486-61053-5.

[30] Dahl (1997:64–78).

[31] Zeeman, P.; Zeeman, P. (1907). "Sir William Crookes, F.R.S". *Nature* **77** (1984): 1–3. Bibcode:1907Natur..77....1C. doi:10.1038/077001a0.

[32] Dahl (1997:99).

[33] Frank Wilczek: "Happy Birthday, Electron" *Scientific American*, June 2012.

[34] Thomson, J.J. (1906). "Nobel Lecture: Carriers of Negative Electricity" (PDF). The Nobel Foundation. Retrieved 2008-08-25.

[35] Trenn, T.J. (1976). "Rutherford on the Alpha-Beta-Gamma Classification of Radioactive Rays". *Isis* **67**(1): 61–75. doi:10.1086/ JSTOR 231134.

[36] Becquerel, H. (1900). "Déviation du Rayonnement du Radium dans un Champ Électrique". *Comptes rendus de l'Académie des sciences* (in French) **130**: 809–815.

[37] Buchwald and Warwick (2001:90–91).

[38] Myers, W.G. (1976). "Becquerel's Discovery of Radioactivity in 1896". *Journal of Nuclear Medicine* **17** (7): 579–582. PMID 775027.

[39] Kikoin, I.K.; Sominskiĭ, I.S. (1961). "Abram Fedorovich Ioffe (on his eightieth birthday)". *Soviet Physics Uspekhi* **3** (5): 798–809. Bibcode:1961SvPhU...3..798K. doi:10.1070/PU1961v003n05ABEH005812. Original publication in Russian: Кикоин, И.К.; Соминский, М.С. (1960). "Академик А.Ф. Иоффе" (PDF). *Успехи Физических Наук* **72** (10): 303–321.

[40] Millikan, R.A. (1911). "The Isolation of an Ion, a Precision Measurement of its Charge, and the Correction of Stokes' Law". *Physical Review* **32** (2): 349–397. Bibcode:1911PhRvI..32..349M. doi:10.1103/PhysRevSeriesI.32.349.

[41] Das Gupta, N.N.; Ghosh, S.K. (1999). "A Report on the Wilson Cloud Chamber and Its Applications in Physics". *Reviews of Modern Physics* **18** (2): 225–290. Bibcode:1946RvMP...18..225G. doi:10.1103/RevModPhys.18.225.

[42] Smirnov, B.M. (2003). *Physics of Atoms and Ions*. Springer. pp. 14–21. ISBN 0-387-95550-X.

[43] Bohr, N. (1922). "Nobel Lecture: The Structure of the Atom" (PDF). The Nobel Foundation. Retrieved 2008-12-03.

[44] Lewis, G.N. (1916). "The Atom and the Molecule". *Journal of the American Chemical Society* **38**(4): 762–786. doi:10.1021/ja0

[45] Arabatzis, T.; Gavroglu, K. (1997). "The chemists' electron". *European Journal of Physics* **18**(3): 150–163. Bibcode:1997EJPh doi:10.1088/0143-0807/18/3/005.

[46] Langmuir, I. (1919). "The Arrangement of Electrons in Atoms and Molecules". *Journal of the American Chemical Society* **41** (6): 868–934. doi:10.1021/ja02227a002.

[47] Scerri, E.R. (2007). *The Periodic Table*. Oxford University Press. pp. 205–226. ISBN 0-19-530573-6.

[48] Massimi, M. (2005). *Pauli's Exclusion Principle, The Origin and Validation of a Scientific Principle*. Cambridge University Press. pp. 7–8. ISBN 0-521-83911-4.

[49] Uhlenbeck, G.E.; Goudsmith, S. (1925). "Ersetzung der Hypothese vom unmechanischen Zwang durch eine Forderung bezüglich des inneren Verhaltens jedes einzelnen Elektrons". *Die Naturwissenschaften* (in German) **13** (47): 953. Bibcode:1925NW.....13. doi:10.1007/BF01558878.

[50] Pauli, W. (1923). "Über die Gesetzmäßigkeiten des anomalen Zeemaneffektes". *Zeitschrift für Physik* (in German) **16** (1): 155–164. Bibcode:1923ZPhy...16..155P. doi:10.1007/BF01327386.

[51] de Broglie, L. (1929). "Nobel Lecture: The Wave Nature of the Electron" (PDF). The Nobel Foundation. Retrieved 2008-08-30.

[52] Falkenburg, B. (2007). *Particle Metaphysics: A Critical Account of Subatomic Reality*. Springer. p. 85. ISBN 3-540-33731-8.

[53] Davisson, C. (1937). "Nobel Lecture: The Discovery of Electron Waves" (PDF). The Nobel Foundation. Retrieved 2008-08-30.

[54] Schrödinger, E. (1926). "Quantisierung als Eigenwertproblem". *Annalen der Physik* (in German) **385**(13): 437–490. Bibcode: doi:10.1002/andp.19263851302.

[55] Rigden, J.S. (2003). *Hydrogen*. Harvard University Press. pp. 59–86. ISBN 0-674-01252-6.

[56] Reed, B.C. (2007). *Quantum Mechanics*. Jones & Bartlett Publishers. pp. 275–350. ISBN 0-7637-4451-4.

[57] Dirac, P.A.M. (1928). "The Quantum Theory of the Electron". *Proceedings of the Royal Society A* **117** (778): 610–624. Bibcode:1928RSPSA.117..610D. doi:10.1098/rspa.1928.0023.

[58] Dirac, P.A.M. (1933). "Nobel Lecture: Theory of Electrons and Positrons" (PDF). The Nobel Foundation. Retrieved 2008-11-01.

[59] "The Nobel Prize in Physics 1965". The Nobel Foundation. Retrieved 2008-11-04.

[60] Panofsky, W.K.H. (1997). "The Evolution of Particle Accelerators & Colliders" (PDF). *Beam Line* (Stanford University) **27** (1): 36–44. Retrieved 2008-09-15.

[61] Elder, F.R.; et al. (1947). "Radiation from Electrons in a Synchrotron". *Physical Review* **71**(11): 829–830. Bibcode:1947PhRv. doi:10.1103/PhysRev.71.829.5.

[62] Hoddeson, L.; et al. (1997). *The Rise of the Standard Model: Particle Physics in the 1960s and 1970s*. Cambridge University Press. pp. 25–26. ISBN 0-521-57816-7.

[63] Bernardini, C. (2004). "AdA: The First Electron–Positron Collider".*Physics in Perspective***6**(2): 156–183.Bibcode:2004PhP... doi:10.1007/s00016-003-0202-y.

[64] "Testing the Standard Model: The LEP experiments". CERN. 2008. Retrieved 2008-09-15.

[65] "LEP reaps a final harvest". *CERN Courier* **40** (10). 2000.

[66] Prati, E.; De Michielis, M.; Belli, M.; Cocco, S.; Fanciulli, M.; Kotekar-Patil, D.; Ruoff, M.; Kern, D. P.; Wharam, D. A.; Verduijn, J.; Tettamanzi, G. C.; Rogge, S.; Roche, B.; Wacquez, R.; Jehl, X.; Vinet, M.; Sanquer, M. (2012). "Few electron limit of n-type metal oxide semiconductor single electron transistors". *Nanotechnology* **23** (21): 215204. arXiv:1203.4811. Bibcode:2012Nanot..23u5204P. doi:10.1088/0957-4484/23/21/215204. PMID 22552118.

[67] Frampton, P.H.; Hung, P.Q.; Sher, Marc (2000). "Quarks and Leptons Beyond the Third Generation". *Physics Reports* **330** (5–6): 263–348. arXiv:hep-ph/9903387. Bibcode:2000PhR...330..263F. doi:10.1016/S0370-1573(99)00095-2.

[68] Raith, W.; Mulvey, T. (2001). *Constituents of Matter: Atoms, Molecules, Nuclei and Particles*. CRC Press. pp. 777–781. ISBN 0-8493-1202-7.

[69] The original source for CODATA is Mohr, P.J.; Taylor, B.N.; Newell, D.B. (2006). "CODATA recommended values of the fundamental physical constants". *Reviews of Modern Physics* **80** (2): 633–730. arXiv:0801.0028. Bibcode:2008RvMP...80..633M. doi:10.1103/RevModPhys.80.633.

    Individual physical constants from the CODATA are available at: "The NIST Reference on Constants, Units and Uncertainty". National Institute of Standards and Technology. Retrieved 2009-01-15.

[70] Zombeck, M.V. (2007). *Handbook of Space Astronomy and Astrophysics* (3rd ed.). Cambridge University Press. p. 14. ISBN 0-521-78242-2.

[71] Murphy, M.T.; et al. (2008). "Strong Limit on a Variable Proton-to-Electron Mass Ratio from Molecules in the Distant Universe". *Science* **320** (5883): 1611–1613. arXiv:0806.3081. Bibcode:2008Sci...320.1611M. doi:10.1126/science.1156352. PMID 18566280.

[72] Zorn, J.C.; Chamberlain, G.E.; Hughes, V.W. (1963). "Experimental Limits for the Electron-Proton Charge Difference and for the Charge of the Neutron".*Physical Review***129**(6): 2566–2576.Bibcode:1963PhRv..129.2566Z.doi:10.1103/PhysRev.129.2

[73] Odom, B.; et al. (2006). "New Measurement of the Electron Magnetic Moment Using a One-Electron Quantum Cyclotron". *Physical Review Letters***97**(3): 030801.Bibcode:2006PhRvL..97c0801O.doi:10.1103/PhysRevLett.97.030801.PMID16907

[74] Anastopoulos, C. (2008). *Particle Or Wave: The Evolution of the Concept of Matter in Modern Physics*. Princeton University Press. pp. 261–262. ISBN 0-691-13512-6.

[75] Gabrielse, G.; et al. (2006). "New Determination of the Fine Structure Constant from the Electron *g* Value and QED". *Physical Review Letters* **97** (3): 030802(1–4). Bibcode:2006PhRvL..97c0802G. doi:10.1103/PhysRevLett.97.030802.

[76] Eduard Shpolsky, Atomic physics (Atomnaia fizika),second edition, 1951

[77] Dehmelt, H. (1988). "A Single Atomic Particle Forever Floating at Rest in Free Space: New Value for Electron Radius". *Physica Scripta* **T22**: 102–10. Bibcode:1988PhST...22..102D. doi:10.1088/0031-8949/1988/T22/016.

[78] Meschede, D. (2004). *Optics, light and lasers: The Practical Approach to Modern Aspects of Photonics and Laser Physics*. Wiley-VCH. p. 168. ISBN 3-527-40364-7.

[79] Steinberg, R.I.; et al. (1999). "Experimental test of charge conservation and the stability of the electron". *Physical Review D* **61** (2): 2582–2586. Bibcode:1975PhRvD..12.2582S. doi:10.1103/PhysRevD.12.2582.

[80] J. Beringer (Particle Data Group); et al. (2012). "Review of Particle Physics: [electron properties]" (PDF). *Physical Review D* **86** (1): 010001. Bibcode:2012PhRvD..86a0001B. doi:10.1103/PhysRevD.86.010001.

[81] Back, H. O.; et al. (2002). "Search for electron decay mode e → γ + ν with prototype of Borexino detector". *Physics Letters B* **525**: 29–40. Bibcode:2002PhLB..525...29B. doi:10.1016/S0370-2693(01)01440-X.

[82] Munowitz, M. (2005). *Knowing, The Nature of Physical Law*. Oxford University Press. ISBN 0-19-516737-6.

[83] Kane, G. (October 9, 2006). "Are virtual particles really constantly popping in and out of existence? Or are they merely a mathematical bookkeeping device for quantum mechanics?". Scientific American. Retrieved 2008-09-19.

[84] Taylor, J. (1989). "Gauge Theories in Particle Physics". In Davies, Paul. *The New Physics*. Cambridge University Press. p. 464. ISBN 0-521-43831-4.

[85] Genz, H. (2001). *Nothingness: The Science of Empty Space*. Da Capo Press. pp. 241–243, 245–247. ISBN 0-7382-0610-5.

[86] Gribbin, J. (January 25, 1997). "More to electrons than meets the eye". *New Scientist*. Retrieved 2008-09-17.

[87] Levine, I.; et al. (1997). "Measurement of the Electromagnetic Coupling at Large Momentum Transfer". *Physical Review Letters* **78** (3): 424–427. Bibcode:1997PhRvL..78..424L. doi:10.1103/PhysRevLett.78.424.

[88] Murayama, H. (March 10–17, 2006). *Supersymmetry Breaking Made Easy, Viable and Generic*. Proceedings of the XLIInd Rencontres de Moriond on Electroweak Interactions and Unified Theories (La Thuile, Italy). arXiv:0709.3041.—lists a 9% mass difference for an electron that is the size of the Planck distance.

[89] Schwinger, J. (1948). "On Quantum-Electrodynamics and the Magnetic Moment of the Electron". *Physical Review* **73** (4): 416–417. Bibcode:1948PhRv...73..416S. doi:10.1103/PhysRev.73.416.

[90] Huang, K. (2007). *Fundamental Forces of Nature: The Story of Gauge Fields*. World Scientific. pp. 123–125. ISBN 981-270-645-3.

[91] Foldy, L.L.; Wouthuysen, S. (1950). "On the Dirac Theory of Spin 1/2 Particles and Its Non-Relativistic Limit". *Physical Review* **78**: 29–36. Bibcode:1950PhRv...78...29F. doi:10.1103/PhysRev.78.29.

[92] Sidharth, B.G. (2008). "Revisiting Zitterbewegung". *International Journal of Theoretical Physics* **48**(2): 497–506.arXiv:0806.0 Bibcode:2009IJTP...48..497S. doi:10.1007/s10773-008-9825-8.

[93] Elliott, R.S. (1978). "The History of Electromagnetics as Hertz Would Have Known It". *IEEE Transactions on Microwave Theory and Techniques* **36** (5): 806–823. Bibcode:1988ITMTT..36..806E. doi:10.1109/22.3600.

[94] Crowell, B. (2000). *Electricity and Magnetism*. Light and Matter. pp. 129–152. ISBN 0-9704670-4-4.

[95] Mahadevan, R.; Narayan, R.; Yi, I. (1996). "Harmony in Electrons: Cyclotron and Synchrotron Emission by Thermal Electrons in a Magnetic Field". *The Astrophysical Journal* **465**: 327–337. arXiv:astro-ph/9601073. Bibcode:1996ApJ...465..327M. doi:10.1086/177422.

[96] Rohrlich, F. (1999). "The Self-Force and Radiation Reaction". *American Journal of Physics* **68**(12): 1109–1112.Bibcode:2000 doi:10.1119/1.1286430.

[97] Georgi, H. (1989). "Grand Unified Theories". In Davies, Paul. *The New Physics*. Cambridge University Press. p. 427. ISBN 0-521-43831-4.

[98] Blumenthal, G.J.; Gould, R. (1970). "Bremsstrahlung, Synchrotron Radiation, and Compton Scattering of High-Energy Electrons Traversing Dilute Gases".*Reviews of Modern Physics***42**(2): 237–270.Bibcode:1970RvMP...42..237B.doi:10.1103/Rev

[99] Staff (2008). "The Nobel Prize in Physics 1927". The Nobel Foundation. Retrieved 2008-09-28.

[100] Chen, S.-Y.; Maksimchuk, A.; Umstadter, D. (1998). "Experimental observation of relativistic nonlinear Thomson scattering". *Nature* **396** (6712): 653–655. arXiv:physics/9810036. Bibcode:1998Natur.396..653C. doi:10.1038/25303.

[101] Beringer, R.; Montgomery, C.G. (1942). "The Angular Distribution of Positron Annihilation Radiation". *Physical Review* **61** (5–6): 222–224. Bibcode:1942PhRv...61..222B. doi:10.1103/PhysRev.61.222.

[102] Buffa, A. (2000). *College Physics* (4th ed.). Prentice Hall. p. 888. ISBN 0-13-082444-5.

[103] Eichler, J. (2005). "Electron–positron pair production in relativistic ion–atom collisions". *Physics Letters A* **347** (1–3): 67–72. Bibcode:2005PhLA..347...67E. doi:10.1016/j.physleta.2005.06.105.

[104] Hubbell, J.H. (2006). "Electron positron pair production by photons: A historical overview". *Radiation Physics and Chemistry* **75** (6): 614–623. Bibcode:2006RaPC...75..614H. doi:10.1016/j.radphyschem.2005.10.008.

[105] Quigg, C. (June 4–30, 2000). *The Electroweak Theory*. TASI 2000: Flavor Physics for the Millennium (Boulder, Colorado). p. 80. arXiv:hep-ph/0204104.

[106]Mulliken, R.S. (1967). "Spectroscopy, Molecular Orbitals, and Chemical Bonding".*Science***157**(3784): 13–24.Bibcode: doi:10.1126/science.157.3784.13. PMID 5338306.

[107]Burhop, E.H.S.(1952).*The Auger Effect and Other Radiationless Transitions.*Cambridge University Press. pp. 2–3. 0-88275-966-3.

[108]Grupen, C. (2000). "Physics of Particle Detection".*AIP Conference Proceedings***536**: 3–34.arXiv:physics/9906063.doi:10

[109] Jiles, D. (1998). *Introduction to Magnetism and Magnetic Materials.* CRC Press. pp. 280–287. ISBN 0-412-79860-3.

[110] Löwdin, P.O.; Erkki Brändas, E.; Kryachko, E.S. (2003). *Fundamental World of Quantum Chemistry: A Tribute to the Memory of Per- Olov Löwdin.* Springer. pp. 393–394. ISBN 1-4020-1290-X.

[111] McQuarrie, D.A.; Simon, J.D. (1997). *Physical Chemistry: A Molecular Approach.* University Science Books. pp. 325–361. ISBN 0-935702-99-7.

[112]Daudel, R.; et al. (1973)."The Electron Pair in Chemistry".*Canadian Journal of Chemistry***52**(8): 1310–1320.doi:10.1139/ 201.

[113] Rakov, V.A.; Uman, M.A. (2007). *Lightning: Physics and Effects.* Cambridge University Press. p. 4. ISBN 0-521-03541-4.

[114] Freeman, G.R.; March, N.H. (1999). "Triboelectricity and some associated phenomena". *Materials Science and Technology* **15** (12): 1454–1458. doi:10.1179/026708399101505464.

[115] Forward, K.M.; Lacks, D.J.; Sankaran, R.M. (2009). "Methodology for studying particle–particle triboelectrification in granular materials". *Journal of Electrostatics* **67** (2–3): 178–183. doi:10.1016/j.elstat.2008.12.002.

[116] Weinberg, S. (2003). *The Discovery of Subatomic Particles.* Cambridge University Press. pp. 15–16. ISBN 0-521-82351-X.

[117] Lou, L.-F. (2003). *Introduction to phonons and electrons.* World Scientific. pp. 162, 164. ISBN 978-981-238-461-4.

[118] Guru, B.S.; Hızıroğlu, H.R. (2004). *Electromagnetic Field Theory.* Cambridge University Press. pp. 138, 276. ISBN 0-521-83016-8.

[119] Achuthan, M.K.; Bhat, K.N. (2007). *Fundamentals of Semiconductor Devices.* Tata McGraw-Hill. pp. 49–67. ISBN 0-07-061220-X.

[120] Ziman, J.M. (2001). *Electrons and Phonons: The Theory of Transport Phenomena in Solids.* Oxford University Press. p. 260. ISBN 0-19-850779-8.

[121] Main, P. (June 12, 1993). "When electrons go with the flow: Remove the obstacles that create electrical resistance, and you get ballistic electrons and a quantum surprise". *New Scientist* **1887**: 30. Retrieved 2008-10-09.

[122] Blackwell, G.R. (2000). *The Electronic Packaging Handbook.* CRC Press. pp. 6.39–6.40. ISBN 0-8493-8591-1.

[123] Durrant, A. (2000). *Quantum Physics of Matter: The Physical World.* CRC Press. pp. 43, 71–78. ISBN 0-7503-0721-8.

[124] Staff (2008). "The Nobel Prize in Physics 1972". The Nobel Foundation. Retrieved 2008-10-13.

[125] Kadin, A.M. (2007). "Spatial Structure of the Cooper Pair". *Journal of Superconductivity and Novel Magnetism* **20** (4): 285–292. arXiv:cond-mat/0510279. doi:10.1007/s10948-006-0198-z.

[126] "Discovery About Behavior Of Building Block Of Nature Could Lead To Computer Revolution". *ScienceDaily*. July 31, 2009. Retrieved 2009-08-01.

[127] Jompol, Y.; et al. (2009). "Probing Spin-Charge Separation in a Tomonaga-Luttinger Liquid". *Science* **325** (5940): 597–601. arXiv:1002.2782. Bibcode:2009Sci...325..597J. doi:10.1126/science.1171769. PMID 19644117.

[128] Staff (2008). "The Nobel Prize in Physics 1958, for the discovery and the interpretation of the Cherenkov effect". The Nobel Foundation. Retrieved 2008-09-25.

[129] Staff (August 26, 2008). "Special Relativity". Stanford Linear Accelerator Center. Retrieved 2008-09-25.

[130] Adams, S. (2000). *Frontiers: Twentieth Century Physics.* CRC Press. p. 215. ISBN 0-7484-0840-1.

[131] Lurquin, P.F. (2003). *The Origins of Life and the Universe.* Columbia University Press. p. 2. ISBN 0-231-12655-7.

[132] Silk, J. (2000). *The Big Bang: The Creation and Evolution of the Universe* (3rd ed.). Macmillan. pp. 110–112, 134–137. ISBN 0-8050-7256-X.

[133] Christianto, V. (2007). "Thirty Unsolved Problems in the Physics of Elementary Particles" (PDF). *Progress in Physics* **4**: 112–114.

[134] Kolb, E.W.; Wolfram, Stephen (1980). "The Development of Baryon Asymmetry in the Early Universe". *Physics Letters B* **91** (2): 217–221. Bibcode:1980PhLB...91..217K. doi:10.1016/0370-2693(80)90435-9.

[135] Sather, E. (Spring–Summer 1996). "The Mystery of Matter Asymmetry" (PDF). *Beam Line*. University of Stanford. Retrieved 2008-11-01.

[136] Burles, S.; Nollett, K.M.; Turner, M.S. (1999). "Big-Bang Nucleosynthesis: Linking Inner Space and Outer Space". arXiv:astro-ph/9903300 [astro-ph].

[137] Boesgaard, A.M.; Steigman, G. (1985). "Big bang nucleosynthesis – Theories and observations". *Annual Review of Astronomy and Astrophysics* **23** (2): 319–378. Bibcode:1985ARA&A..23..319B. doi:10.1146/annurev.aa.23.090185.001535.

[138] Barkana, R. (2006). "The First Stars in the Universe and Cosmic Reionization". *Science* **313** (5789): 931–934. arXiv:astro-ph/0608450. Bibcode:2006Sci...313..931B. doi:10.1126/science.1125644. PMID 16917052.

[139] Burbidge, E.M.; et al. (1957). "Synthesis of Elements in Stars". *Reviews of Modern Physics* **29**(4): 548–647. Bibcode:1957RvM doi:10.1103/RevModPhys.29.547.

[140] Rodberg, L.S.; Weisskopf, V. (1957). "Fall of Parity: Recent Discoveries Related to Symmetry of Laws of Nature". *Science* **125** (3249): 627–633. Bibcode:1957Sci...125..627R. doi:10.1126/science.125.3249.627. PMID 17810563.

[141] Fryer, C.L. (1999). "Mass Limits For Black Hole Formation". *The Astrophysical Journal* **522** (1): 413–418. arXiv:astro-ph/9902315. Bibcode:1999ApJ...522..413F. doi:10.1086/307647.

[142] Parikh, M.K.; Wilczek, F. (2000). "Hawking Radiation As Tunneling". *Physical Review Letters* **85** (24): 5042–5045. arXiv:hep-th/9907001. Bibcode:2000PhRvL..85.5042P. doi:10.1103/PhysRevLett.85.5042. PMID 11102182.

[143] Hawking, S.W. (1974). "Black hole explosions?". *Nature* **248**(5443): 30–31. Bibcode:1974Natur.248...30H. doi:10.1038/2480

[144] Halzen, F.; Hooper, D. (2002). "High-energy neutrino astronomy: the cosmic ray connection". *Reports on Progress in Physics* **66** (7): 1025–1078. arXiv:astro-ph/0204527. Bibcode:2002astro.ph..4527H. doi:10.1088/0034-4885/65/7/201.

[145] Ziegler, J.F. (1998). "Terrestrial cosmic ray intensities". *IBM Journal of Research and Development* **42**(1): 117–139. doi:10.114

[146] Sutton, C. (August 4, 1990). "Muons, pions and other strange particles". *New Scientist*. Retrieved 2008-08-28.

[147] Wolpert, S. (July 24, 2008). "Scientists solve 30-year-old aurora borealis mystery". University of California. Retrieved 2008-10-11.

[148] Gurnett, D.A.; Anderson, R. (1976). "Electron Plasma Oscillations Associated with Type III Radio Bursts". *Science* **194** (4270): 1159–1162. Bibcode:1976Sci...194.1159G. doi:10.1126/science.194.4270.1159. PMID 17790910.

[149] Martin, W.C.; Wiese, W.L. (2007). "Atomic Spectroscopy: A Compendium of Basic Ideas, Notation, Data, and Formulas". National Institute of Standards and Technology. Retrieved 2007-01-08.

[150] Fowles, G.R. (1989). *Introduction to Modern Optics*. Courier Dover. pp. 227–233. ISBN 0-486-65957-7.

[151] Staff (2008). "The Nobel Prize in Physics 1989". The Nobel Foundation. Retrieved 2008-09-24.

[152] Ekstrom, P.; Wineland, David (1980). "The isolated Electron" (PDF). *Scientific American* **243**(2): 91–101. doi:10.1038/scienti 104. Retrieved 2008-09-24.

[153] Mauritsson, J. "Electron filmed for the first time ever" (PDF). Lund University. Archived from the original (PDF) on March 25, 2009. Retrieved 2008-09-17.

[154] Mauritsson, J.; et al. (2008). "Coherent Electron Scattering Captured by an Attosecond Quantum Stroboscope". *Physical Review Letters* **100** (7): 073003. arXiv:0708.1060. Bibcode:2008PhRvL.100g3003M. doi:10.1103/PhysRevLett.100.073003. PMID 18352546.

[155] Damascelli, A. (2004). "Probing the Electronic Structure of Complex Systems by ARPES". *Physica Scripta* **T109**: 61–74. arXiv:cond-mat/0307085. Bibcode:2004PhST..109...61D. doi:10.1238/Physica.Topical.109a00061.

[156] Staff (April 4, 1975). "Image # L-1975-02972". Langley Research Center, NASA. Retrieved 2008-09-20.

[157] Elmer, J. (March 3, 2008). "Standardizing the Art of Electron-Beam Welding". Lawrence Livermore National Laboratory. Retrieved 2008-10-16.

[158] Schultz, H. (1993). *Electron Beam Welding*. Woodhead Publishing. pp. 2–3. ISBN 1-85573-050-2.

[159] Benedict, G.F. (1987). *Nontraditional Manufacturing Processes*. Manufacturing engineering and materials processing **19**. CRC Press. p. 273. ISBN 0-8247-7352-7.

[160] Ozdemir, F.S. (June 25–27, 1979). *Electron beam lithography*. Proceedings of the 16th Conference on Design automation (San Diego, CA, USA: IEEE Press). pp. 383–391. Retrieved 2008-10-16.

[161] Madou, M.J. (2002). *Fundamentals of Microfabrication: the Science of Miniaturization* (2nd ed.). CRC Press. pp. 53–54. ISBN 0-8493-0826-7.

[162] Jongen, Y.; Herer, A. (May 2–5, 1996). *Electron Beam Scanning in Industrial Applications*. APS/AAPT Joint Meeting (American Physical Society). Bibcode:1996APS..MAY.H9902J.

[163] Mobus G. et al. (2010). Journal of Nuclear Materials, v. 396, 264–271, doi:10.1016/j.jnucmat.2009.11.020

[164] Beddar, A.S.; Domanovic, Mary Ann; Kubu, Mary Lou; Ellis, Rod J.; Sibata, Claudio H.; Kinsella, Timothy J. (2001). "Mobile linear accelerators for intraoperative radiation therapy". *AORN Journal* **74** (5): 700. doi:10.1016/S0001-2092(06)61769-9.

[165] Gazda, M.J.; Coia, L.R. (June 1, 2007). "Principles of Radiation Therapy" (PDF). Retrieved 2013-10-31.

[166] Chao, A.W.; Tigner, M. (1999). *Handbook of Accelerator Physics and Engineering*. World Scientific. pp. 155, 188. ISBN 981-02-3500-3.

[167] Oura, K.; et al. (2003). *Surface Science: An Introduction*. Springer. pp. 1–45. ISBN 3-540-00545-5.

[168] Ichimiya, A.; Cohen, P.I. (2004). *Reflection High-energy Electron Diffraction*. Cambridge University Press. p. 1. ISBN 0-521-45373-9.

[169] Heppell, T.A. (1967). "A combined low energy and reflection high energy electron diffraction apparatus". *Journal of Scientific Instruments* **44** (9): 686–688. Bibcode:1967JScI...44..686H. doi:10.1088/0950-7671/44/9/311.

[170] McMullan, D. (1993). "Scanning Electron Microscopy: 1928–1965". University of Cambridge. Retrieved 2009-03-23.

[171] Slayter, H.S. (1992). *Light and electron microscopy*. Cambridge University Press. p. 1. ISBN 0-521-33948-0.

[172] Cember, H. (1996). *Introduction to Health Physics*. McGraw-Hill Professional. pp. 42–43. ISBN 0-07-105461-8.

[173] Erni, R.; et al. (2009). "Atomic-Resolution Imaging with a Sub-50-pm Electron Probe". *Physical Review Letters* **102** (9): 096101. Bibcode:2009PhRvL.102i6101E. doi:10.1103/PhysRevLett.102.096101. PMID 19392535.

[174] Bozzola, J.J.; Russell, L.D. (1999). *Electron Microscopy: Principles and Techniques for Biologists*. Jones & Bartlett Publishers. pp. 12, 197–199. ISBN 0-7637-0192-0.

[175] Flegler, S.L.; Heckman Jr., J.W.; Klomparens, K.L. (1995). *Scanning and Transmission Electron Microscopy: An Introduction* (Reprint ed.). Oxford University Press. pp. 43–45. ISBN 0-19-510751-9.

[176] Bozzola, J.J.; Russell, L.D. (1999). *Electron Microscopy: Principles and Techniques for Biologists* (2nd ed.). Jones & Bartlett Publishers. p. 9. ISBN 0-7637-0192-0.

[177] Freund, H.P.; Antonsen, T. (1996). *Principles of Free-Electron Lasers*. Springer. pp. 1–30. ISBN 0-412-72540-1.

[178] Kitzmiller, J.W. (1995). *Television Picture Tubes and Other Cathode-Ray Tubes: Industry and Trade Summary*. DIANE Publishing. pp. 3–5. ISBN 0-7881-2100-6.

[179] Sclater, N. (1999). *Electronic Technology Handbook*. McGraw-Hill Professional. pp. 227–228. ISBN 0-07-058048-0.

[180] Staff (2008). "The History of the Integrated Circuit". The Nobel Foundation. Retrieved 2008-10-18.

23.9 External links

- "The Discovery of the Electron". American Institute of Physics, Center for History of Physics.

- "Particle Data Group". University of California.

- Bock, R.K.; Vasilescu, A. (1998). *The Particle Detector BriefBook* (14th ed.). Springer. ISBN 3-540-64120-3.

- Copeland, Ed. "Spherical Electron". *Sixty Symbols*. Brady Haran for the University of Nottingham.

Robert Millikan

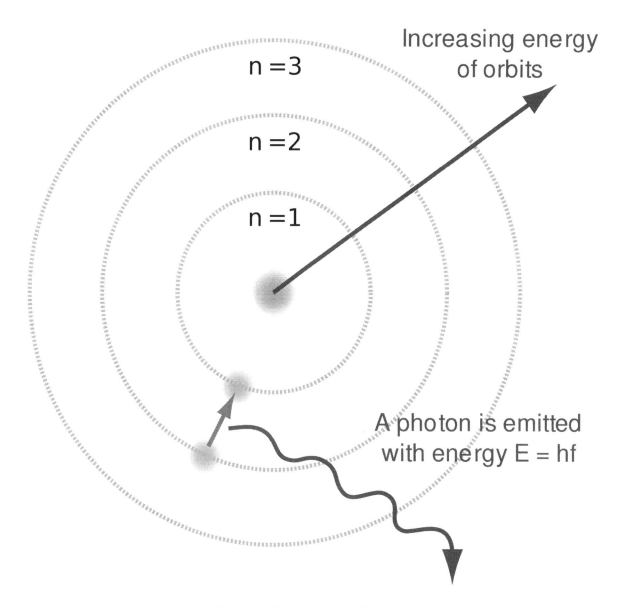

The Bohr model of the atom, showing states of electron with energy quantized by the number n. An electron dropping to a lower orbit emits a photon equal to the energy difference between the orbits.

206

CHAPTER 23. ELECTRON

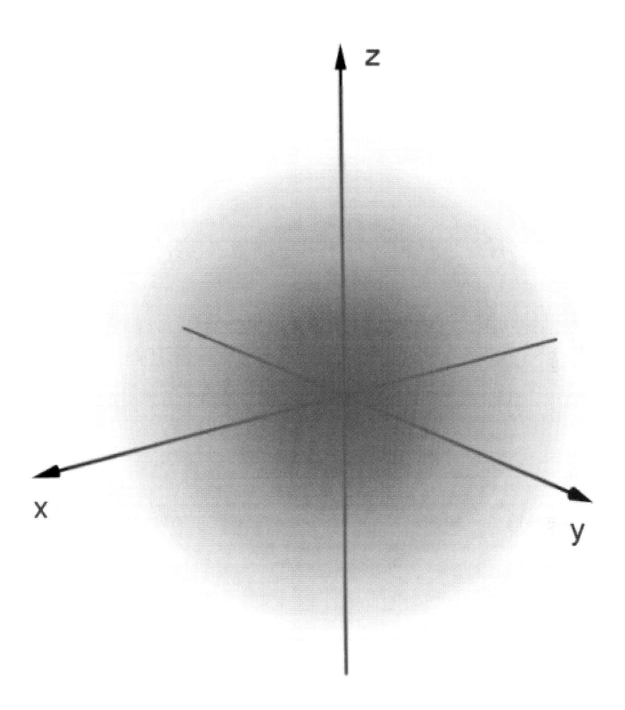

Orbital s ($\ell = 0$, $m_\ell = 0$)

In quantum mechanics, the behavior of an electron in an atom is described by an orbital, which is a probability distribution rather than an orbit. In the figure, the shading indicates the relative probability to "find" the electron, having the energy corresponding to the given quantum numbers, at that point.

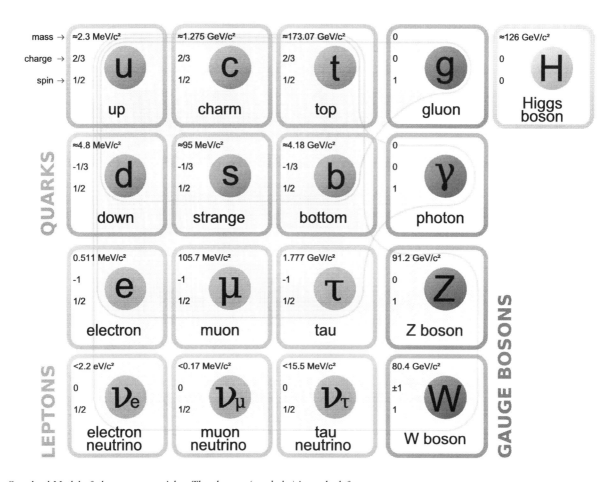

Standard Model of elementary particles. The electron (symbol e) is on the left.

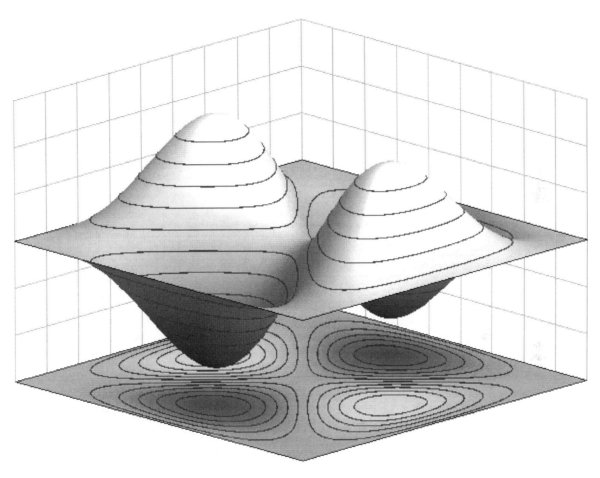

Example of an antisymmetric wave function for a quantum state of two identical fermions in a 1-dimensional box. If the particles swap position, the wave function inverts its sign.

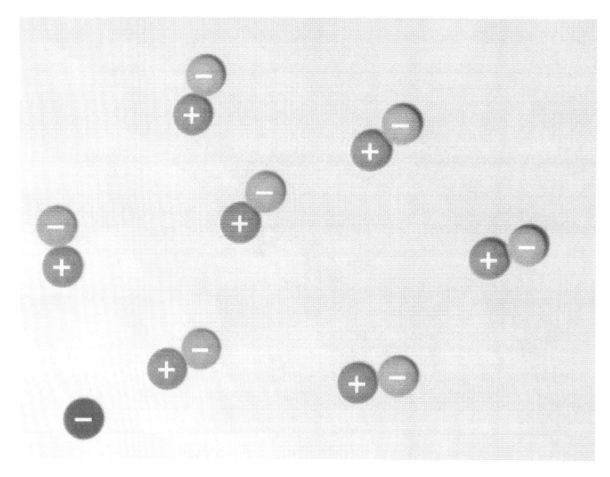

A schematic depiction of virtual electron–positron pairs appearing at random near an electron (at lower left)

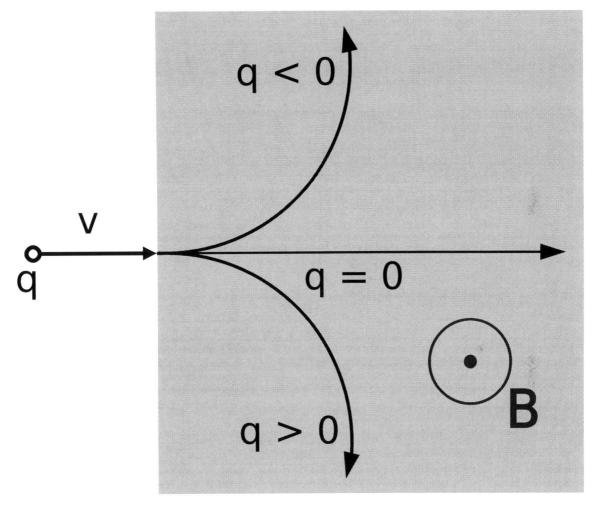

A particle with charge q *(at left) is moving with velocity* v *through a magnetic field* B *that is oriented toward the viewer. For an electron,* q *is negative so it follows a curved trajectory toward the top.*

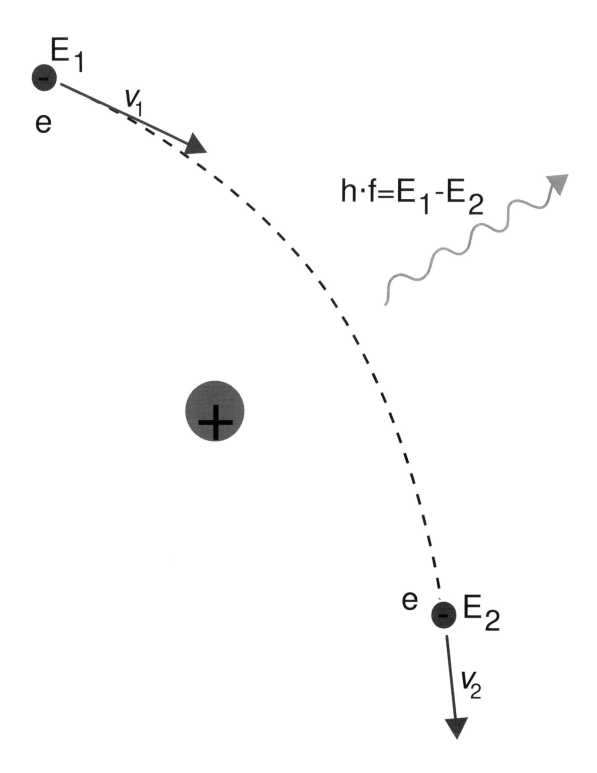

Here, Bremsstrahlung is produced by an electron e *deflected by the electric field of an atomic nucleus. The energy change* $E_2 - E_1$
determines the frequency f *of the emitted photon.*

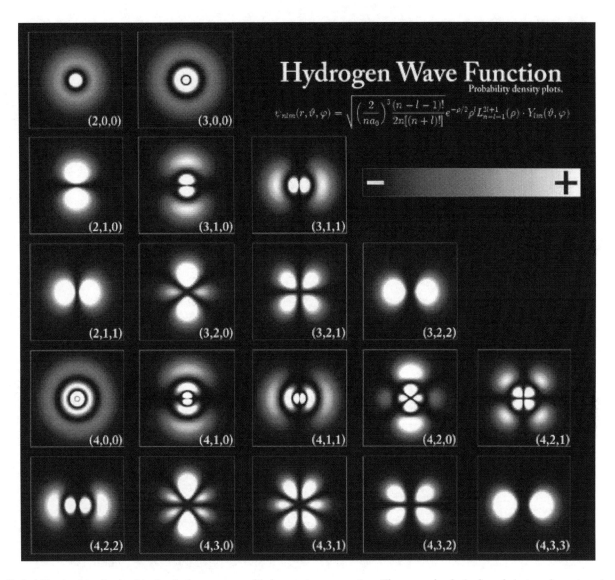

Probability densities for the first few hydrogen atom orbitals, seen in cross-section. The energy level of a bound electron determines the orbital it occupies, and the color reflects the probability of finding the electron at a given position.

A lightning discharge consists primarily of a flow of electrons.[113] *The electric potential needed for lightning may be generated by a triboelectric effect.*[114][115]

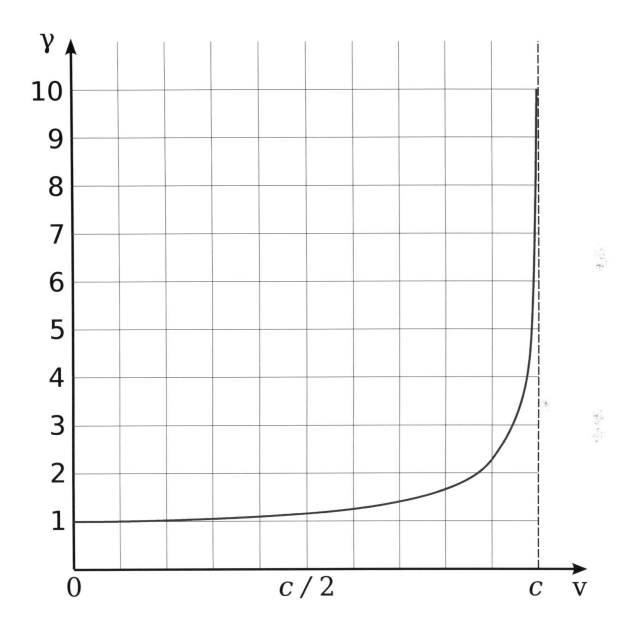

Lorentz factor as a function of velocity. It starts at value 1 and goes to infinity as v *approaches* c.

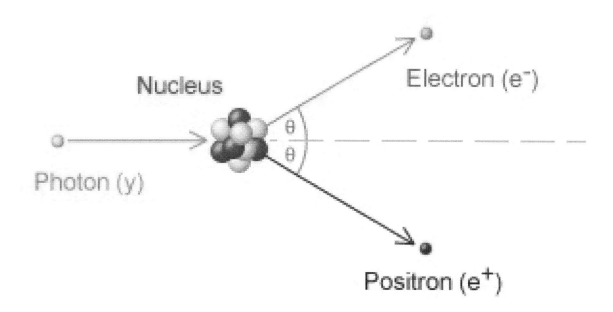

Pair production caused by the collision of a photon with an atomic nucleus

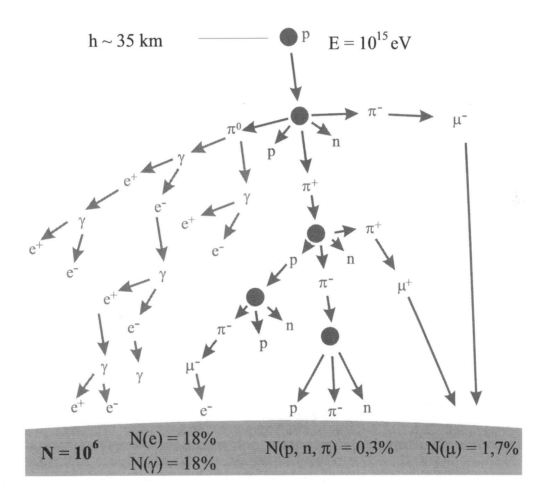

An extended air shower generated by an energetic cosmic ray striking the Earth's atmosphere

Aurorae are mostly caused by energetic electrons precipitating into the atmosphere.[147]

During a NASA wind tunnel test, a model of the Space Shuttle is targeted by a beam of electrons, simulating the effect of ionizing gases during re-entry.[156]

Chapter 24

Electron neutrino

The **electron neutrino** (ν
e) is a subatomic lepton elementary particle which has no net electric charge. Together with the electron it forms the first generation of leptons, hence its name *electron neutrino*. It was first hypothesized by Wolfgang Pauli in 1930, to account for missing momentum and missing energy in beta decay, and was discovered in 1956 by a team led by Clyde Cowan and Frederick Reines (see Cowan–Reines neutrino experiment).[1]

24.1 Proposal

In the early 1900s, theories predicted that the electrons resulting from beta decay should have been emitted at a specific energy. However, in 1914, James Chadwick showed that electrons were instead emitted in a continuous spectrum.[1]

$$n0 \rightarrow p+ + e-$$

The early understanding of beta decay

In 1930, Wolfgang Pauli theorized that an undetected particle was carrying away the observed difference between the energy, momentum, and angular momentum of the initial and final particles.[nb 1][2]

$$n0 \rightarrow p+ + e- + \nu0$$
e

Pauli's version of beta decay

24.1.1 Pauli's letter

On 4 December 1930, Pauli wrote a letter to the Physical Institute of the Federal Institute of Technology, Zürich, in which he proposed the electron neutrino as a potential solution to solve the problem of the continuous beta decay spectrum. An excerpt of the letter reads:[1]

> Dear radioactive ladies and gentlemen,
> As the bearer of these lines [...] will explain more exactly, considering the 'false' statistics of N-14 and Li-6 nuclei, as well as the continuous β-spectrum, I have hit upon a desperate remedy to save the "exchange theorem" of statistics and the energy theorem. Namely [there is] the possibility that there could exist in the nuclei electrically neutral particles that I wish to call neutrons,[nb 2] which have spin 1/2 and obey the exclusion principle, and additionally differ from light quanta in that they do not travel with the velocity of light: The

mass of the neutron must be of the same order of magnitude as the electron mass and, in any case, not larger than 0.01 proton mass. The continuous β-spectrum would then become understandable by the assumption that in β decay a neutron is emitted together with the electron, in such a way that the sum of the energies of neutron and electron is constant.

[...]

But I don't feel secure enough to publish anything about this idea, so I first turn confidently to you, dear radioactives, with a question as to the situation concerning experimental proof of such a neutron, if it has something like about 10 times the penetrating capacity of a γ ray.

I admit that my remedy may appear to have a small *a priori* probability because neutrons, if they exist, would probably have long ago been seen. However, only those who wager can win, and the seriousness of the situation of the continuous β-spectrum can be made clear by the saying of my honored predecessor in office, Mr. Debye, [...] "One does best not to think about that at all, like the new taxes." [...] So, dear radioactives, put it to test and set it right. [...]

With many greetings to you, also to Mr. Back, your devoted servant,

W. Pauli

A translated reprint of the full letter can be found in the September 1978 issue of *Physics Today*.[3]

24.2 Discovery

Main article: Cowan–Reines neutrino experiment

The electron neutrino was discovered by Clyde Cowan and Frederick Reines in 1956.[1][4]

24.3 Name

Pauli originally named his proposed light particle a *neutron*. When James Chadwick discovered a much more massive nuclear particle in 1932 and also named it a neutron, this left the two particles with the same name. Enrico Fermi, who developed the theory of beta decay, coined the term *neutrino* in 1934 to resolve the confusion. It was a pun on *neutrone*, the Italian equivalent of *neutron*: the *-one* ending can be an augmentative in Italian, so *neutrone* could be read as the "large neutral thing"; *-ino* replaces the augmentative suffix with a diminutive one. [5]

Upon the prediction and discovery of a second neutrino, it became important to distinguish between different types of neutrinos. Pauli's neutrino is now identified as the *electron neutrino*, while the second neutrino is identified as the *muon neutrino*.

24.4 Electron antineutrino

Like all fermions, the electron neutrino has a corresponding antiparticle, the electron antineutrino (ν e), which differs only in that some of its properties have equal magnitude but opposite sign. The process of beta decay produces both beta particles and electron antineutrinos. Wolfgang Pauli proposed the existence of these particles, in 1930, to ensure that beta decay conserved energy (the electrons in beta decay have a continuum of energies and momentum (the momentum of the electron and recoil nucleus – in beta decay – do not add up to zero).

24.5 Notes

[1] Niels Bohr was notably opposed to this interpretation of beta decay and was ready to accept that energy, momentum and angular momentum were not conserved quantities.

[2] See *Name*.

24.6 See also

- Muon neutrino

- PMNS matrix

- Tau neutrino

24.7 References

[1] "The Reines-Cowan Experiments: Detecting the Poltergeist" (PDF). *Los Alamos Science* **25**: 3. 1997. Retrieved 2010-02-10.

[2] K. Riesselmann (2007). "Logbook: Neutrino Invention". *Symmetry Magazine* **4** (2).

[3] L.M. Brown (1978). "The idea of the neutrino". *Physics Today* **31** (9): 23. Bibcode:1978PhT....31i..23B. doi:10.1063/1.2995181.

[4] F. Reines, C.L. Cowan, Jr. (1956). "The Neutrino". *Nature* **178** (4531): 446. Bibcode:1956Natur.178..446R. doi:10.1038/1784

[5] M.F. L'Annunziata (2007). *Radioactivity*. Elsevier. p. 100. ISBN 978-0-444-52715-8.

24.8 Further reading

- F. Reines, C.L. Cowan, Jr. (1956). "The Neutrino". *Nature* **178** (4531): 446. Bibcode:1956Natur.178..446R. doi:10.1038/178446a0.

- C.L. Cowan, Jr., F. Reines, F.B. Harrison, H.W. Kruse, A.D. McGuire (1956). "Detection of the Free Neutrino: A Confirmation". *Science* **124** (3212): 103–4. Bibcode:1956Sci...124..103C. doi:10.1126/science.124.3212.103. PMID 17796274.

Chapter 25

Muon

The **muon** (/ˈmjuːɒn/; from the Greek letter mu (μ) used to represent it) is an elementary particle similar to the electron, with electric charge of −1 e and a spin of $\frac{1}{2}$, but with a much greater mass (105.7 MeV/c^2). It is classified as a lepton, together with the electron (mass 0.511 MeV/c^2), the tau (mass 1776.82 MeV/c^2), and the three neutrinos (electron neutrino ν

e, muon neutrino ν

μ and tau neutrino ν

τ). As is the case with other leptons, the muon is not believed to have any sub-structure—that is, it is not thought to be composed of any simpler particles.

The muon is an unstable subatomic particle with a mean lifetime of 2.2 μs. Among all known unstable subatomic particles, only the neutron (lasting around 15 minutes) and some atomic nuclei have a longer decay lifetime; others decay significantly faster. The decay of the muon (as well as of the neutron, the longest-lived unstable baryon), is mediated by the weak interaction exclusively. Muon decay always produces at least three particles, which must include an electron of the same charge as the muon and two neutrinos of different types.

Like all elementary particles, the muon has a corresponding antiparticle of opposite charge (+1 e) but equal mass and spin: the **antimuon** (also called a *positive muon*). Muons are denoted by μ− and antimuons by μ+. Muons were previously called **mu mesons**, but are not classified as mesons by modern particle physicists (see § History), and that name is no longer used by the physics community.

Muons have a mass of 105.7 MeV/c^2, which is about 207 times that of the electron. Due to their greater mass, muons are not as sharply accelerated when they encounter electromagnetic fields, and do not emit as much bremsstrahlung (deceleration radiation). This allows muons of a given energy to penetrate far more deeply into matter than electrons, since the deceleration of electrons and muons is primarily due to energy loss by the bremsstrahlung mechanism. As an example, so-called "secondary muons", generated by cosmic rays hitting the atmosphere, can penetrate to the Earth's surface, and even into deep mines.

Because muons have a very large mass and energy compared with the decay energy of radioactivity, they are never produced by radioactive decay. They are, however, produced in copious amounts in high-energy interactions in normal matter, in certain particle accelerator experiments with hadrons, or naturally in cosmic ray interactions with matter. These interactions usually produce pi mesons initially, which most often decay to muons.

As with the case of the other charged leptons, the muon has an associated muon neutrino, denoted by ν

μ, which is not the same particle as the electron neutrino, and does not participate in the same nuclear reactions.

25.1 History

Muons were discovered by Carl D. Anderson and Seth Neddermeyer at Caltech in 1936, while studying cosmic radiation. Anderson had noticed particles that curved differently from electrons and other known particles when passed through a magnetic field. They were negatively charged but curved less sharply than electrons, but more sharply than protons, for particles of the same velocity. It was assumed that the magnitude of their negative electric charge was equal to that of the electron, and so to account for the difference in curvature, it was supposed that their mass was greater than an electron but smaller than a proton. Thus Anderson initially called the new particle a *mesotron*, adopting the prefix *meso-* from the Greek word for "mid-". The existence of the muon was confirmed in 1937 by J. C. Street and E. C. Stevenson's cloud chamber experiment.[2]

A particle with a mass in the meson range had been predicted before the discovery of any mesons, by theorist Hideki Yukawa:[3]

> It seems natural to modify the theory of Heisenberg and Fermi in the following way. The transition of a heavy particle from neutron state to proton state is not always accompanied by the emission of light particles. The transition is sometimes taken up by another heavy particle.

Because of its mass, the mu meson was initially thought to be Yukawa's particle, but it later proved to have the wrong properties. Yukawa's predicted particle, the pi meson, was finally identified in 1947 (again from cosmic ray interactions), and shown to differ from the earlier-discovered mu meson by having the correct properties to be a particle which mediated the nuclear force.

With two particles now known with the intermediate mass, the more general term *meson* was adopted to refer to any such particle within the correct mass range between electrons and nucleons. Further, in order to differentiate between the two different types of mesons after the second meson was discovered, the initial mesotron particle was renamed the *mu meson* (the Greek letter μ (*mu*) corresponds to *m*), and the new 1947 meson (Yukawa's particle) was named the pi meson.

As more types of mesons were discovered in accelerator experiments later, it was eventually found that the mu meson significantly differed not only from the pi meson (of about the same mass), but also from all other types of mesons. The difference, in part, was that mu mesons did not interact with the nuclear force, as pi mesons did (and were required to do, in Yukawa's theory). Newer mesons also showed evidence of behaving like the pi meson in nuclear interactions, but not like the mu meson. Also, the mu meson's decay products included both a neutrino and an antineutrino, rather than just one or the other, as was observed in the decay of other charged mesons.

In the eventual Standard Model of particle physics codified in the 1970s, all mesons other than the mu meson were understood to be hadrons—that is, particles made of quarks—and thus subject to the nuclear force. In the quark model, a *meson* was no longer defined by mass (for some had been discovered that were very massive—more than nucleons), but instead were particles composed of exactly two quarks (a quark and antiquark), unlike the baryons, which are defined as particles composed of three quarks (protons and neutrons were the lightest baryons). Mu mesons, however, had shown themselves to be fundamental particles (leptons) like electrons, with no quark structure. Thus, mu mesons were not mesons at all, in the new sense and use of the term *meson* used with the quark model of particle structure.

With this change in definition, the term *mu meson* was abandoned, and replaced whenever possible with the modern term *muon*, making the term mu meson only historical. In the new quark model, other types of mesons sometimes continued to be referred to in shorter terminology (e.g., *pion* for pi meson), but in the case of the muon, it retained the shorter name and was never again properly referred to by older "mu meson" terminology.

The eventual recognition of the "mu meson" muon as a simple "heavy electron" with no role at all in the nuclear interaction, seemed so incongruous and surprising at the time, that Nobel laureate I. I. Rabi famously quipped, "Who ordered that?"

In the Rossi–Hall experiment (1941), muons were used to observe the time dilation (or alternatively, length contraction) predicted by special relativity, for the first time.

25.2 Muon sources

On Earth, most naturally occurring muons are created by quasars and supernovas, which consist mostly of protons, many arriving from deep space at very high energy[4]

> About 10,000 muons reach every square meter of the earth's surface a minute; these charged particles form as by-products of cosmic rays colliding with molecules in the upper atmosphere. Traveling at relativistic

speeds, muons can penetrate tens of meters into rocks and other matter before attenuating as a result of absorption or deflection by other atoms.[5]

When a cosmic ray proton impacts atomic nuclei in the upper atmosphere, pions are created. These decay within a relatively short distance (meters) into muons (their preferred decay product), and muon neutrinos. The muons from these high energy cosmic rays generally continue in about the same direction as the original proton, at a velocity near the speed of light. Although their lifetime *without* relativistic effects would allow a half-survival distance of only about 456 m (2,197 μs×ln(2) × 0,9997×c) at most (as seen from Earth) the time dilation effect of special relativity (from the viewpoint of the Earth) allows cosmic ray secondary muons to survive the flight to the Earth's surface, since in the Earth frame, the muons have a longer half life due to their velocity. From the viewpoint (inertial frame) of the muon, on the other hand, it is the length contraction effect of special relativity which allows this penetration, since in the muon frame, its lifetime is unaffected, but the length contraction causes distances through the atmosphere and Earth to be far shorter than these distances in the Earth rest-frame. Both effects are equally valid ways of explaining the fast muon's unusual survival over distances.

Since muons are unusually penetrative of ordinary matter, like neutrinos, they are also detectable deep underground (700 meters at the Soudan 2 detector) and underwater, where they form a major part of the natural background ionizing radiation. Like cosmic rays, as noted, this secondary muon radiation is also directional.

The same nuclear reaction described above (i.e. hadron-hadron impacts to produce pion beams, which then quickly decay to muon beams over short distances) is used by particle physicists to produce muon beams, such as the beam used for the muon $g - 2$ experiment.[6]

25.3   Muon decay

See also: Michel parameters

Muons are unstable elementary particles and are heavier than electrons and neutrinos but lighter than all other matter particles. They decay via the weak interaction. Because lepton numbers must be conserved, one of the product neutrinos of muon decay must be a muon-type neutrino and the other an electron-type antineutrino (antimuon decay produces the corresponding antiparticles, as detailed below). Because charge must be conserved, one of the products of muon decay is always an electron of the same charge as the muon (a positron if it is a positive muon). Thus all muons decay to at least an electron, and two neutrinos. Sometimes, besides these necessary products, additional other particles that have no net charge and spin of zero (e.g., a pair of photons, or an electron-positron pair), are produced.

The dominant muon decay mode (sometimes called the Michel decay after Louis Michel) is the simplest possible: the muon decays to an electron, an electron antineutrino, and a muon neutrino. Antimuons, in mirror fashion, most often decay to the corresponding antiparticles: a positron, an electron neutrino, and a muon antineutrino. In formulaic terms, these two decays are:

μ– → e– + ν
e + ν
μ
μ+ → e+ + ν
e + ν
μ

The mean lifetime, τ = 1/Γ, of the (positive) muon is (2.1969811±0.0000022) μs.[1] The equality of the muon and antimuon lifetimes has been established to better than one part in 10^4.

The muon decay width which follows from Fermi's golden rule follows Sargent's law of fifth-power dependence on $m\mu$,

$$\Gamma = \frac{G_F^2 m_\mu^5}{192\pi^3} I\left(\frac{m_e^2}{m_\mu^2}\right),$$

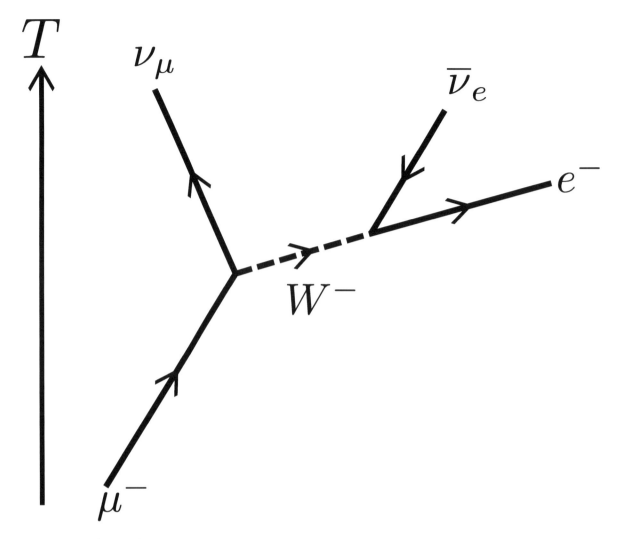

The most common decay of the muon

where $I(x) = 1 - 8x - 12x^2 \ln x + 8x^3 - x^4$, G_F is the Fermi coupling constant and $x = 2E_e/m_\mu c^2$ is the fraction of the maximum energy transmitted to the electron.

The decay distributions of the electron in muon decays have been parameterised using the so-called Michel parameters. The values of these four parameters are predicted unambiguously in the Standard Model of particle physics, thus muon decays represent a good test of the space-time structure of the weak interaction. No deviation from the Standard Model predictions has yet been found.

For the decay of the muon, the expected decay distribution for the Standard Model values of Michel parameters is

$$\frac{d^2\Gamma}{dx\,d\cos\theta} \sim x^2[(3 - 2x) + P_\mu \cos\theta(1 - 2x)]$$

where θ is the angle between the muon's polarization vector $\mathbf{P}_\mu$ and the decay-electron momentum vector, and $P_\mu = |\mathbf{P}_\mu|$ is the fraction of muons that are forward-polarized. Integrating this expression over electron energy gives the angular distribution of the daughter electrons:

$$\frac{d\Gamma}{d\cos\theta} \sim 1 - \frac{1}{3}P_\mu \cos\theta.$$

The electron energy distribution integrated over the polar angle (valid for $x < 1$) is

$$\frac{d\Gamma}{dx} \sim (3x^2 - 2x^3).$$

Due to the muons decaying by the weak interaction, parity conservation is violated. Replacing the $\cos\theta$ term in the expected decay values of the Michel Parameters with a $\cos\omega t$ term, where ω is the Larmor frequency from Larmor precession of the muon in a uniform magnetic field, given by:

$\omega = \frac{egB}{2m}$

where m is mass of the muon, e is charge, g is the muon g-factor and B is applied field.

A change in the electron distribution computed using the standard, unprecessional, Michel Parameters can be seen displaying a periodicity of π radians. This can be shown to physically correspond to a phase change of π, introduced in the electron distribution as the angular momentum is changed by the action of the charge conjugation operator, which is conserved by the weak interaction.

The observation of Parity violation in muon decay can be compared to the concept of violation of parity in weak interactions in general as an extension of The Wu Experiment, as well as the change of angular momentum introduced by a phase change of π corresponding to the charge-parity operator being invariant in this interaction. This fact is true for all lepton interactions in The Standard Model.

Certain neutrino-less decay modes are kinematically allowed but forbidden in the Standard Model. Examples forbidden by lepton flavour conservation are:

μ− → e− + γ and

μ− → e− + e+ + e− .

Observation of such decay modes would constitute clear evidence for theories beyond the Standard Model. Upper limits for the branching fractions of such decay modes were measured in many experiments starting more than 50 years ago. The current upper limit for the μ+ → e+ + γ branching fraction was measured 2013 in the MEG experiment and is 5.7 × 10^{-13}.[7]

25.4 Muonic atoms

The muon was the first elementary particle discovered that does not appear in ordinary atoms. Negative muons can, however, form muonic atoms (also called mu-mesic atoms), by replacing an electron in ordinary atoms. Muonic hydrogen atoms are much smaller than typical hydrogen atoms because the much larger mass of the muon gives it a much more localized ground-state wavefunction than is observed for the electron. In multi-electron atoms, when only one of the electrons is replaced by a muon, the size of the atom continues to be determined by the other electrons, and the atomic size is nearly unchanged. However, in such cases the orbital of the muon continues to be smaller and far closer to the nucleus than the atomic orbitals of the electrons.

Muonic helium is created by substituting a muon for one of the electrons in helium-4. The muon orbits much closer to the nucleus, so muonic helium can therefore be regarded like an isotope of helium whose nucleus consists of two neutrons, two protons and a muon, with a single electron outside. Colloquially, it could be called "helium 4.1", since the mass of the muon is roughly 0.1 amu. Chemically, muonic helium, possessing an unpaired valence electron, can bond with other atoms, and behaves more like a hydrogen atom than an inert helium atom.[8][9][10]

A positive muon, when stopped in ordinary matter, can also bind an electron and form an exotic atom known as muonium (Mu) atom, in which the muon acts as the nucleus. The positive muon, in this context, can be considered a pseudo-isotope of hydrogen with one ninth of the mass of the proton. Because the reduced mass of muonium, and hence its Bohr radius, is very close to that of hydrogen, this short-lived "atom" behaves chemically — to a first approximation — like hydrogen, deuterium and tritium.

25.5 Use in measurement of the proton charge radius

The experimental technique that is expected to provide the most precise determination of the root-mean-square charge radius of the proton is the measurement of the frequency of photons (precise "color" of light) emitted or absorbed by atomic transitions in muonic hydrogen. This form of hydrogen atom is composed of a negatively charged muon bound to a proton. The muon is particularly well suited for this purpose because its much larger mass results in a much more compact bound state and hence a larger probability for it to be found inside the proton in muonic hydrogen compared to the electron in atomic hydrogen.[11] The Lamb shift in muonic hydrogen was measured by driving the muon from a 2s state up to an excited 2p state using a laser. The frequency of the photons required to induce two such (slightly different) transitions were reported in 2014 to be 50 and 55 THz which, according to present theories of quantum electrodynamics, yield an appropriately averaged value of 0.84087±0.00039 fm for the charge radius of the proton.[12]

The internationally accepted value of the proton's charge radius is based on a suitable average of results from older measurements of effects caused by the nonzero size of the proton on scattering of electrons by nuclei and the light spectrum (photon energies) from excited atomic hydrogen. The official value updated in 2014 is 0.8751±0.0061 fm (see orders of magnitude for comparison to other sizes).[13] The expected precision of this result is inferior to that from muonic hydrogen by about a factor of fifteen, yet they disagree by about 5.6 times the nominal uncertainty in the difference (a discrepancy called 5.6σ in scientific notation). A conference of the world experts on this topic led to the decision to exclude the muon result from influencing the official 2014 value, in order to avoid hiding the mysterious discrepancy.[14] This "proton radius puzzle" remained unresolved as of late 2015, and has attracted much attention, in part because of the possibility that both measurements are valid, which would imply the influence of some "new physics".[15]

25.6 Anomalous magnetic dipole moment

The anomalous magnetic dipole moment is the difference between the experimentally observed value of the magnetic dipole moment and the theoretical value predicted by the Dirac equation. The measurement and prediction of this value is very important in the precision tests of QED (quantum electrodynamics). The E821 experiment[16] at Brookhaven National Laboratory (BNL) studied the precession of muon and anti-muon in a constant external magnetic field as they circulated in a confining storage ring. E821 reported the following average value[17] in 2006:

$$a = \frac{g-2}{2} = 0.00116592080(54)(33)$$

where the first errors are statistical and the second systematic.

The prediction for the value of the muon anomalous magnetic moment includes three parts:

$$a\mu^{SM} = a\mu^{QED} + a\mu^{EW} + a\mu^{had}.$$

The difference between the g-factors of the muon and the electron is due to their difference in mass. Because of the muon's larger mass, contributions to the theoretical calculation of its anomalous magnetic dipole moment from Standard Model weak interactions and from contributions involving hadrons are important at the current level of precision, whereas these effects are not important for the electron. The muon's anomalous magnetic dipole moment is also sensitive to contributions from new physics beyond the Standard Model, such as supersymmetry. For this reason, the muon's anomalous magnetic moment is normally used as a probe for new physics beyond the Standard Model rather than as a test of QED.[18] A new experiment at Fermilab using the E821 magnet will improve the precision of this measurement.[19]

25.7 Muon radiography and tomography

Main article: Muon tomography

Since muons are much more deeply penetrating than X-rays or gamma rays, muon imaging can be used with much thicker material or, with cosmic ray sources, larger objects. One example is commercial muon tomography used to image entire cargo containers to detect shielded nuclear material, as well as explosives or other contraband.[20]

The technique of muon transmission radiography based on cosmic ray sources was first used in the 1950s to measure the depth of the overburden of a tunnel in Australia[21] and in the 1960s to search for possible hidden chambers in the Pyramid of Chephren in Giza.[22]

In 2003, the scientists at Los Alamos National Laboratory developed a new imaging technique: **muon scattering tomography**. With muon scattering tomography, both incoming and outgoing trajectories for each particle are reconstructed, such as with sealed aluminum drift tubes.[23] Since the development of this technique, several companies have started to use it.

In August 2014, Decision Sciences International Corporation announced it had been awarded a contract by Toshiba for use of its muon tracking detectors in reclaiming the Fukushima nuclear complex.[24] The Fukushima Daiichi Tracker (FDT) was proposed to make a few months of muon measurements to show the distribution of the reactor cores.

In December 2014, Tepco reported that they would be using two different muon imaging techniques at Fukushima, "Muon Scanning Method" on Unit 1 (the most badly damaged, where the fuel may have left the reactor vessel) and "Muon Scattering Method" on Unit 2.[25]

The International Research Institute for Nuclear Decommissioning IRID in Japan and the High Energy Accelerator Research Organization KEK call the method they developed for Unit 1 the **muon permeation method**; 1,200 optical fibers for wavelength conversion light up when muons come into contact with them.[26] After a month of data collection, it is hoped to reveal the location and amount of fuel debris still inside the reactor. The measurements began in February 2015.[27]

25.8   See also

- Muonic atoms

- Muon spin spectroscopy

- Muon-catalyzed fusion

- Muon Tomography

- Mu2e, an experiment to detect neutrinoless conversion of muons to electrons

- List of particles

25.9   References

[1] J. Beringer et al. (Particle Data Group) (2012). "PDGLive Particle Summary 'Leptons (e, mu, tau, ... neutrinos ...)'" (PDF). Particle Data Group. Retrieved 2013-01-12.

[2] New Evidence for the Existence of a Particle Intermediate Between the Proton and Electron", Phys. Rev. 52, 1003 (1937).

[3] Yukaya Hideka, On the Interaction of Elementary Particles 1, Proceedings of the Physico-Mathematical Society of Japan (3) 17, 48, pp 139–148 (1935). (Read 17 November 1934)

[4] S. Carroll (2004). *Spacetime and Geometry: An Introduction to General Relativity*. Addison Wesley. p. 204

[5] Mark Wolverton (September 2007). "Muons for Peace: New Way to Spot Hidden Nukes Gets Ready to Debut". *Scientific American* **297** (3): 26–28. doi:10.1038/scientificamerican0907-26.

[6] "Physicists Announce Latest Muon g-2 Measurement" (Press release). Brookhaven National Laboratory. 30 July 2002. Retrieved 2009-11-14.

[7] J. Adam (MEG Collaboration); et al. (2013). "New Constraint on the Existence of the mu+ -> e+ gamma Decay". *Physical Review Letters* **110** (20): 201801. arXiv:1303.0754. Bibcode:2013PhRvL.110t1801A. doi:10.1103/PhysRevLett.110.201801.

[8] Fleming, D. G.; Arseneau, D. J.; Sukhorukov, O.; Brewer, J. H.; Mielke, S. L.; Schatz, G. C.; Garrett, B. C.; Peterson, K. A.; Truhlar, D. G. (28 Jan 2011). "Kinetic Isotope Effects for the Reactions of Muonic Helium and Muonium with H2". *Science* **331** (6016): 448–450. Bibcode:2011Sci...331..448F. doi:10.1126/science.1199421. PMID 21273484.

[9] Moncada, F.; Cruz, D.; Reyes, A. "Muonic alchemy: Transmuting elements with the inclusion of negative muons". *Chemical Physics Letters* **539**: 209–213. Bibcode:2012CPL...539..209M. doi:10.1016/j.cplett.2012.04.062.

[10] Moncada, F.; Cruz, D.; Reyes, A (10 May 2013). "Electronic properties of atoms and molecules containing one and two negative muons". *Chemical Physics Letters* **570**: 16–21. Bibcode:2013CPL...570...16M. doi:10.1016/j.cplett.2013.03.004.

[11] TRIUMF Muonic Hydrogen collaboration. "A brief description of Muonic Hydrogen research". Retrieved 2010-11-7

[12] Antognini, A.; Nez, F.; Schuhmann, K.; Amaro, F. D.; Biraben, F.; Cardoso, J. M. R.; Covita, D. S.; Dax, A.; Dhawan, S.; Diepold, M.; Fernandes, L. M. P.; Giesen, A.; Gouvea, A. L.; Graf, T.; Hänsch, T. W.; Indelicato, P.; Julien, L.; Kao, C. -Y.; Knowles, P.; Kottmann, F.; Le Bigot, E. -O.; Liu, Y. -W.; Lopes, J. A. M.; Ludhova, L.; Monteiro, C. M. B.; Mulhauser, F.; Nebel, T.; Rabinowitz, P.; Dos Santos, J. M. F.; Schaller, L. A. (2013). "Proton Structure from the Measurement of 2S-2P Transition Frequencies of Muonic Hydrogen". *Science* **339** (6118): 417–420. doi:10.1126/science.1230016. PMID 23349284.

[13] Mohr, Peter J.; Newell, David B.; Taylor, Barry N. (2015). "CODATA recommended values of the fundamental physical constants: 2014". *Zenodo.* arXiv:1507.07956. doi:10.5281/zenodo.22826.

[14] Wood, B. (3–4 November 2014). "Report on the Meeting of the CODATA Task Group on Fundamental Constants" (PDF). BIPM. p. 7.

[15] Carlson, Carl E. (May 2015). "The Proton Radius Puzzle". *Progress in Particle and Nuclear Physics* **82**: 59–77. arXiv:1502.05314. doi:10.1016/j.ppnp.2015.01.002.

[16] "The Muon g-2 Experiment Home Page". G-2.bnl.gov. 2004-01-08. Retrieved 2012-01-06.

[17] "(from the July 2007 review by Particle Data Group)" (PDF). Retrieved 2012-01-06.

[18] Hagiwara, K; Martin, A; Nomura, D; Teubner, T (2007). "Improved predictions for g−2g−2 of the muon and αQED(MZ2)". *Physics Letters B* **649** (2–3): 173. arXiv:hep-ph/0611102. Bibcode:2007PhLB..649..173H. doi:10.1016/j.physletb.2007.04.012.

[19] "Revolutionary muon experiment to begin with 3,200-mile move of 50-foot-wide particle storage ring". May 8, 2013. Retrieved Mar 16, 2015.

[20] "Decision Sciences Corp".

[21] George, E.P. (July 1, 1955). "Cosmic rays measure overburden of tunnel". *Commonwealth Engineer*: 455.

[22] Alvarez, L.W. (1970). "Search for hidden chambers in the pyramids using cosmic rays". *Science* **167**: 832. Bibcode:1970Sci... doi:10.1126/science.167.3919.832.

[23] Konstantin N. Borozdin, Gary E. Hogan, Christopher Morris, William C. Priedhorsky, Alexander Saunders, Larry J. Schultz & Margaret E. Teasdale. "Radiographic imaging with cosmic-ray muons". Nature.

[24] http://www.decisionsciencescorp.com/ds-awarded-toshiba-contract-fukushima-daiichi-nuclear-project/

[25] Tepco to start "scanning" inside of Reactor 1 in early February by using muon Fukushima Diary

[26] "Muon measuring instrument production for "muon permeation method" and its review by international experts". IRID.or.jp.

[27] Muon Scans Begin At Fukushima Daiichi - SimplyInfo

- S.H. Neddermeyer, C.D. Anderson; Anderson (1937). "Note on the Nature of Cosmic-Ray Particles". *Physical Review* **51** (10): 884–886. Bibcode:1937PhRv...51..884N. doi:10.1103/PhysRev.51.884.

- J.C. Street, E.C. Stevenson; Stevenson (1937). "New Evidence for the Existence of a Particle of Mass Intermediate Between the Proton and Electron". *Physical Review* **52** (9): 1003–1004. Bibcode:1937PhRv...52.1003S. doi:10.1103/PhysRev.52.1003.

- G. Feinberg, S. Weinberg; Weinberg (1961). "Law of Conservation of Muons". *Physical Review Letters* **6** (7): 381–383. Bibcode:1961PhRvL...6..381F. doi:10.1103/PhysRevLett.6.381.

- Serway & Faughn (1995). *College Physics* (4th ed.). Saunders. p. 841.

- M. Knecht (2003). "The Anomalous Magnetic Moments of the Electron and the Muon". In B. Duplantier, V. Rivasseau. *Poincaré Seminar 2002: Vacuum Energy – Renormalization*. Progress in Mathematical Physics **30**. Birkhäuser Verlag. p. 265. ISBN 3-7643-0579-7.

- E. Derman (2004). *My Life As A Quant*. Wiley. pp. 58–62.

25.10   External links

- Muon anomalous magnetic moment and supersymmetry

- g-2 (muon anomalous magnetic moment) experiment

- muLan (Measurement of the Positive Muon Lifetime) experiment

- The Review of Particle Physics

- The TRIUMF Weak Interaction Symmetry Test

- The MEG Experiment (Search for the decay Muon → Positron + Gamma)

- King, Philip. "Making Muons". *Backstage Science*. Brady Haran.

Chapter 26

Muon neutrino

The **muon neutrino** is a subatomic lepton elementary particle which has the symbol ν_μ and no net electric charge. Together with the muon it forms the second generation of leptons, hence its name *muon neutrino*. It was first hypothesized in the early 1940s by several people, and was discovered in 1962 by Leon Lederman, Melvin Schwartz and Jack Steinberger. The discovery was rewarded with the 1988 Nobel Prize in Physics.

26.1 Discovery

In 1962 Leon M. Lederman, Melvin Schwartz and Jack Steinberger established by performing an experiment at the Brookhaven National Laboratory[1] that more than one type of neutrino exists by first detecting interactions of the muon neutrino (already hypothesised with the name *neutretto*[2]), which earned them the 1988 Nobel Prize.[3]

26.2 Speed

Main article: Faster-than-light neutrino anomaly

In September 2011, OPERA researchers reported that muon neutrinos were apparently traveling at faster than light speed. This result was confirmed again in a second experiment in November 2011. These results have been viewed skeptically by the scientific community at large, and more experiments have/are investigating the phenomenon. In March 2012, the ICARUS team published results directly contradicting the results of OPERA.[4]

Later in July 2012 the apparent anomalous super-luminous propagation of neutrinos was traced to a faulty element of the fibre optic timing system in Gran-Sasso. After it was corrected the neutrinos appeared to travel with the speed of light within the errors of the experiment.[5]

26.3 See also

- Electron neutrino

- Neutrino oscillation

- PMNS matrix

- Tau neutrino

26.4   References

[1] G. Danby, J.-M. Gaillard, K. Goulianos, L. M. Lederman, N. B. Mistry, M. Schwartz, J. Steinberger (1962). "Observation of high-energy neutrino reactions and the existence of two kinds of neutrinos". *Physical Review Letters* **9**: 36. Bibcode:1962PhRvL doi:10.1103/PhysRevLett.9.36.

[2] I.V. Anicin (2005). "The Neutrino – Its Past, Present and Future". *SFIN (Institute of Physics, Belgrade) year XV, Series A: Conferences, No. A* **2**: 3–59. arXiv:physics/0503172. Bibcode:2005physics...3172A.

[3] "The Nobel Prize in Physics 1988". The Nobel Foundation. Retrieved 2010-02-11.

[4] M. Antonello et at. (2012). "Measurement of the neutrino velocity with the ICARUS detector at the CNGS beam". http://arxiv.org/abs/1203.3433v3

[5] "OPERA experiment reports anomaly in flight time of neutrinos from CERN to Gran Sasso (UPDATE 8 June 2012)". *CERN press office*. 8 June 2012. Retrieved 19 April 2013.

26.5   Further reading

- Leon M. Lederman (1988). "Observations in Particle Physics from Two Neutrinos to the Standard Model" (PDF). *Nobel Lectures*. The Nobel Foundation. Retrieved 2010-02-11.

- Melvin Schwartz (1988). "The First High Energy Neutrino Experiment" (PDF). *Nobel Lectures*. The Nobel Foundation. Retrieved 2010-02-11.

- Jack Steinberger (1988). "Experiments with High-Energy Neutrino Beams" (PDF). *Nobel Lectures*. The Nobel Foundation. Retrieved 2010-02-11.

Chapter 27

Tau (particle)

Not to be confused with the τ^+ of the τ–θ puzzle, which is now identified as a kaon.

The **tau** (τ), also called the **tau lepton**, **tau particle**, or **tauon**, is an elementary particle similar to the electron, with negative electric charge and a spin of $\frac{1}{2}$. Together with the electron, the muon, and the three neutrinos, it is a lepton. Like all elementary particles with half-integral spin, the tau has a corresponding antiparticle of opposite charge but equal mass and spin, which in the tau's case is the **antitau** (also called the *positive tau*). Tau particles are denoted by τ– and the antitau by τ+.

Tau leptons have a lifetime of 2.9×10^{-13} s and a mass of 1776.82 MeV/c^2 (compared to 105.7 MeV/c^2 for muons and 0.511 MeV/c^2 for electrons). Since their interactions are very similar to those of the electron, a tau can be thought of as a much heavier version of the electron. Because of their greater mass, tau particles do not emit as much bremsstrahlung radiation as electrons; consequently they are potentially highly penetrating, much more so than electrons. However, because of their short lifetime, the range of the tau is mainly set by their decay length, which is too small for bremsstrahlung to be noticeable: their penetrating power appears only at ultra high energy (above PeV energies).[4]

As with the case of the other charged leptons, the tau has an associated tau neutrino, denoted by ν
τ.

27.1 History

The tau was detected in a series of experiments between 1974 and 1977 by Martin Lewis Perl with his colleagues at the SLAC-LBL group.[2] Their equipment consisted of SLAC's then-new e+–e– colliding ring, called SPEAR, and the LBL magnetic detector. They could detect and distinguish between leptons, hadrons and photons. They did not detect the tau directly, but rather discovered anomalous events:

"We have discovered 64 events of the form

e+ + e– → e± + μ∓ + at least two undetected particles

for which we have no conventional explanation."

The need for at least two undetected particles was shown by the inability to conserve energy and momentum with only one. However, no other muons, electrons, photons, or hadrons were detected. It was proposed that this event was the production and subsequent decay of a new particle pair:

e+ + e– → τ+ + τ– → e± + μ∓ + 4ν

This was difficult to verify, because the energy to produce the τ+τ– pair is similar to the threshold for D meson production.

Work done at DESY-Hamburg, and with the Direct Electron Counter (DELCO) at SPEAR, subsequently established the mass and spin of the tau.

The symbol τ was derived from the Greek τρίτον (*triton*, meaning "third" in English), since it was the third charged lepton discovered.[5]

Martin Perl shared the 1995 Nobel Prize in Physics with Frederick Reines. The latter was awarded his share of the prize for experimental discovery of the neutrino.

27.2   Tau decay

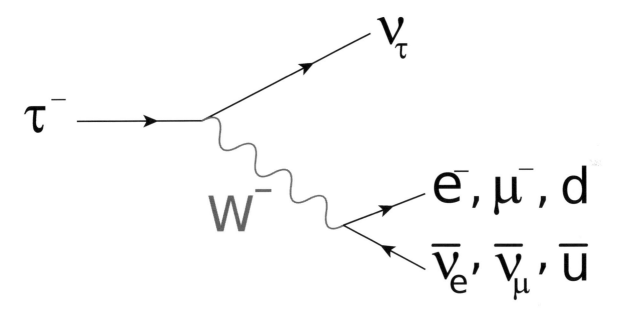

Feynman diagram of the common decays of the tau by emission of a W boson.

The tau is the only lepton that can decay into hadrons – the other leptons do not have the necessary mass. Like the other decay modes of the tau, the hadronic decay is through the weak interaction.[6]

The branching ratio of the dominant hadronic tau decays are:[3]

- 25.52% for decay into a charged pion, a neutral pion, and a tau neutrino;

- 10.83% for decay into a charged pion and a tau neutrino;

- 9.30% for decay into a charged pion, two neutral pions, and a tau neutrino;

- 8.99% for decay into three charged pions (of which two have the same electrical charge) and a tau neutrino;

- 2.70% for decay into three charged pions (of which two have the same electrical charge), a neutral pion, and a tau neutrino;

- 1.05% for decay into three neutral pions, a charged pion, and a tau neutrino.

In total, the tau lepton will decay hadronically approximately 64.79% of the time.

Since the tauonic lepton number is conserved in weak decays, a tau neutrino is always created when a tau decays.[6]

The branching ratio of the common purely leptonic tau decays are:[3]

- 17.82% for decay into a tau neutrino, electron and electron antineutrino;

- 17.39% for decay into a tau neutrino, muon and muon antineutrino.

The similarity of values of the two branching ratios is a consequence of lepton universality.

27.3 Exotic atoms

The tau lepton is predicted to form exotic atoms like other charged subatomic particles. One of such, called **tauonium** by the analogy to muonium, consists in antitauon and an electron: τ+e−.[7]

Another one is an onium atom τ+τ− called *true tauonium* and is difficult to detect due to tau's extremely short lifetime at low (non-relativistic) energies needed to form this atom. Its detection is important for quantum electrodynamics.[7]

27.4 See also

- Koide formula

27.5 References

[1] L. B. Okun (1980). *Leptons and Quarks*. V.I. Kisin (trans.). North-Holland Publishing. p. 103. ISBN 978-0444869241.

[2] Perl, M. L.; Abrams, G.; Boyarski, A.; Breidenbach, M.; Briggs, D.; Bulos, F.; Chinowsky, W.; Dakin, J.; et al. (1975). "Evidence for Anomalous Lepton Production in e+e− Annihilation". *Physical Review Letters* **35** (22): 1489. Bibcode:1975PhRvL..35.1489P.doi:10.1103/PhysRevLett.35.1489.

[3] J. Beringer *et al.* (Particle Data Group) (2012). "Review of Particle Physics". *Journal of Physics G* **86** (1): 581–651. Bibcode:2012PhRvD..86a0001B. doi:10.1103/PhysRevD.86.010001. |chapter= ignored (help)

[4] D. Fargion, P.G. De Sanctis Lucentini, M. De Santis, M. Grossi (2004). "Tau Air Showers from Earth". *The Astrophysical Journal* **613** (2): 1285. arXiv:hep-ph/0305128. Bibcode:2004ApJ...613.1285F. doi:10.1086/423124.

[5] M.L. Perl (1977). "Evidence for, and properties of, the new charged heavy lepton" (PDF). In T. Thanh Van (ed.). *Proceedings of the XII Rencontre de Moriond*. SLAC-PUB-1923.

[6] Riazuddin (2009). "Non-standard interactions" (PDF). *NCP 5th Particle Physics Sypnoisis* (Islamabad,: Riazuddin, Head of High-Energy Theory Group at National Center for Physics) **1** (1): 1–25.

[7] Brodsky, Stanley J.; Lebed, Richard F. (2009). "Production of the Smallest QED Atom: True Muonium ($\mu^+\mu^-$)". *Physical Review Letters* **102** (21): 213401. arXiv:0904.2225. Bibcode:2009PhRvL.102u3401B. doi:10.1103/PhysRevLett.102.213401.

27.6 External links

- Nobel Prize in Physics 1995

- Perl's logbook showing tau discovery

- A Tale of Three Papers gives the covers of the three original papers announcing the discovery.

Chapter 28

Tau neutrino

The **tau neutrino** or **tauon neutrino** is a subatomic elementary particle which has the symbol ν
τ and no net electric charge. Together with the tau, it forms the third generation of leptons, hence its name *tau neutrino*. Its existence was immediately implied after the tau particle was detected in a series of experiments between 1974 and 1977 by Martin Lewis Perl with his colleagues at the SLAC–LBL group.[1] The discovery of the tau neutrino was announced in July 2000 by the DONUT collaboration.[2][3]

28.1   Discovery

Main article: DONUT

The tau neutrino is last of the leptons, and is the second most recent particle of the Standard Model to be discovered. The DONUT experiment (which stands for *Direct Observation of the Nu Tau*) from Fermilab was built during the 1990s to specifically detect the tau neutrino. These efforts came to fruition in July 2000, when the DONUT collaboration reported its detection.[2][3]

28.2   See also

- Electron neutrino

- Muon neutrino

- PMNS matrix

28.3   References

[1] M. L. Perl; Abrams, G.; Boyarski, A.; Breidenbach, M.; Briggs, D.; Bulos, F.; Chinowsky, W.; Dakin, J.; Feldman, G.; Friedberg, C.; Fryberger, D.; Goldhaber, G.; Hanson, G.; Heile, F.; Jean-Marie, B.; Kadyk, J.; Larsen, R.; Litke, A.; Lüke, D.; Lulu, B.; Lüth, V.; Lyon, D.; Morehouse, C.; Paterson, J.; Pierre, F.; Pun, T.; Rapidis, P.; Richter, B.; Sadoulet, B.; et al. (1975). "Evidence for Anomalous Lepton Production in e+e− Annihilation". *Physical Review Letters* **35** (22): 1489. Bibcode:1975PhRvL..35.1489P. doi:10.1103/PhysRevLett.35.1489.

[2] "Physicists Find First Direct Evidence for Tau Neutrino at Fermilab" (Press release). Fermilab. 20 July 2000.

[3] K. Kodama *et al.* (DONUT Collaboration; Kodama; Ushida; Andreopoulos; Saoulidou; Tzanakos; Yager; Baller; Boehnlein; Freeman; Lundberg; Morfin; Rameika; Yun; Song; Yoon; Chung; Berghaus; Kubantsev; Reay; Sidwell; Stanton; Yoshida; Aoki;

Hara; Rhee; Ciampa; Erickson; Graham; et al. (2001). "Observation of tau neutrino interactions". *Physics Letters B* **504** (3): 218. arXiv:hep-ex/0012035. Bibcode:2001PhLB..504..218D. doi:10.1016/S0370-2693(01)00307-0.

Chapter 29

Antiparticle

Corresponding to most kinds of particles, there is an associated antimatter **antiparticle** with the same mass and opposite charge (including electric charge). For example, the antiparticle of the electron is the positively charged positron, which is produced naturally in certain types of radioactive decay.

The laws of nature are very nearly symmetrical with respect to particles and antiparticles. For example, an antiproton and a positron can form an antihydrogen atom, which is believed to have the same properties as a hydrogen atom. This leads to the question of why the formation of matter after the Big Bang resulted in a universe consisting almost entirely of matter, rather than being a half-and-half mixture of matter and antimatter. The discovery of Charge Parity violation helped to shed light on this problem by showing that this symmetry, originally thought to be perfect, was only approximate.

Particle-antiparticle pairs can annihilate each other, producing photons; since the charges of the particle and antiparticle are opposite, total charge is conserved. For example, the positrons produced in natural radioactive decay quickly annihilate themselves with electrons, producing pairs of gamma rays, a process exploited in positron emission tomography.

Antiparticles are produced naturally in beta decay, and in the interaction of cosmic rays in the Earth's atmosphere.

Because charge is conserved, it is not possible to create an antiparticle without either destroying a particle of the same charge (as in beta decay, when a proton (positive charge) is destroyed, a neutron created and a positron (positive charge, antiparticle) is also created and emitted) or by creating a particle of the opposite charge. The latter is seen in many processes in which both a particle and its antiparticle are created simultaneously, as in particle accelerators. This is the inverse of the particle-antiparticle annihilation process.

Although particles and their antiparticles have opposite charges, electrically neutral particles need not be identical to their antiparticles. The neutron, for example, is made out of quarks, the antineutron from antiquarks, and they are distinguishable from one another because neutrons and antineutrons annihilate each other upon contact. However, other neutral particles are their own antiparticles, such as photons, hypothetical gravitons, and some WIMPs.

29.1 History

29.1.1 Experiment

In 1932, soon after the prediction of positrons by Paul Dirac, Carl D. Anderson found that cosmic-ray collisions produced these particles in a cloud chamber— a particle detector in which moving electrons (or positrons) leave behind trails as they move through the gas. The electric charge-to-mass ratio of a particle can be measured by observing the radius of curling of its cloud-chamber track in a magnetic field. Positrons, because of the direction that their paths curled, were at first mistaken for electrons travelling in the opposite direction. Positron paths in a cloud-chamber trace the same helical path as an electron but rotate in the opposite direction with respect to the magnetic field direction due to their having the same magnitude of charge-to-mass ratio but with opposite charge and, therefore, opposite signed charge-to-mass ratios.

The antiproton and antineutron were found by Emilio Segrè and Owen Chamberlain in 1955 at the University of Califor-

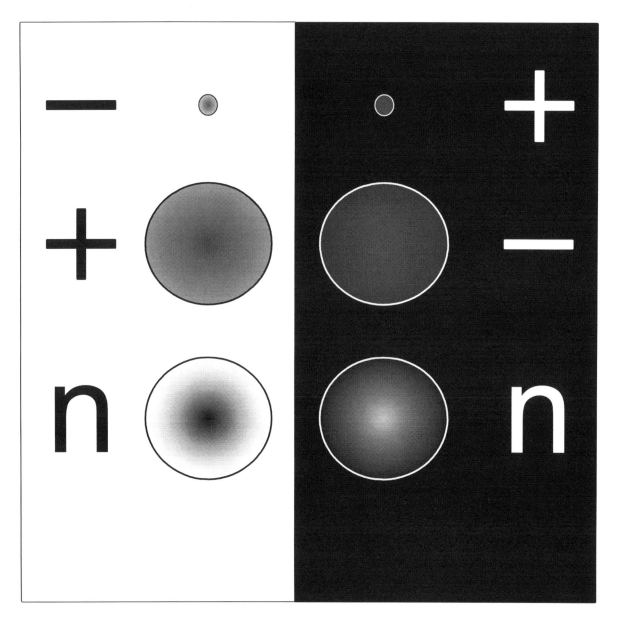

Illustration of electric charge of particles (left) and antiparticles (right). From top to bottom; electron/positron, proton/antiproton, neutron/antineutron.

nia, Berkeley. Since then, the antiparticles of many other subatomic particles have been created in particle accelerator experiments. In recent years, complete atoms of antimatter have been assembled out of antiprotons and positrons, collected in electromagnetic traps.[1]

29.1.2 Dirac Hole theory

... the development of quantum field theory made the interpretation of antiparticles as holes unnecessary, even though it lingers on in many textbooks.

Steven Weinberg[2]

Solutions of the Dirac equation contained negative energy quantum states. As a result, an electron could always radiate energy and fall into a negative energy state. Even worse, it could keep radiating infinite amounts of energy because there were infinitely many negative energy states available. To prevent this unphysical situation from happening, Dirac proposed that a "sea" of negative-energy electrons fills the universe, already occupying all of the lower-energy states so that, due to the Pauli exclusion principle, no other electron could fall into them. Sometimes, however, one of these negative-energy particles could be lifted out of this Dirac sea to become a positive-energy particle. But, when lifted out, it would leave behind a *hole* in the sea that would act exactly like a positive-energy electron with a reversed charge. These he interpreted as "negative-energy electrons" and attempted to identify them with protons in his 1930 paper *A Theory of Electrons and Protons*[3] However, these "negative-energy electrons" turned out to be positrons, and not protons.

This picture implied an infinite negative charge for the universe—a problem of which Dirac was aware. Dirac tried to argue that we would perceive this as the normal state of zero charge. Another difficulty was the difference in masses of the electron and the proton. Dirac tried to argue that this was due to the electromagnetic interactions with the sea, until Hermann Weyl proved that hole theory was completely symmetric between negative and positive charges. Dirac also predicted a reaction e− + p+ → γ + γ, where an electron and a proton annihilate to give two photons. Robert Oppenheimer and Igor Tamm proved that this would cause ordinary matter to disappear too fast. A year later, in 1931, Dirac modified his theory and postulated the positron, a new particle of the same mass as the electron. The discovery of this particle the next year removed the last two objections to his theory.

However, the problem of infinite charge of the universe remains. Also, as we now know, bosons also have antiparticles, but since bosons do not obey the Pauli exclusion principle (only fermions do), hole theory does not work for them. A unified interpretation of antiparticles is now available in quantum field theory, which solves both these problems.

29.2   Particle-antiparticle annihilation

Main article: Annihilation
If a particle and antiparticle are in the appropriate quantum states, then they can annihilate each other and produce

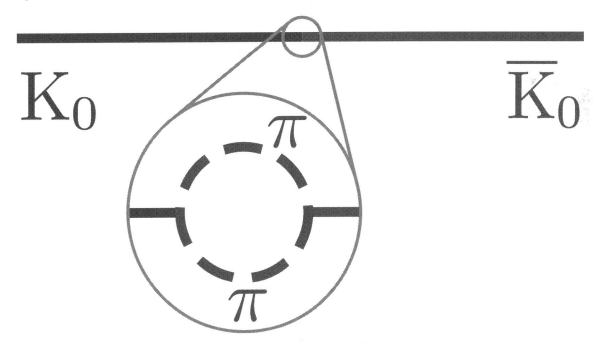

An example of a virtual pion pair that influences the propagation of a kaon, causing a neutral kaon to mix with the antikaon. This is an example of renormalization in quantum field theory— the field theory being necessary because of the change in particle number.

other particles. Reactions such as e− + e+ → γ + γ (the two-photon annihilation of an electron-positron pair) are an example. The single-photon annihilation of an electron-positron pair, e− + e+ → γ, cannot occur in free space because it is impossible to conserve energy and momentum together in this process. However, in the Coulomb field of a nucleus the translational invariance is broken and single-photon annihilation may occur.[4] The reverse reaction (in free space, without an atomic nucleus) is also impossible for this reason. In quantum field theory, this process is allowed only as an intermediate quantum state for times short enough that the violation of energy conservation can be accommodated by the uncertainty principle. This opens the way for virtual pair production or annihilation in which a one particle quantum state may *fluctuate* into a two particle state and back. These processes are important in the vacuum state and renormalization of a quantum field theory. It also opens the way for neutral particle mixing through processes such as the one pictured here, which is a complicated example of mass renormalization.

29.3 Properties of antiparticles

Quantum states of a particle and an antiparticle can be interchanged by applying the charge conjugation (**C**), parity (**P**), and time reversal (**T**) operators. If $|p, \sigma, n\rangle$ denotes the quantum state of a particle (**n**) with momentum **p**, spin **J** whose component in the z-direction is σ, then one has

$$CPT \, |p, \sigma, n\rangle \; = \; (-1)^{J-\sigma} \, |p, -\sigma, n^c\rangle,$$

where **n^c** denotes the charge conjugate state, *i.e.*, the antiparticle. This behaviour under **CPT** is the same as the statement that the particle and its antiparticle lie in the same irreducible representation of the Poincaré group. Properties of antiparticles can be related to those of particles through this. If **T** is a good symmetry of the dynamics, then

$$T \, |p, \sigma, n\rangle \; \propto \; |-p, -\sigma, n\rangle,$$
$$CP \, |p, \sigma, n\rangle \; \propto \; |-p, \sigma, n^c\rangle,$$
$$C \, |p, \sigma, n\rangle \; \propto \; |p, \sigma, n^c\rangle,$$

where the proportionality sign indicates that there might be a phase on the right hand side. In other words, particle and antiparticle must have

- the same mass **m**
- the same spin state **J**
- opposite electric charges **q** and **-q**.

29.4 Quantum field theory

This section draws upon the ideas, language and notation of canonical quantization of a quantum field theory.

One may try to quantize an electron field without mixing the annihilation and creation operators by writing

$$\psi(x) = \sum_k u_k(x) a_k e^{-iE(k)t},$$

where we use the symbol k to denote the quantum numbers p and σ of the previous section and the sign of the energy, $E(k)$, and ak denotes the corresponding annihilation operators. Of course, since we are dealing with fermions, we have to have the operators satisfy canonical anti-commutation relations. However, if one now writes down the Hamiltonian

$$H = \sum_k E(k) a_k^\dagger a_k,$$

then one sees immediately that the expectation value of H need not be positive. This is because $E(k)$ can have any sign whatsoever, and the combination of creation and annihilation operators has expectation value 1 or 0.

So one has to introduce the charge conjugate *antiparticle* field, with its own creation and annihilation operators satisfying the relations

$$b_{k\prime} = a_k^\dagger \text{ and } b_{k\prime}^\dagger = a_k,$$

where k has the same p, and opposite σ and sign of the energy. Then one can rewrite the field in the form

$$\psi(x) = \sum_{k_+} u_k(x) a_k e^{-iE(k)t} + \sum_{k_-} u_k(x) b_k^\dagger e^{-iE(k)t},$$

where the first sum is over positive energy states and the second over those of negative energy. The energy becomes

$$H = \sum_{k_+} E_k a_k^\dagger a_k + \sum_{k_-} |E(k)| b_k^\dagger b_k + E_0,$$

where E_0 is an infinite negative constant. The vacuum state is defined as the state with no particle or antiparticle, *i.e.*, $a_k|0\rangle = 0$ and $b_k|0\rangle = 0$. Then the energy of the vacuum is exactly E_0. Since all energies are measured relative to the vacuum, **H** is positive definite. Analysis of the properties of *ak* and *bk* shows that one is the annihilation operator for particles and the other for antiparticles. This is the case of a fermion.

This approach is due to Vladimir Fock, Wendell Furry and Robert Oppenheimer. If one quantizes a real scalar field, then one finds that there is only one kind of annihilation operator; therefore, real scalar fields describe neutral bosons. Since complex scalar fields admit two different kinds of annihilation operators, which are related by conjugation, such fields describe charged bosons.

29.4.1 Feynman–Stueckelberg interpretation

By considering the propagation of the negative energy modes of the electron field backward in time, Ernst Stueckelberg reached a pictorial understanding of the fact that the particle and antiparticle have equal mass **m** and spin **J** but opposite charges **q**. This allowed him to rewrite perturbation theory precisely in the form of diagrams. Richard Feynman later gave an independent systematic derivation of these diagrams from a particle formalism, and they are now called Feynman diagrams. Each line of a diagram represents a particle propagating either backward or forward in time. This technique is the most widespread method of computing amplitudes in quantum field theory today.

Since this picture was first developed by Ernst Stueckelberg, and acquired its modern form in Feynman's work, it is called the *Feynman-Stueckelberg interpretation* of antiparticles to honor both scientists.

As a consequence of this interpretation, Villata argued that the assumption of antimatter as CPT-transformed matter would imply that the gravitational interaction between matter and antimatter is repulsive.[5]

29.5 See also

- Gravitational interaction of antimatter

- Parity, charge conjugation and time reversal symmetry.

- CP violations and the baryon asymmetry of the universe.

- Quantum field theory and the list of particles

- Baryogenesis

29.6 References

[1] http://news.nationalgeographic.com/news/2010/11/101118-antimatter-trapped-engines-bombs-nature-science-cern/

[2] Weinberg, Steve. *The quantum theory of fields, Volume 1 : Foundations.* p. 14. ISBN 0-521-55001-7.

[3] Dirac, Paul (1930). "A Theory of Electrons and Protons". *Proceedings of the Royal Society A* **126**(801): 360–365.Bibcode: doi:10.1098/rspa.1930.0013.

[4] Sodickson, L.; W. Bowman; J. Stephenson (1961). "Single-Quantum Annihilation of Positrons". *Physical Review* **124** (6): 1851–1861. Bibcode:1961PhRv..124.1851S. doi:10.1103/PhysRev.124.1851.

[5] M. Villata, CPT symmetry and antimatter gravity in general relativity, 2011, EPL (Europhysics Letters) 94, 20001

- Feynman, R. P. (1987). "The reason for antiparticles". In R. P. Feynman and S. Weinberg. *The 1986 Dirac memorial lectures.* Cambridge University Press. ISBN 0-521-34000-4.

- Weinberg, S. (1995). *The Quantum Theory of Fields, Volume 1: Foundations.* Cambridge University Press. ISBN 0-521-55001-7.

Chapter 30

Weyl semimetal

Weyl fermions are massless chiral fermions that play an important role in quantum field theory and the standard model. They may be thought of as a building block for fermions in quantum field theory, and were predicted from a solution to the Dirac equation derived by Hermann Weyl.[1] For example, one-half of a charged Dirac fermion of a definite chirality is a Weyl fermion.[2] They have not been observed as a fundamental particle in nature. Weyl fermions may be realized as emergent quasiparticles in a low-energy condensed matter system.[3][4]

30.1 Experimental discovery

A **Weyl semimetal** is a solid state crystal whose low energy excitations are Weyl fermions.[6][7] A Weyl semimetal enables the first-ever realization of Weyl fermions.[8] It is a topologically nontrivial phase of matter that broadens the topological classification beyond topological insulators.[4] The Weyl fermions at zero energy correspond to points of bulk band degeneracy, the Weyl nodes that are separated in momentum space. Weyl fermions have distinct chiralities, either left handed or right handed. In a Weyl semimetal crystal, the chiralities associated with the Weyl nodes can be understood as topological charges, leading to monopoles and anti-monopoles of Berry curvature in momentum space, which (the splitting) serve as the topological invariant of this phase.[6] Comparing to the Dirac fermions in graphene or on the surface of topological insulators, Weyl fermions in a Weyl semimetal are the most robust electrons and do not depend on symmetries except the translation symmetry of the crystal lattice. Hence the Weyl fermion quasiparticles in a Weyl semimetal possess a high degree of mobility. Due to the nontrivial topology, a Weyl semimetal is expected to demonstrate Fermi arc electron states on its surface.[6][8] These arcs are discontinuous or disjoint segments of a two dimensional Fermi contour, which are terminated onto the projections of the Weyl fermion nodes on the surface.

On July 16, 2015 the first experimental observations of Weyl fermion semimetal in an inversion symmetry-breaking single crystal material tantalum arsenide (TaAs) were made.[8] Both Weyl fermions and Fermi arc surface states were observed, which established its topological character.[8] This discovery was built upon previous theoretical predictions proposed in November 2014.[9][10] Weyl points were also observed in a non-fermionic system, a photonic crystal. [11] [12]

30.2 Applications

The Weyl fermions in the bulk and the Fermi arcs on the surface are of interest in physics and materials technology.[1][13] Their high mobility may find use in electronics and computing.

30.3 Further reading

- "Weyl fermions are spotted at long last". *Physics World*. Retrieved 23 July 2015.

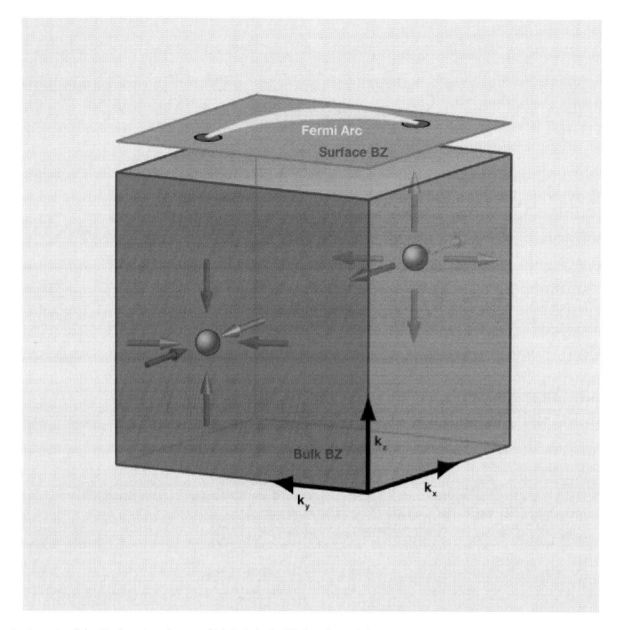

A schematic of the Weyl semimetal state, which include the Weyl nodes and the Fermi arcs. The Weyl nodes are momentum space monopoles and anti-monopoles. The sketch is adapted from Ref.[5]

- "Weyl fermions: Massless yet real". *Nature Materials*. Retrieved 20 August 2015.

- "Where the Weyl Things Are". *APS Physics*. Retrieved 8 September 2015.

30.4 References

[1] Johnston, Hamish (2015). "Weyl fermions are spotted at long last". *Physics World*.

[2] Weyl, H. (1929). "Elektron und gravitation". *I. Z. Phys.* **56**: 330–352. doi:10.1007/bf01339504.

[3] Herring, C. (1937). "Accidental Degeneracy in the Energy Bands of Crystals". *Phys. Rev.* **52**: 365. doi:10.1103/physrev.52.365.

[4] Murakami, S. (2007). "Phase transition between the quantum spin Hall and insulator phases in 3D: emergence of a topological gapless phase". *New J. Phys.* **9**: 356. doi:10.1088/1367-2630/9/9/356.

[5] Balents, L. (2011). "Weyl electrons kiss". *Physics* **4**: 36. doi:10.1103/physics.4.36.

[6] Wan, X.; Turner, A. M.; Vishwanath, A.; Savrasov, S. Y. (2011). "Topological Semimetal and Fermi-arc surface states in the electronic structure of pyrochlore iridates". *Phys. Rev. B* **83**: 205101. doi:10.1103/physrevb.83.205101.

[7] Burkov, A. A.; Balents, L. (2011). "Weyl Semimetal in a Topological Insulator Multilayer". *Phys. Rev. Lett.* **107**: 127205. doi:10.1103/physrevlett.107.127205.

[8] Xu, S.-Y.; Belopolski, I.; Alidoust, N.; Neupane, M.; Bian, G.; Zhang, C.; Sankar, R.; Chang, G.; Yuan, Z.; Lee, C.-C.; Huang, S.-M.; Zheng, H.; Ma, J.; Sanchez, D. S.; Wang, B. K.; Bansil, A.; Chou, F.-C.; Shibayev, P. P.; Lin, H.; Jia, S.; Hasan, M. Z. (2015). "Discovery of a Weyl Fermion semimetal and topological Fermi arcs". *Science.* doi:10.1126/science.aaa9297.

[9] Huang, S.-M.; Xu, S.-Y.; Belopolski, I.; Lee, C.-C.; Chang, G.; Wang, B. K.; Alidoust, N.; Bian, G.; Neupane, M.; Zhang, C.; Jia, S.; Bansil, A.; Lin, H.; Hasan, M. Z. (2015). "A Weyl Fermion semimetal with surface Fermi arcs in the transition metal monopnictide TaAs class". *Nature Commun.* **6**: 7373. doi:10.1038/ncomms8373.

[10] Weng, H.; Fang, C.; Fang, Z.; Bernevig, A.; Dai, X. (2015). "Weyl semimetal phase in non-centrosymmetric transition metal monophosphides". *Phys. Rev. X* **5**: 011029.

[11] Lu, L.; Fu, L.; Joannopoulos, J.; Soljačić, M. (2013). "Weyl points and line nodes in gyroid photonic crystals". *Nature Photonics* **7**: 294. doi:10.1038/nphoton.2013.42.

[12] Lu, L.; Wang, Z.; Ye, D.; Fu, L.; Joannopoulos, J.; Soljačić, M. (2015). "Experimental observation of Weyl points". *Science* **349**: 622. doi:10.1126/science.aaa9273.

[13] Shekhar, C.; et al. (2015). "Extremely large magnetoresistance and ultrahigh mobility in the topological Weyl semimetal candidate NbP". *Nature Physics* **11**: 645. doi:10.1038/nphys3372.

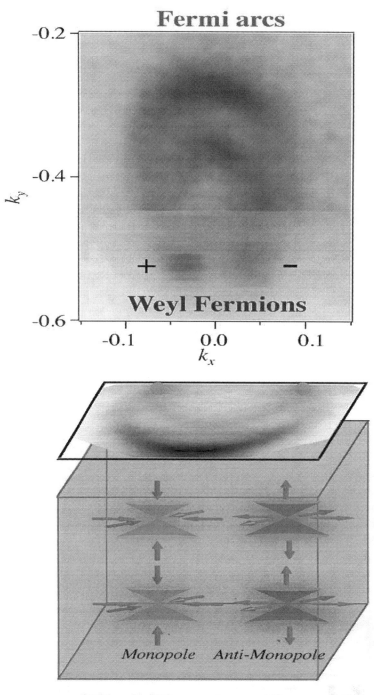

A detector image (top) signals the existence of Weyl fermion nodes and the Fermi arcs.[8] The plus and minus signs note the particle's chirality. A schematic (bottom) shows the way Weyl fermions inside a crystal can be thought as monopole and antimonopole in momentum space. (Image art by Su-Yang Xu and M. Zahid Hasan)

Chapter 31

Dirac fermion

In physics, a **Dirac fermion** is a fermion which is not its own antiparticle. It is named for Paul Dirac. They can be modelled with the Dirac equation.

The alternative to a Dirac fermion is a Majorana fermion.

In particle physics, all fermions in the standard model, except possibly neutrinos, are Dirac fermions.

In condensed matter physics, low-energy excitations in graphene and topological insulators, among others, are fermionic quasiparticles described by a pseudo-relativistic Dirac equation.

31.1   See also

- Dirac spinor, a wavefunction-like description of a Dirac fermion

- Spinor, mathematical details

- Dirac sea, a conceptual metaphor

Chapter 32

Majorana fermion

Not to be confused with Majoron.

A **Majorana fermion** (/maɪəˈrɒnə ˈfɛərmiːɒn/[1]), also referred to as a **Majorana particle**, is a fermion that is its own antiparticle. They were hypothesized by Ettore Majorana in 1937. The term is sometimes used in opposition to a Dirac fermion, which describes fermions that are not their own antiparticles.

All of the Standard Model fermions except the neutrino behave as Dirac fermions at low energy (after electroweak symmetry breaking), but the (massive) nature of the neutrino is not settled and it may be either Dirac or Majorana. In condensed matter physics, Majorana fermions exist as quasiparticle excitations in superconductors and can be used to form Majorana bound states governed by non-abelian statistics.

32.1 Theory

The concept goes back to Majorana's suggestion in 1937[2] that neutral spin$-1/2$ particles can be described by a real wave equation (the Majorana equation), and would therefore be identical to their antiparticle (because the wave functions of particle and antiparticle are related by complex conjugation).

The difference between Majorana fermions and Dirac fermions can be expressed mathematically in terms of the creation and annihilation operators of second quantization. The creation operator $\gamma_j^\dagger$ creates a fermion in quantum state j (described by a *real* wave function), whereas the annihilation operator γ_j annihilates it (or, equivalently, creates the corresponding antiparticle). For a Dirac fermion the operators $\gamma_j^\dagger$ and γ_j are distinct, whereas for a Majorana fermion they are identical.

The ordinary fermionic operators in Majorana representation present as follows:

1. $\gamma_1 = \frac{(f+f^\dagger)}{\sqrt{2}}$

2. $\gamma_2 = -i\frac{(f-f^\dagger)}{\sqrt{2}}$

Where f and $f^\dagger$ are respectively fermion annihilation and creation operators.

In supersymmetry models, neutralinos--superpartners of gauge bosons and Higgs bosons--are Majorana.

32.2 Elementary particle

Because particles and antiparticles have opposite conserved charges, in order to be a Majorana fermion, namely, it is its own antiparticle, it is necessarily uncharged. All of the elementary fermions of the Standard Model have gauge charges,

Ettore Majorana hypothesised the existence of Majorana fermions in 1937

so they cannot have fundamental Majorana masses. However, the right-handed sterile neutrinos introduced to explain neutrino oscillation could have Majorana masses. If they do, then at low energy (after electroweak symmetry breaking), by the seesaw mechanism, the neutrino fields would naturally behave as six Majorana fields, with three expected to have very high masses (comparable to the GUT scale) and the other three expected to have very low masses (comparable to 1 eV). If right-handed neutrinos exist but do not have a Majorana mass, the neutrinos would instead behave as three Dirac fermions and their antiparticles with masses coming directly from the Higgs interaction, like the other Standard Model fermions.

The seesaw mechanism is appealing because it would naturally explain why the observed neutrino masses are so small. However, if the neutrinos are Majorana then they violate the conservation of lepton number and even B − L.

Neutrinoless double beta decay, which can be viewed as two beta decay events with the produced antineutrinos immediately annihilating with one another, is only possible if neutrinos are their own antiparticles.[3] Experiments are underway to search for this type of decay.[4]

The high-energy analog of the neutrinoless double beta decay process is the production of same sign charged lepton pairs at hadron colliders;[5] it is being searched for by both the ATLAS and CMS experiments at the Large Hadron Collider. In theories based on left–right symmetry, there is a deep connection between these processes.[6] In the most accepted explanation of the smallness of neutrino mass, the seesaw mechanism, the neutrino is naturally a Majorana fermion.

Majorana fermions cannot possess intrinsic electric or magnetic moments, only toroidal moments.[7][8][9] Such minimal interaction with electromagnetic fields makes them potential candidates for cold dark matter.[10][11]

32.3 Majorana bound states

In superconducting materials, Majorana fermions can emerge as (non-fundamental) quasiparticles (which are more commonly referred as Bogoliubov quasiparticles in condensed matter.). This becomes possible because a quasiparticle in a superconductor is its own antiparticle. Majorana fermions (i.e. the Bogoliubov quasiparticles) in superconductors were observed by many experiments many years ago.

Mathematically, the superconductor imposes electron hole "symmetry" on the quasiparticle excitations, relating the creation operator $\gamma(E)$ at energy E to the annihilation operator $\gamma^{\dagger}(-E)$ at energy $-E$. Majorana fermions can be bound to a defect at zero energy, and then the combined objects are called Majorana bound states or Majorana zero modes.[12] This name is more appropriate than Majorana fermion (although the distinction is not always made in the literature), because the statistics of these objects is no longer fermionic. Instead, the Majorana bound states are an example of non-abelian anyons: interchanging them changes the state of the system in a way that depends only on the order in which the exchange was performed. The non-abelian statistics that Majorana bound states possess allows them to be used as a building block for a topological quantum computer.[13]

A quantum vortex in certain superconductors or superfluids can trap midgap states, so this is one source of Majorana bound states.[14][15][16] Shockley states at the end points of superconducting wires or line defects are an alternative, purely electrical, source.[17] An altogether different source uses the fractional quantum Hall effect as a substitute for the superconductor.[18]

32.3.1 Experiments in superconductivity

In 2008, Fu and Kane provided a groundbreaking development by theoretically predicting that Majorana bound states can appear at the interface between topological insulators and superconductors.[19][20] Many proposals of a similar spirit soon followed, where it was shown that Majorana bound states can appear even without any topological insulator. An intense search to provide experimental evidence of Majorana bound states in superconductors[21][22] first produced some positive results in 2012.[23][24] A team from the Kavli Institute of Nanoscience at Delft University of Technology in the Netherlands reported an experiment involving indium antimonide nanowires connected to a circuit with a gold contact at one end and a slice of superconductor at the other. When exposed to a moderately strong magnetic field the apparatus showed a peak electrical conductance at zero voltage that is consistent with the formation of a pair of Majorana bound states, one at either end of the region of the nanowire in contact with the superconductor.[25] This type of bounded state with zero energy was soon detected by several other groups in similar hybrid devices.[26][27][28][29]

This experiment from Delft marks a possible verification of independent 2010 theoretical proposals from two groups[30][31] predicting the solid state manifestation of Majorana bound states in semiconducting wires. However, it was also pointed out that some other trivial non-topological bounded states[32] could highly mimic the zero voltage conductance peak of Majorana bound state.

In 2014, evidence of Majorana bound states was observed using a low-temperature scanning tunneling microscope, by scientists at Princeton University.[33][34] It was suggested that Majorana bound states appeared at the edges of a chain of iron atoms formed on the surface of superconducting lead. Physicist Jason Alicea of California Institute of Technology, not involved in the research, said the study offered "compelling evidence" for Majorana fermions but that "we should keep in mind possible alternative explanations—even if there are no immediately obvious candidates".[35]

32.4   References

[1] "Quantum Computation possible with Majorana Fermions" on YouTube, uploaded 19 April 2013, retrieved 5 October 2014; and also based on the physicist's name's pronunciation.

[2] Majorana, Ettore; Maiani, Luciano (2006). "A symmetric theory of electrons and positrons". In Bassani, Giuseppe Franco. *Ettore Majorana Scientific Papers*. pp. 201–33. doi:10.1007/978-3-540-48095-2_10. ISBN 978-3-540-48091-4. Translated from: Majorana, Ettore (1937). "Teoria simmetrica dell'elettrone e del positrone". *Il Nuovo Cimento* (in Italian) **14** (4): 171–84. doi:10.1007/bf02961314.

[3] Schechter, J.; Valle, J.W.F. (1982). "Neutrinoless Double beta Decay in SU(2) x U(1) Theories". *Physical Review D* **25** (11): 2951. Bibcode:1982PhRvD..25.2951S. doi:10.1103/PhysRevD.25.2951. (subscription required (help)).

[4] Rodejohann, Werner (2011). "Neutrino-less Double Beta Decay and Particle Physics". *International Journal of Modern Physics* **E20** (9): 1833. arXiv:1106.1334. Bibcode:2011IJMPE..20.1833R. doi:10.1142/S0218301311020186. (registration required (help)).

[5] Keung, Wai-Yee; Senjanović, Goran (1983). "Majorana Neutrinos and the Production of the Right-Handed Charged Gauge Boson". *Physical Review Letters* **50** (19): 1427. Bibcode:1983PhRvL..50.1427K. doi:10.1103/PhysRevLett.50.1427. (subscription required (help)).

[6] Tello, Vladimir; Nemevšek, Miha; Nesti, Fabrizio; Senjanović, Goran; Vissani, Francesco (2011). "Left-Right Symmetry: from LHC to Neutrinoless Double Beta Decay". *Physical Review Letters* **106** (15): 151801. arXiv:1011.3522. Bibcode:2011PhRvL.1 06o1801T.doi:10.1103/PhysRevLett.106.151801. (subscription required(help)).

[7] Kayser, Boris; Goldhaber, Alfred S. (1983). "CPT and CP properties of Majorana particles, and the consequences". *Physical Review D* **28** (9): 2341–2344. Bibcode:1983PhRvD..28.2341K. doi:10.1103/PhysRevD.28.2341. (subscription required (help)).

[8] Radescu, E. E. (1985). "On the electromagnetic properties of Majorana fermions". *Physical Review D* **32** (5): 1266–1268. Bibcode:1985PhRvD..32.1266R. doi:10.1103/PhysRevD.32.1266. (subscription required (help)).

[9] Boudjema, F.; Hamzaoui, C.; Rahal, V.; Ren, H. C. (1989). "Electromagnetic Properties of Generalized Majorana Particles". *Physical Review Letters* **62** (8): 852–854. Bibcode:1989PhRvL..62..852B. doi:10.1103/PhysRevLett.62.852. (subscription required (help)).

[10] Pospelov, Maxim; ter Veldhuis, Tonnis (2000). "Direct and indirect limits on the electro-magnetic form factors of WIMPs". *Physics Letters B* **480**: 181–186. arXiv:hep-ph/0003010. Bibcode:2000PhLB..480..181P. doi:10.1016/S0370-2693(00)00358-0.

[11] Ho, Chiu Man; Scherrer, Robert J. (2013). "Anapole Dark Matter". *Physics Letters B* **722** (8): 341–346. arXiv:1211.0503. Bibcode:2013PhLB..722..341H. doi:10.1016/j.physletb.2013.04.039.

[12] Wilczek, Frank (2009)."Majorana returns"(PDF).*Nature Physics***5**(9): 614–618.Bibcode:2009NatPh...5..614W.doi:10.103

[13] Nayak, Chetan; Simon, Steven H.; Stern, Ady; Freedman, Michael; Das Sarma, Sankar (2008). "Non-Abelian anyons and topological quantum computation". *Reviews of Modern Physics* **80** (3): 1083. arXiv:0707.1889. Bibcode:2008RvMP...80.1083N. doi:10.1103/RevModPhys.80.1083.

[14] N.B. Kopnin; M.M. Salomaa (1991). "Mutual friction in superfluid ^{3}He: Effects of bound states in the vortex core". *Physical Review B* **44** (17): 9667. Bibcode:1991PhRvB..44.9667K. doi:10.1103/PhysRevB.44.9667.

[15] Volovik, G. E. (1999). "Fermion zero modes on vortices in chiral superconductors". *JETP Letters* **70** (9): 609–614. arXiv:cond-mat/9909426. Bibcode:1999JETPL..70..609V. doi:10.1134/1.568223.

[16] Read, N.; Green, Dmitry (2000). "Paired states of fermions in two dimensions with breaking of parity and time-reversal symmetries and the fractional quantum Hall effect". *Physical Review B* **61** (15): 10267. arXiv:cond-mat/9906453. Bibcode:2000PhRvB ..6110267R.doi:10.1103/PhysRevB.61.10267.

[17] Kitaev, A. Yu (2001). "Unpaired Majorana fermions in quantum wires". *Physics-Uspekhi (supplement)* **44** (131): 131. arXiv:cond-mat/0010440. Bibcode:2001PhyU...44..131K. doi:10.1070/1063-7869/44/10S/S29.

[18] Moore, Gregory; Read, Nicholas (August 1991). "Nonabelions in the fractional quantum Hall effect". *Nuclear Physics B* **360** (2–3): 362. Bibcode:1991NuPhB.360..362M. doi:10.1016/0550-3213(91)90407-O.

[19] Fu, Liang; Kane, Charles L. (2008). "Superconducting Proximity Effect and Majorana Fermions at the Surface of a Topological Insulator".*Physical Review Letters***10**(9): 096407.arXiv:0707.1692.Bibcode:2008PhRvL.100i6407F.doi:10.1103/PhysRev

[20] Fu, Liang; Kane, Charles L. (2009). "Josephson current and noise at a superconductor/quantum-spin-Hall-insulator/superconductor junction". *Physical Review B* **79** (16): 161408. arXiv:0804.4469. Bibcode:2009PhRvB..79p1408F. doi:10.1103/PhysRevB.79 .161408.(subscription required(help)).

[21] Alicea, Jason (2012). "New directions in the pursuit of Majorana fermions in solid state systems". *Reports on Progress in Physics* **75** (7): 076501. arXiv:1202.1293. Bibcode:2012RPPh...75g6501A. doi:10.1088/0034-4885/75/7/076501. PMID 22790778. (subscription required (help)).

[22] Beenakker, C. W. J. (April 2013). "Search for Majorana fermions in superconductors". *Annual Review of Condensed Matter Physics* **4** (113): 113–136. arXiv:1112.1950. Bibcode:2013ARCMP...4..113B. doi:10.1146/annurev-conmatphys-030212-184337. (subscription required (help)).

[23] Reich, Eugenie Samuel (28 February 2012). "Quest for quirky quantum particles may have struck gold". *Nature News.* doi:10.1038/nature.2012.10124.

[24] Amos, Jonathan (13 April 2012). "Majorana particle glimpsed in lab". *BBC News.* Retrieved 15 April 2012.

[25] Mourik, V.; Zuo, K.; Frolov, S. M.; Plissard, S. R.; Bakkers, E. P. A. M.; Kouwenhoven, L. P. (12 April 2012). "Signatures of Majorana fermions in hybrid superconductor-semiconductor nanowire devices". *Science* **336** (6084): 1003–1007. arXiv:1204.2792. Bibcode:2012Sci...336.1003M. doi:10.1126/science.1222360.

[26] Deng, M.T.; Yu, C.L.; Huang, G.Y.; Larsson, M.; Caroff, P.; Xu, H.Q. (28 November 2012). "Anomalous zero-bias conductance peak in a Nb-InSb nanowire-Nb hybrid device". *Nano Letters* **12** (12): 6414–6419. Bibcode:2012NanoL..12.6414D. doi:10.1021/nl303758w.

[27] Das, A.; Ronen, Y.; Most, Y.; Oreg, Y.; Heiblum, M.; Shtrikman, H. (11 November 2012). "Zero-bias peaks and splitting in an Al-InAs nanowire topological superconductor as a signature of Majorana fermions.". *Nature Physics* **8** (12): 887–895. arXiv:1205.7073. Bibcode:2012NatPh...8..887D. doi:10.1038/nphys2479.

[28] Churchill, H. O. H.; Fatemi, V.; Grove-Rasmussen, K.; Deng, M.T.; Caroff, P.; Xu, H.Q.; Marcus, C.M. (6 June 2013). "Superconductor-nanowire devices from tunneling to the multichannel regime: Zero-bias oscillations and magnetoconductance crossover".*PHYSICAL REVIEW B***87**(24):241401(R).arXiv:1303.2407.Bibcode:2013PhRvB..87x1401C.doi:10.1103/Phys

[29] Deng, M.T.; Yu, C.L.; Huang, G.Y.; Larsson, Marcus; Caroff, P.; Xu, H.Q. (11 November 2014). "Parity independence of the zero-bias conductance peak in a nanowire based topological superconductor-quantum dot hybrid device". *Scientific Reports* **4**: 7261. arXiv:1406.4435. Bibcode:2014NatSR...4E7261D. doi:10.1038/srep07261.

[30] Lutchyn, Roman M.; Sau, Jay D.; Das Sarma, S. (August 2010). "Majorana Fermions and a Topological Phase Transition in Semiconductor-Superconductor Heterostructures". *Physical Review Letters* **105** (7): 077001. arXiv:1002.4033. Bibcode:2010Ph RvL.105g7001L.doi:10.1103/PhysRevLett.105.077001.

[31] Oreg, Yuval; Refael, Gil; von Oppen, Felix (October 2010). "Helical Liquids and Majorana Bound States in Quantum Wires". *Physical Review Letters***105**(17): 177002.arXiv:1003.1145.Bibcode:2010PhRvL.105q7002O.doi:10.1103/PhysRevLett.105

[32] Lee, E. J. H.; Jiang, X.; Houzet, M.; Aguado, R.; Lieber, C.M.; Franceschi, S.D. (15 December 2013). "Spin-resolved Andreev levels and parity crossings in hybrid superconductor–semiconductor nanostructures". *Nature Nanotechnology* **9**: 79–84. arXiv:1302.2611. Bibcode:2014NatNa...9...79L. doi:10.1038/nnano.2013.267.

[33] Nadj-Perge, Stevan; Drozdov, Ilya K.; Li, Jian; Chen, Hua; Jeon, Sangjun; Seo, Jungpil; MacDonald, Allan H.; Bernevig, B. Andrei; Yazdani, Ali (2 October 2014). "Observation of Majorana fermions in ferromagnetic atomic chains on a superconductor". Science. arXiv:1410.3453. Bibcode:2014Sci...346..602N. doi:10.1126/science.1259327. (subscription required (help)).

[34] "Majorana fermion: Physicists observe elusive particle that is its own antiparticle". Phys.org. October 2, 2014. Retrieved 3 October 2014.

[35] "New Particle Is Both Matter and Antimatter". *Scientific American*. October 2, 2014. Retrieved 3 October 2014.

32.5   Further reading

- Pal, Palash B. (2011) [12 October 2010]. "Dirac, Majorana and Weyl fermions". *American Journal of Physics* **79** (5): 485. arXiv:1006.1718. Bibcode:2011AmJPh..79..485P. doi:10.1119/1.3549729. (subscription required (help)).

Chapter 33

Skyrmion

In particle theory, the **skyrmion** (/ˈskɜrmi.ɒn/) is a hypothetical particle related originally[1] to baryons. It was described by Tony Skyrme and consists of a quantum superposition of baryons and resonance states.[2] It could be predicted from some nuclear matter properties.[3]

Skyrmions as topological objects are important in solid state physics, especially in the emerging technology of spintronics. A two-dimensional magnetic skyrmion, as a topological object, is formed, e.g., from a 3D effective-spin "hedgehog" (in the field of micromagnetics: out of a so-called "Bloch point" singularity of homotopy degree +1) by a stereographic projection, whereby the positive north-pole spin is mapped onto a far-off edge circle of a 2D-disk, while the negative south-pole spin is mapped onto the center of the disk.

33.1 Mathematical definition

In field theory, skyrmions are homotopically non-trivial classical solutions of a nonlinear sigma model with a non-trivial target manifold topology – hence, they are topological solitons. An example occurs in chiral models[4] of mesons, where the target manifold is a homogeneous space of the structure group

$$\left(\frac{SU(N)_L \times SU(N)_R}{SU(N)_{\text{diag}}} \right)$$

where $SU(N)L$ and $SU(N)R$ are the left and right parts of the $SU(N)$ matrix, and $SU(N)_{\text{diag}}$ is the diagonal subgroup.

If spacetime has the topology $S^3 \times \mathbf{R}$, then classical configurations can be classified by an integral winding number[5] because the third homotopy group

$$\pi_3 \left(\frac{SU(N)_L \times SU(N)_R}{SU(N)_{\text{diag}}} \cong SU(N) \right)$$

is equivalent to the ring of integers, with the congruence sign referring to homeomorphism.

A topological term can be added to the chiral Lagrangian, whose integral depends only upon the homotopy class; this results in superselection sectors in the quantised model. A skyrmion can be approximated by a soliton of the Sine-Gordon equation; after quantisation by the Bethe ansatz or otherwise, it turns into a fermion interacting according to the massive Thirring model.

Skyrmions have been reported, but not conclusively proven, to be in Bose-Einstein condensates,[6] superconductors,[7] thin magnetic films[8] and in chiral nematic liquid crystals.[9]

33.2 Magnetic materials/data storage

One particular form of skyrmions is found in magnetic materials that break the inversion symmetry and where the Dzyaloshinskii-Moriya interaction plays an important role. They form "domains" as small as a 1 nm (e.g. in Fe on

Ir(111)).[10] The small size and low energy consumption of magnetic skyrmions make them a good candidate for future data storage solutions and other spintronics devices.[11][12][13] Researchers could read and write skyrmions using scanning tunneling microscopy.[14] The topological charge, representing the existence and non-existence of skyrmions, can represent the bit states "1" and "0". Room temperature skyrmions were reported.[15][16]

Skyrmions operate at magnetic fields that are several orders of magnitude weaker than conventional magnetic devices. In 2015 a practical way to create and access magnetic skyrmions, under ambient room-temperature conditions was announced. The device used arrays of magnetized cobalt disks as artificial Bloch skyrmion lattices atop a thin film of cobalt and palladium. Asymmetric magnetic nanodots were patterned with controlled circularity on an underlayer with perpendicular magnetic anisotropy (PMA). Polarity is controlled by a tailored magnetic field sequence and demonstrated in magnetometry measurements. The vortex structure is imprinted into the underlayer's interfacial region via suppressing the PMA by a critical ion-irradiation step. The lattices are identified with polarized neutron reflectometry and confirmed by magnetoresistance measurements.[17][18]

33.3 External links

- Developments in Magnetic Skyrmions Come in Bunches, IEEE Spectrum 2015 web article

33.4 References

[1] At later stages the model was also related to mesons.

[2] Wong, Stephen (2002). "What exactly is a Skyrmion?". arXiv:hep-ph/0202250 [hep/ph].

[3] M.R.Khoshbin-e-Khoshnazar,"Correlated Quasiskyrmions as Alpha Particles",*Eur.Phys.J.A* **14**,207-209 (2002).

[4] Chiral models stress the difference between "left-handedness" and "right-handedness".

[5] The same classification applies to the mentioned effective-spin "hedgehog" singularity": spin upwards at the northpole, but downward at the southpole.
See also Döring, W. (1968). "Point Singularities in Micromagnetism". *Journal of Applied Physics* **39** (2): 1006. Bibcode:1968JAP ....39.1006D.doi:10.1063/1.1656144.

[6] Al Khawaja, Usama; Stoof, Henk (2001). "Skyrmions in a ferromagnetic Bose–Einstein condensate". *Nature* **411** (6840): 918–20. Bibcode:2001Natur.411..918A. doi:10.1038/35082010. PMID 11418849.

[7] Baskaran, G. (2011). "Possibility of Skyrmion Superconductivity in Doped Antiferromagnet $K_2Fe_4Se_5$". arXiv:1108.3562 [cond-mat.supr-con].

[8] Kiselev, N. S.; Bogdanov, A. N.; Schäfer, R.; Rößler, U. K. (2011). "Chiral skyrmions in thin magnetic films: New objects for magnetic storage technologies?". *Journal of Physics D: Applied Physics* **44** (39): 392001. arXiv:1102.2726. Bibcode:2011JPhD. ..44M2001K.doi:10.1088/0022-3727/44/39/392001.

[9] Fukuda, J.-I.; Žumer, S. (2011). "Quasi-two-dimensional Skyrmion lattices in a chiral nematic liquid crystal". *Nature Communications* **2**: 246. Bibcode:2011NatCo...2E.246F. doi:10.1038/ncomms1250. PMID 21427717.

[10] Heinze, Stefan; Von Bergmann, Kirsten; Menzel, Matthias; Brede, Jens; Kubetzka, André; Wiesendanger, Roland; Bihlmayer, Gustav; Blügel, Stefan (2011). "Spontaneous atomic-scale magnetic skyrmion lattice in two dimensions". *Nature Physics* **7** (9): 713–718. Bibcode:2011NatPh...7..713H. doi:10.1038/NPHYS2045. Lay summary (Jul 31, 2011).

[11] A. Fert, V. Cros, and J. Sampaio (2013). "Skyrmions on the track".*Nature Nanotechnology***8**: 152–156.doi:10.1038/nnano.

[12] Y. Zhou, E. Iacocca, A.A. Awad, R.K. Dumas, F.C. Zhang, H.B. Braun and J. Akerman (2015). "Dynamically stabilized magnetic skyrmions". *Nature Communications* **6**: 8193. doi:10.1038/ncomms9193.

[13] X.C. Zhang, M. Ezawa, Y. Zhou (2014). "Magnetic skyrmion logic gates: conversion, duplication and merging of skyrmions". *Scientific Reports* **5**: 9400. doi:10.1038/srep09400.

[14] Romming, N.; Hanneken, C.; Menzel, M.; Bickel, J. E.; Wolter, B.; Von Bergmann, K.; Kubetzka, A.; Wiesendanger, R. (2013). "Writing and Deleting Single Magnetic Skyrmions". *Science* **341** (6146): 636–9. Bibcode:2013Sci...341..636R. doi:10.1126/science.1240573. PMID 23929977. Lay summary – *phys.org* (Aug 8, 2013).

[15] Jiang, Wanjun; Upadhyaya, Pramey; Zhang, Wei; Yu, Guoqiang; Jungfleisch, M. Benjamin; Fradin, Frank Y.; Pearson, John E.; Tserkovnyak, Yaroslav; Wang, Kang L. (2015-07-17). "Blowing magnetic skyrmion bubbles". *Science* **349** (6245): 283–286. doi:10.1126/science.aaa1442. ISSN 0036-8075. PMID 26067256.

[16] D.A. Gilbert, B.B. Maranville, A.L. Balk, B.J. Kirby, P. Fischer, D.T. Pierce, J. Unguris, J.A. Borchers, K. Liu (8 October 2015). "Realization of ground state artificial skyrmion lattices at room temperature". *Nature Communications* **6**: 8462. doi:10.1038/ncomms9462. Lay summary – *NIST*.

[17] Gilbert, Dustin A.; Maranville, Brian B.; Balk, Andrew L.; Kirby, Brian J.; Fischer, Peter; Pierce, Daniel T.; Unguris, John; Borchers, Julie A.; Liu, Kai (2015-10-08). "Realization of ground-state artificial skyrmion lattices at room temperature". *Nature Communications* **6**. doi:10.1038/ncomms9462.

[18] "A new way to create spintronic magnetic information storage | KurzweilAI". *www.kurzweilai.net*. October 9, 2015. Retrieved 2015-10-14.

Chapter 34

Soliton

For other uses, see Soliton (disambiguation).

In mathematics and physics, a **soliton** is a self-reinforcing solitary wave (a wave packet or pulse) that maintains its shape

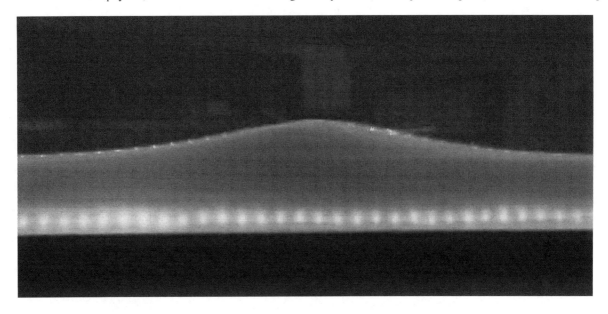

Solitary wave in a laboratory wave channel

while it propagates at a constant velocity. Solitons are caused by a cancellation of nonlinear and dispersive effects in the medium. (The term "dispersive effects" refers to a property of certain systems where the speed of the waves varies according to frequency). Solitons are the solutions of a widespread class of weakly nonlinear dispersive partial differential equations describing physical systems.

The soliton phenomenon was first described in 1834 by John Scott Russell (1808–1882) who observed a solitary wave in the Union Canal in Scotland. He reproduced the phenomenon in a wave tank and named it the "Wave of Translation".

34.1   Definition

A single, consensus definition of a soliton is difficult to find. Drazin & Johnson (1989, p. 15) ascribe three properties to solitons:

   1. They are of permanent form;

2. They are localized within a region;

3. They can interact with other solitons, and emerge from the collision unchanged, except for a phase shift.

More formal definitions exist, but they require substantial mathematics. Moreover, some scientists use the term *soliton* for phenomena that do not quite have these three properties (for instance, the 'light bullets' of nonlinear optics are often called solitons despite losing energy during interaction).[1] The shape of a soliton is described by the hyperbolic secant function.

34.2 Explanation

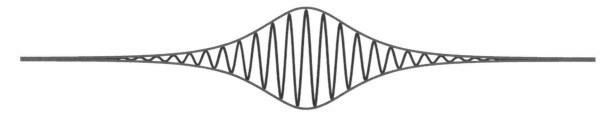

A hyperbolic secant (sech) envelope soliton for water waves. The blue line is the carrier signal, while the red line is the envelope soliton.

Dispersion and non-linearity can interact to produce permanent and localized wave forms. Consider a pulse of light traveling in glass. This pulse can be thought of as consisting of light of several different frequencies. Since glass shows dispersion, these different frequencies will travel at different speeds and the shape of the pulse will therefore change over time. However, there is also the non-linear Kerr effect: the refractive index of a material at a given frequency depends on the light's amplitude or strength. If the pulse has just the right shape, the Kerr effect will exactly cancel the dispersion effect, and the pulse's shape will not change over time: a soliton. See soliton (optics) for a more detailed description.

Many exactly solvable models have soliton solutions, including the Korteweg–de Vries equation, the nonlinear Schrödinger equation, the coupled nonlinear Schrödinger equation, and the sine-Gordon equation. The soliton solutions are typically obtained by means of the inverse scattering transform and owe their stability to the integrability of the field equations. The mathematical theory of these equations is a broad and very active field of mathematical research.

Some types of tidal bore, a wave phenomenon of a few rivers including the River Severn, are 'undular': a wavefront followed by a train of solitons. Other solitons occur as the undersea internal waves, initiated by seabed topography, that propagate on the oceanic pycnocline. Atmospheric solitons also exist, such as the Morning Glory Cloud of the Gulf of Carpentaria, where pressure solitons traveling in a temperature inversion layer produce vast linear roll clouds. The recent and not widely accepted soliton model in neuroscience proposes to explain the signal conduction within neurons as pressure solitons.

A topological soliton, also called a topological defect, is any solution of a set of partial differential equations that is stable against decay to the "trivial solution". Soliton stability is due to topological constraints, rather than integrability of the field equations. The constraints arise almost always because the differential equations must obey a set of boundary conditions, and the boundary has a non-trivial homotopy group, preserved by the differential equations. Thus, the differential equation solutions can be classified into homotopy classes.

There is no continuous transformation that will map a solution in one homotopy class to another. The solutions are truly distinct, and maintain their integrity, even in the face of extremely powerful forces. Examples of topological solitons include the screw dislocation in a crystalline lattice, the Dirac string and the magnetic monopole in electromagnetism, the Skyrmion and the Wess–Zumino–Witten model in quantum field theory, the magnetic skyrmion in condensed matter physics, and cosmic strings and domain walls in cosmology.

34.3 History

In 1834, John Scott Russell describes his *wave of translation*.[nb 1] The discovery is described here in Scott Russell's own words:[nb 2]

> I was observing the motion of a boat which was rapidly drawn along a narrow channel by a pair of horses, when the boat suddenly stopped – not so the mass of water in the channel which it had put in motion; it accumulated round the prow of the vessel in a state of violent agitation, then suddenly leaving it behind, rolled forward with great velocity, assuming the form of a large solitary elevation, a rounded, smooth and well-defined heap of water, which continued its course along the channel apparently without change of form or diminution of speed. I followed it on horseback, and overtook it still rolling on at a rate of some eight or nine miles an hour, preserving its original figure some thirty feet long and a foot to a foot and a half in height. Its height gradually diminished, and after a chase of one or two miles I lost it in the windings of the channel. Such, in the month of August 1834, was my first chance interview with that singular and beautiful phenomenon which I have called the Wave of Translation.[2]

Scott Russell spent some time making practical and theoretical investigations of these waves. He built wave tanks at his home and noticed some key properties:

- The waves are stable, and can travel over very large distances (normal waves would tend to either flatten out, or steepen and topple over)

- The speed depends on the size of the wave, and its width on the depth of water.

- Unlike normal waves they will never merge – so a small wave is overtaken by a large one, rather than the two combining.

- If a wave is too big for the depth of water, it splits into two, one big and one small.

Scott Russell's experimental work seemed at odds with Isaac Newton's and Daniel Bernoulli's theories of hydrodynamics. George Biddell Airy and George Gabriel Stokes had difficulty accepting Scott Russell's experimental observations because they could not be explained by the existing water wave theories. Their contemporaries spent some time attempting to extend the theory but it would take until the 1870s before Joseph Boussinesq and Lord Rayleigh published a theoretical treatment and solutions.[nb 3] In 1895 Diederik Korteweg and Gustav de Vries provided what is now known as the Korteweg–de Vries equation, including solitary wave and periodic cnoidal wave solutions.[3][nb 4]

In 1965 Norman Zabusky of Bell Labs and Martin Kruskal of Princeton University first demonstrated soliton behavior in media subject to the Korteweg–de Vries equation (KdV equation) in a computational investigation using a finite difference approach. They also showed how this behavior explained the puzzling earlier work of Fermi, Pasta and Ulam.[5]

In 1967, Gardner, Greene, Kruskal and Miura discovered an inverse scattering transform enabling analytical solution of the KdV equation.[6] The work of Peter Lax on Lax pairs and the Lax equation has since extended this to solution of many related soliton-generating systems.

Note that solitons are, by definition, unaltered in shape and speed by a collision with other solitons.[7] So solitary waves on a water surface are *near*-solitons, but not exactly – after the interaction of two (colliding or overtaking) solitary waves, they have changed a bit in amplitude and an oscillatory residual is left behind.[8]

34.4 Solitons in fibre optics

See also: Soliton (optics)

Much experimentation has been done using solitons in fibre optics applications. Solitons in a fibre optic system are described by the Manakov equations. Solitons' inherent stability make long-distance transmission possible without the use of repeaters, and could potentially double transmission capacity as well.[9]

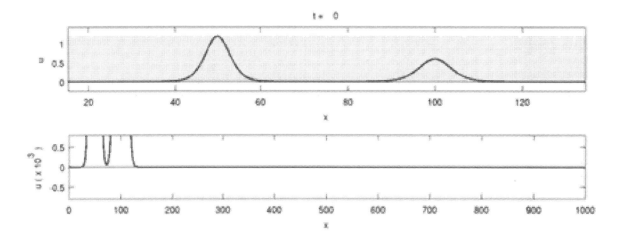

An animation of the overtaking of two solitary waves according to the Benjamin–Bona–Mahony equation – or BBM equation, a model equation for (among others) long surface gravity waves. The wave heights of the solitary waves are 1.2 and 0.6, respectively, and their velocities are 1.4 and 1.2.

The upper graph is for a frame of reference moving with the average velocity of the solitary waves.

The lower graph (with a different vertical scale and in a stationary frame of reference) shows the oscillatory tail produced by the interaction.[4] Thus, the solitary wave solutions of the BBM equation are not solitons.

34.5   Solitons in biology

Solitons may occur in proteins[13] and DNA.[14] Solitons are related to the low-frequency collective motion in proteins and DNA.[15]

A recently developedmodel in neuroscienceproposes that signals are conducted within neurons in the form of solitons.

34.6   Solitons in magnets

In magnets, there also exist different types of solitons and other nonlinear waves.[19] These magnetic solitons are an exact solution of classical nonlinear differential equations — magnetic equations, e.g. the Landau–Lifshitz equation, continuum Heisenberg model, Ishimori equation, nonlinear Schrödinger equation and others.

34.7   Bions

The bound state of two solitons is known as a *bion,* or in systems where the bound state periodically oscillates, a *"breather."*

In field theory *Bion* usually refers to the solution of the Born–Infeld model. The name appears to have been coined by G. W. Gibbons in order to distinguish this solution from the conventional soliton, understood as a *regular*, finite-energy (and usually stable) solution of a differential equation describing some physical system.[20] The word *regular* means a smooth solution carrying no sources at all. However, the solution of the Born–Infeld model still carries a source in the form of a Dirac-delta function at the origin. As a consequence it displays a singularity in this point (although the electric field is everywhere regular). In some physical contexts (for instance string theory) this feature can be important, which motivated the introduction of a special name for this class of solitons.

On the other hand, when gravity is added (i.e. when considering the coupling of the Born–Infeld model to general relativity) the corresponding solution is called *EBIon*, where "E" stands for Einstein.

34.8   See also

- Compacton, a soliton with compact support

- Freak waves may be a Peregrine soliton related phenomenon involving breather waves which exhibit concentrated localized energy with non-linear properties.[21]

- Nematicons

- Oscillons

- Peakon, a soliton with a non-differentiable peak

- Soliton (topological)

- Non-topological soliton, in Quantum Field Theory

- Q-ball a non-topological soliton

- Soliton model of nerve impulse propagation

- Topological quantum number

- Sine-Gordon equation

- Nonlinear Schrödinger equation

- Vector soliton

- Soliton distribution

- Soliton hypothesis for ball lightning, by David Finkelstein

- Pattern formation

34.9   Notes

[1] "Translation" here means that there is real mass transport, although it is not the same water which is transported from one end of the canal to the other end by this "Wave of Translation". Rather, a fluid parcel acquires momentum during the passage of the solitary wave, and comes to rest again after the passage of the wave. But the fluid parcel has been displaced substantially forward during the process – by Stokes drift in the wave propagation direction. And a net mass transport is the result. Usually there is little mass transport from one side to another side for ordinary waves.

[2] This passage has been repeated in many papers and books on soliton theory.

[3] Lord Rayleigh published a paper in Philosophical Magazine in 1876 to support John Scott Russell's experimental observation with his mathematical theory. In his 1876 paper, Lord Rayleigh mentioned Scott Russell's name and also admitted that the first theoretical treatment was by Joseph Valentin Boussinesq in 1871. Joseph Boussinesq mentioned Russell's name in his 1871 paper. Thus Scott Russell's observations on solitons were accepted as true by some prominent scientists within his own lifetime of 1808–1882.

[4] Korteweg and de Vries did not mention John Scott Russell's name at all in their 1895 paper but they did quote Boussinesq's paper of 1871 and Lord Rayleigh's paper of 1876. The paper by Korteweg and de Vries in 1895 was not the first theoretical treatment of this subject but it was a very important milestone in the history of the development of soliton theory.

34.10   References

[1] "Light bullets".

[2] Scott Russell, J. (1844). "Report on waves". *Fourteenth meeting of the British Association for the Advancement of Science.*

[3] Korteweg, D. J.; de Vries, G. (1895). "On the Change of Form of Long Waves advancing in a Rectangular Canal and on a New Type of Long Stationary Waves". *Philosophical Magazine* **39**: 422–443. doi:10.1080/14786449508620739.

[4] Bona, J. L.; Pritchard, W. G.; Scott, L. R. (1980). "Solitary-wave interaction".*Physics of Fluids***23**(3): 438–441.Bibcode:1980 doi:10.1063/1.863011.

[5] Zabusky & Kruskal (1965)

[6] Gardner, Clifford S.; Greene, John M.; Kruskal, Martin D.; Miura, Robert M. (1967). "Method for Solving the Korteweg–deVries Equation".*Physical Review Letters***19**(19): 1095–1097.Bibcode:1967PhRvL..19.1095G.doi:10.1103/PhysRevLett.

[7] Remoissenet, M. (1999). *Waves called solitons: Concepts and experiments.* Springer. p. 11. ISBN 9783540659198.

[8] See e.g.:
   • Maxworthy, T. (1976). "Experiments on collisions between solitary waves". *Journal of Fluid Mechanics* **76** (1): 177–186. Bibcode:1976JFM....76..177M. doi:10.1017/S0022112076003194.
   • Fenton, J.D.; Rienecker, M.M. (1982). "A Fourier method for solving nonlinear water-wave problems: application to solitary-wave interactions". *Journal of Fluid Mechanics* **118**: 411–443. Bibcode:1982JFM...118..411F. doi:10.1017/S0022112082001141.
   • Craig, W.; Guyenne, P.; Hammack, J.; Henderson, D.; Sulem, C. (2006). "Solitary water wave interactions". *Physics of Fluids* **18** (057106): 25 pp. Bibcode:2006PhFl...18e7106C. doi:10.1063/1.2205916.

[9] "Photons advance on two fronts". EETimes.com. October 24, 2005. Retrieved 2011-02-15.

[10] Fred Tappert (January 29, 1998). "Reminiscences on Optical Soliton Research with Akira Hasegawa" (PDF).

[11] Cundiff, S. T.; Collings, B. C.; Akhmediev, N. N.; Soto-Crespo, J. M.; Bergman, K.; Knox, W. H. (1999). "Observation of Polarization-Locked Vector Solitons in an Optical Fibre". *Physical Review Letters* **82** (20): 3988. Bibcode:1999PhRvL..82.3988C. doi:10.1103/PhysRevLett.82.3988.

[12] Tang, D. Y.; Zhang, H.; Zhao, L. M.; Wu, X. (2008). "Observation of high-order polarization-locked vector solitons in a fibre laser". *Physical Review Letters* **101** (15): 153904. Bibcode:2008PhRvL.101o3904T. doi:10.1103/PhysRevLett.101.153904. PMID 18999601.

[13] Davydov, Aleksandr S. (1991). *Solitons in molecular systems.* Mathematics and its applications (Soviet Series) **61** (2nd ed.). Kluwer Academic Publishers. ISBN 0-7923-1029-2.

[14] Yakushevich, Ludmila V. (2004). *Nonlinear physics of DNA* (2nd revised ed.). Wiley-VCH. ISBN 3-527-40417-1.

[15] Sinkala, Z. (August 2006). "Soliton/exciton transport in proteins".*J. Theor. Biol.***241**(4): 919–27.doi:10.1016/j.jtbi.2006.01 PMID 16516929.

[16] Heimburg, T., Jackson, A.D. (12 July 2005). "On soliton propagation in biomembranes and nerves". *Proc. Natl. Acad. Sci. U.S.A.* **102** (2): 9790. Bibcode:2005PNAS..102.9790H. doi:10.1073/pnas.0503823102.

[17] Heimburg, T., Jackson, A.D. (2007). "On the action potential as a propagating density pulse and the role of anesthetics". *Biophys. Rev. Lett.* **2**: 57–78. arXiv:physics/0610117. Bibcode:2006physics..10117H. doi:10.1142/S179304800700043X.

[18] Andersen, S.S.L., Jackson, A.D., Heimburg, T. (2009). "Towards a thermodynamic theory of nerve pulse propagation". *Progr. Neurobiol.* **88** (2): 104–113. doi:10.1016/j.pneurobio.2009.03.002.

[19] Kosevich, A. M.; Gann, V. V.; Zhukov, A. I.; Voronov, V. P. (1998). "Magnetic soliton motion in a nonuniform magnetic field". *Journal of Experimental and Theoretical Physics* **87** (2): 401–407. Bibcode:1998JETP...87..401K. doi:10.1134/1.558674.

[20] Gibbons, G. W. (1998). "Born–Infeld particles and Dirichlet *p*-branes". *Nuclear Physics B* **514** (3): 603–639. arXiv:hep-th/9709027. Bibcode:1998NuPhB.514..603G. doi:10.1016/S0550-3213(97)00795-5.

[21] Powell, Devin (20 May 2011). "Rogue Waves Captured". Science News. Retrieved 24 May 2011.

34.11   Further reading

- Zabusky, N. J.; Kruskal, M. D. (1965). "Interaction of 'solitons' in a collisionless plasma and the recurrence of initial states". *Phys. Rev. Lett.* **15** (6): 240–243. Bibcode:1965PhRvL..15..240Z. doi:10.1103/PhysRevLett.15.240.

- Hasegawa, A.; Tappert, F. (1973). "Transmission of stationary nonlinear optical pulses in dispersive dielectric fibers. I. Anomalous dispersion". *Appl. Phys. Lett.* **23** (3): 142–144. Bibcode:1973ApPhL..23..142H. doi:10.1063/1.1654836.

- Emplit, P.; Hamaide, J. P.; Reynaud, F.; Froehly, C.; Barthelemy, A. (1987). "Picosecond steps and dark pulses through nonlinear single mode fibers". *Optics Comm.* **62** (6): 374–379. Bibcode:1987OptCo..62..374E. doi:10.1016/0030-4018(87)90003-4.

- Tao, Terence (2009). "Why are solutions stable?". *Bull. Am. Math. Soc.* **46** (1): 1–33. MR 2457070.

- Drazin, P. G.; Johnson, R. S. (1989). *Solitons: an introduction* (2nd ed.). Cambridge University Press. ISBN 0-521-33655-4.

- Dunajski, M. (2009). *Solitons, Instantons and Twistors.* Oxford University Press. ISBN 978-0-19-857063-9.

- Jaffe, A.; Taubes, C. H. (1980). *Vortices and monopoles.* Birkhauser. ISBN 0-8176-3025-2.

- Manton, N.; Sutcliffe, P. (2004). *Topological solitons.* Cambridge University Press. ISBN 0-521-83836-3.

- Mollenauer, Linn F.; Gordon, James P. (2006). *Solitons in optical fibers.* Elsevier Academic Press. ISBN 0-12-504190-X.

- Rajaraman, R. (1982). *Solitons and instantons.* North-Holland. ISBN 0-444-86229-3.

- Yang, Y. (2001). *Solitons in field theory and nonlinear analysis.* Springer-Verlag. ISBN 0-387-95242-X.

34.12   External links

Related to John Scott Russell

- John Scott Russell and the solitary wave

- John Scott Russell biography

- Photograph of soliton on the Scott Russell Aqueduct

Other

- Heriot–Watt University soliton page

- The many faces of solitons

- Helmholtz solitons, Salford University

- Soliton in electrical engineering

- Miura's home page

- Short didactic review on optical solitons

- Solitons & nonlinear wave equations

- Star Trek's solitons are real

Chapter 35

Fermionic condensate

A **fermionic condensate** is a superfluid phase formed by fermionic particles at low temperatures. It is closely related to the Bose–Einstein condensate, a superfluid phase formed by bosonic atoms under similar conditions. Unlike the Bose–Einstein condensates, fermionic condensates are formed using fermions instead of bosons. The earliest recognized fermionic condensate described the state of electrons in a superconductor; the physics of other examples including recent work with fermionic atoms is analogous. The first atomic fermionic condensate was created by a team led by Deborah S. Jin in 2003. A **chiral condensate** is an example of a fermionic condensate that appears in theories of massless fermions with chiral symmetry breaking.

35.1 Background

35.1.1 Superfluidity

Fermionic condensates are attained at temperatures lower than Bose–Einstein condensates. Fermionic condensates are a type of superfluid. As the name suggests, a superfluid possesses fluid properties similar to those possessed by ordinary liquids and gases, such as the lack of a definite shape and the ability to flow in response to applied forces. However, superfluids possess some properties that do not appear in ordinary matter. For instance, they can flow at low velocities without dissipating any energy—i.e. zero viscosity. At higher velocities, energy is dissipated by the formation of quantized vortices, which act as "holes" in the medium where superfluidity breaks down.

Superfluidity was originally discovered in liquid helium-4, in 1938, by Pyotr Kapitsa, John Allen and Don Misener. Superfluidity in helium-4, which occurs at temperatures below 2.17 kelvins (K), has long been understood to result from Bose condensation, the same mechanism that produces the Bose–Einstein condensates. The primary difference between superfluid helium and a Bose–Einstein condensate is that the former is condensed from a liquid while the latter is condensed from a gas.

35.1.2 Fermionic superfluids

It is far more difficult to produce a fermionic superfluid than a bosonic one, because the Pauli exclusion principle prohibits fermions from occupying the same quantum state. However, there is a well-known mechanism by which a superfluid may be formed from fermions. This is the BCS transition, discovered in 1957 by John Bardeen, Leon Cooper and Robert Schrieffer for describing superconductivity. These authors showed that, below a certain temperature, electrons (which are fermions) can pair up to form bound pairs now known as Cooper pairs. As long as collisions with the ionic lattice of the solid do not supply enough energy to break the Cooper pairs, the electron fluid will be able to flow without dissipation. As a result, it becomes a superfluid, and the material through which it flows a superconductor.

The BCS theory was phenomenally successful in describing superconductors. Soon after the publication of the BCS paper, several theorists proposed that a similar phenomenon could occur in fluids made up of fermions other than electrons, such as helium-3 atoms. These speculations were confirmed in 1971, when experiments performed by Douglas D. Osheroff showed that helium-3 becomes a superfluid below 0.0025 K. It was soon verified that the superfluidity of helium-3 arises from a BCS-like mechanism. (The theory of superfluid helium-3 is a little more complicated than the BCS theory of superconductivity. These complications arise because helium atoms repel each other much more strongly than electrons, but the basic idea is the same.)

35.1.3   Creation of the first fermionic condensates

When Eric Cornell and Carl Wieman produced a Bose–Einstein condensate from rubidium atoms in 1995, there naturally arose the prospect of creating a similar sort of condensate made from fermionic atoms, which would form a superfluid by the BCS mechanism. However, early calculations indicated that the temperature required for producing Cooper pairing in atoms would be too cold to achieve. In 2001, Murray Holland at JILA suggested a way of bypassing this difficulty. He speculated that fermionic atoms could be coaxed into pairing up by subjecting them to a strong magnetic field.

In 2003, working on Holland's suggestion, Deborah Jin at JILA, Rudolf Grimm at the University of Innsbruck, and Wolfgang Ketterle at MIT managed to coax fermionic atoms into forming molecular bosons, which then underwent Bose–Einstein condensation. However, this was not a true fermionic condensate. On December 16, 2003, Jin managed to produce a condensate out of fermionic atoms for the first time. The experiment involved 500,000 potassium−40 atoms cooled to a temperature of 5×10^{-8} K, subjected to a time-varying magnetic field. The findings were published in the online edition of *Physical Review Letters* on January 24, 2004.

35.2   Examples

35.2.1   BCS theory

The BCS theory of superconductivity has a fermion condensate. A pair of electrons in a metal, with opposite spins can form a scalar bound state called a Cooper pair. Then, the bound states themselves form a condensate. Since the Cooper pair has electric charge, this fermion condensate breaks the electromagnetic gauge symmetry of a superconductor, giving rise to the wonderful electromagnetic properties of such states.

35.2.2   QCD

In quantum chromodynamics (QCD) the chiral condensate is also called the **quark condensate**. This property of the QCD vacuum is partly responsible for giving masses to hadrons (along with other condensates like the gluon condensate).

In an approximate version of QCD, which has vanishing quark masses for N quark flavours, there is an exact chiral SU(N) $\times$ SU(N) symmetry of the theory. The QCD vacuum breaks this symmetry to SU(N) by forming a quark condensate. The existence of such a fermion condensate was first shown explicitly in the lattice formulation of QCD. The quark condensate is therefore an order parameter of transitions between several phases of quark matter in this limit.

This is very similar to the BCS theory of superconductivity. The Cooper pairs are analogous to the pseudoscalar mesons. However, the vacuum carries no charge. Hence all the gauge symmetries are unbroken. Corrections for the masses of the quarks can be incorporated using chiral perturbation theory.

35.2.3   Helium-3 superfluid

A helium-3 atom is a fermion and at very low temperatures, they form two-atom Cooper pairs which are bosonic and condense into a superfluid. These Cooper pairs are substantially larger than the interatomic separation.

35.3 References

- Guenault, Tony (2003). *Basic superfluids*. Taylor & Francis. ISBN 0-7484-0892-4.

- University of Colorado (January 28, 2004). *NIST/University of Colorado Scientists Create New Form of Matter: A Fermionic Condensate*. Press Release.

- Rodgers, Peter & Dumé, Bell (January 28, 2004). *Fermionic condensate makes its debut*. PhysicWeb.

- Haegler, Philipp, "Hadron Structure from Lattice Quantum Chromodynamics", Physics Reports 490, 49-175 (2010) [DOI 10.1016/j.physrep.2009.12.008]

Chapter 36

Fermionic field

In quantum field theory, a **fermionic field** is a quantum field whose quanta are fermions; that is, they obey Fermi–Dirac statistics. Fermionic fields obey canonical anticommutation relations rather than the canonical commutation relations of bosonic fields.

The most prominent example of a fermionic field is the Dirac field, which describes fermions with spin$-1/2$: electrons, protons, quarks, etc. The Dirac field can be described as either a 4-component spinor or as a pair of 2-component Weyl spinors. Spin-1/2 Majorana fermions, such as the hypothetical neutralino, can be described as either a dependent 4-component Majorana spinor or a single 2-component Weyl spinor. It is not known whether the neutrino is a Majorana fermion or a Dirac fermion (see also Neutrinoless double-beta decay for experimental efforts to determine this).

36.1 Basic properties

Free (non-interacting) fermionic fields obey canonical anticommutation relations, i.e., involve the anticommutators $\{a,b\}$ = $ab + ba$ rather than the commutators $[a,b] = ab - ba$ of bosonic or standard quantum mechanics. Those relations also hold for interacting fermionic fields in the interaction picture, where the fields evolve in time as if free and the effects of the interaction are encoded in the evolution of the states.

It is these anticommutation relations that imply Fermi–Dirac statistics for the field quanta. They also result in the Pauli exclusion principle: two fermionic particles cannot occupy the same state at the same time.

36.2 Dirac fields

The prominent example of a spin-1/2 fermion field is the **Dirac field** (named after Paul Dirac), and denoted by $\psi(x)$. The equation of motion for a free field is the Dirac equation,

$$(i\gamma^\mu \partial_\mu - m)\psi(x) = 0.$$

where γ^μ are gamma matrices and m is the mass. The simplest possible solutions to this equation are plane wave solutions, $\psi_1(x) = u(p)e^{-ip\cdot x}$ and $\psi_2(x) = v(p)e^{ip\cdot x}$. These plane wave solutions form a basis for the Fourier components of $\psi(x)$, allowing for the general expansion of the Dirac field as follows,

$$\psi(x) = \int \frac{d^3 p}{(2\pi)^3} \frac{1}{\sqrt{2E_p}} \sum_s \left(a_{\mathbf{p}}^s u^s(p)e^{-ip\cdot x} + b_{\mathbf{p}}^{s\dagger} v^s(p)e^{ip\cdot x} \right).$$

u and v are spinors, labelled by spin, s. For the electron, a spin 1/2 particle, $s = +1/2$ or s=$-1/2$. The energy factor is the result of having a Lorentz invariant integration measure. Since $\psi(x)$ can be thought of as an operator, the coefficients

268

of its Fourier modes must be operators too. Hence, $a_{\mathbf{p}}^s$ and $b_{\mathbf{p}}^{s\dagger}$ are operators. The properties of these operators can be discerned from the properties of the field. $\psi(x)$ and $\psi(y)^\dagger$ obey the anticommutation relations:

$$\{\psi_a(\mathbf{x}), \psi_b^\dagger(\mathbf{y})\} = \delta^{(3)}(\mathbf{x} - \mathbf{y})\delta_{ab},$$

where a and b are spinor indices. We impose an anticommutator relation (as opposed to a commutation relation as we do for the bosonic field) in order to make the operators compatible with Fermi–Dirac statistics. By putting in the expansions for $\psi(x)$ and $\psi(y)$, the anticommutation relations for the coefficients can be computed.

$$\{a_{\mathbf{p}}^r, a_{\mathbf{q}}^{s\dagger}\} = \{b_{\mathbf{p}}^r, b_{\mathbf{q}}^{s\dagger}\} = (2\pi)^3\delta^3(\mathbf{p} - \mathbf{q})\delta^{rs},$$

In a manner analogous to non-relativistic annihilation and creation operators and their commutators, these algebras lead to the physical interpretation that $a_{\mathbf{p}}^{s\dagger}$ creates a fermion of momentum $\mathbf{p}$ and spin s, and $b_{\mathbf{q}}^{r\dagger}$ creates an antifermion of momentum $\mathbf{q}$ and spin r. The general field $\psi(x)$ is now seen to be a weighed (by the energy factor) summation over all possible spins and momenta for creating fermions and antifermions. Its conjugate field, $\bar{\psi} \overset{\text{def}}{=} \psi^\dagger \gamma^0$, is the opposite, a weighted summation over all possible spins and momenta for annihilating fermions and antifermions.

With the field modes understood and the conjugate field defined, it is possible to construct Lorentz invariant quantities for fermionic fields. The simplest is the quantity $\bar{\psi}\psi$. This makes the reason for the choice of $\bar{\psi} = \psi^\dagger \gamma^0$ clear. This is because the general Lorentz transform on ψ is not unitary so the quantity $\psi^\dagger \psi$ would not be invariant under such transforms, so the inclusion of γ^0 is to correct for this. The other possible non-zero Lorentz invariant quantity, up to an overall conjugation, constructible from the fermionic fields is $\bar{\psi}\gamma^\mu \partial_\mu \psi$.

Since linear combinations of these quantities are also Lorentz invariant, this leads naturally to the Lagrangian density for the Dirac field by the requirement that the Euler–Lagrange equation of the system recover the Dirac equation.

$$\mathcal{L}_D = \bar{\psi}(i\gamma^\mu \partial_\mu - m)\psi$$

Such an expression has its indices suppressed. When reintroduced the full expression is

$$\mathcal{L}_D = \bar{\psi}_a(i\gamma_{ab}^\mu \partial_\mu - m\mathbb{I}_{ab})\psi_b$$

The Hamiltonian (energy) density can also be constructed by first defining the momentum canonically conjugate to $\psi(x)$, called $\Pi(x)$:

$$\Pi \overset{\text{def}}{=} \frac{\partial \mathcal{L}_D}{\partial(\partial_0 \psi)} = -\bar{\psi}\gamma^0.$$

With that definition of Π, the Hamiltonian density is:

$$\mathcal{H}_D = \Pi\gamma^0[\vec{\gamma} \cdot \vec{\nabla} + m]\psi,$$

where $\vec{\nabla}$ is the standard gradient of the space-like coordinates, and $\vec{\gamma}$ is a vector of the space-like γ matrices. It is surprising that the Hamiltonian density doesn't depend on the time derivative of ψ, directly, but the expression is correct.

Given the expression for $\psi(x)$ we can construct the Feynman propagator for the fermion field:

$$D_F(x - y) = \langle 0|T(\psi(x)\bar{\psi}(y))|0\rangle$$

we define the time-ordered product for fermions with a minus sign due to their anticommuting nature

$$T(\psi(x)\bar{\psi}(y)) \overset{\text{def}}{=} \theta(x^0 - y^0)\psi(x)\bar{\psi}(y) - \theta(y^0 - x^0)\bar{\psi}(y)\psi(x).$$

Plugging our plane wave expansion for the fermion field into the above equation yields:

$$D_F(x - y) = \int \frac{d^4 p}{(2\pi)^4} \frac{i(\not{p} + m)}{p^2 - m^2 + i\epsilon} e^{-ip \cdot (x - y)}$$

where we have employed the Feynman slash notation. This result makes sense since the factor

$$\frac{i(\not{p} + m)}{p^2 - m^2}$$

is just the inverse of the operator acting on $\psi(x)$ in the Dirac equation. Note that the Feynman propagator for the Klein–Gordon field has this same property. Since all reasonable observables (such as energy, charge, particle number, etc.) are built out of an even number of fermion fields, the commutation relation vanishes between any two observables at spacetime points outside the light cone. As we know from elementary quantum mechanics two simultaneously commuting observables can be measured simultaneously. We have therefore correctly implemented Lorentz invariance for the Dirac field, and preserved causality.

More complicated field theories involving interactions (such as Yukawa theory, or quantum electrodynamics) can be analyzed too, by various perturbative and non-perturbative methods.

Dirac fields are an important ingredient of the Standard Model.

36.3 See also

- Dirac equation

- Einstein–Maxwell–Dirac equations

- Spin-statistics theorem

- Spinor

36.4 References

- Edwards, D. (1981). "The Mathematical Foundations of Quantum Field Theory: Fermions, Gauge Fields, and Super-symmetry, Part I: Lattice Field Theories". *International J. of Theor. Phys.* **20** (7): 503–517. Bibcode:1981IJTP...20..503E.doi:10.1007/BF00669437.

- Peskin, M and Schroeder, D. (1995). *An Introduction to Quantum Field Theory*, Westview Press. (See pages 35–63.)

- Srednicki, Mark (2007). *Quantum Field Theory*, Cambridge University Press, ISBN 978-0-521-86449-7.

- Weinberg, Steven (1995). *The Quantum Theory of Fields*, (3 volumes) Cambridge University Press.

Chapter 37

Kogut–Susskind fermion

Kogut–Susskind fermions are a lattice version of Kähler–Dirac fermions, which obey a first-order differential equation by taking an alternative square root of the Laplacian than that used by Dirac.[1] They are named after John Kogut and Leonard Susskind.

37.1 Notes

[1] Stone, M. (2000). *The Physics of Quantum Fields*. Springer New York. p. 200. ISBN 9780387989099. Retrieved 2015-04-08.

37.2 References

- Kogut, J.; Susskind, L. (1975), "Hamiltonian formulation of Wilson's lattice gauge theories", *Phys. Rev. D* **11**: 395, Bibcode:1975PhRvD..11..395K, doi:10.1103/PhysRevD.11.395

37.3 Text and image sources, contributors, and licenses

37.3.1 Text

- **Fermion** *Source:* https://en.wikipedia.org/wiki/Fermion?oldid=695996537 *Contributors:* AxelBoldt, Chenyu, Derek Ross, CYD, Mav, Bryan Derksen, The Anome, Ben-Zin~enwiki, Alan Peakall, Dominus, Dcljr, Looxix~enwiki, Glenn, Nikai, Andres, Wikiborg, David Latapie, Phys, Bevo, Stormie, Olathe, Donarreiskoffer, Robbot, Merovingian, Rorro, Wikibot, HaeB, Giftlite, Fropuff, Xerxes314, Vivektewary, JoJan, Karol Langner, Tothebarricades.tk, Icairns, Hidaspal, Vsmith, Laurascudder, Lysdexia, Ashlux, Graham87, Magister Mathematicae, Kbdank71, Syndicate, Strait, Protez, Drrngrvy, FlaBot, Srleffler, Chobot, YurikBot, RobotE, Jimp, Bhny, Captaindan, SpuriousQ, Salsb, Lomn, Enormousdude, CharlesHBennett, Federalist51, Tom Lougheed, Unyoyega, Jrockley, MK8, BabuBhatt, Complexica, Zachorious, Shalom Yechiel, QFT, Garry Denke, Daniel.Cardenas, SashatoBot, Flipperinu, Dan Gluck, LearningKnight, Happy-melon, Paulfriedman7, Cydebot, Meno25, Zalgo, Thijs!bot, Mbell, Headbomb, Nick Number, Orionus, Shlomi Hillel, CosineKitty, NE2, Mwarren us, ZPM, Vanished user ty12kl89jq10, Joshua Davis, R'n'B, Tensegrity, Rod57, Dgiraffes, Alpvax, VolkovBot, TXiKiBoT, Red Act, Anonymous Dissident, Abdullais4u, בל יכול, Tanhueiming, Antixt, Haiviet~enwiki, EmxBot, Kbrose, SieBot, Likebox, Jojalozzo, Dhatfield, Oxymoron83, TubularWorld, ClueBot, Seervoitek, Rodhullandemu, Jorisverbiest, Feebas factor, ChandlerMapBot, Nilradical, Wikeepedian, Stephen Poppitt, Addbot, Vectorboson, Luckas-bot, Yobot, Planlips, Dickdock, AnomieBOT, Icalanise, Materialscientist, Xqbot, Br77rino, Balaonair, 豆豆, Paine Ellsworth, Blackoutjack, Kikeku, Rameshngbot, Tom.Reding, RedBot, Alarichus, Michael9422, Silicon-28, TjBot, EmausBot, WikitanvirBot, Quazar121, Solomonfromfinland, JSquish, Fimin, Quondum, AManWithNoPlan, EdoBot, ClueBot NG, PBot1, EthanChant, Bibcode Bot, BG19bot, Petermahlzahn, KingKhan85, ChrisGualtieri, BoethiusUK, DerekWinters, Tentinator, JNrgbKLM, Mohit rajpal, KasparBot, Jiswin1992, Even This Is Taken, Wulframm, Chemistry1111 and Anonymous: 120

- **Particle physics** *Source:* https://en.wikipedia.org/wiki/Particle_physics?oldid=698160676 *Contributors:* AxelBoldt, Chenyu, Matthew Woodcraft, Trelvis, The Epopt, Sodium, Lee Daniel Crocker, CYD, Eloquence, Mav, Gareth Owen, Larry Sanger, XJaM, Roadrunner, SimonP, Ark~enwiki, Hfastedge, Bdesham, Patrick, Boud, Michael Hardy, Ixfd64, Fruge~enwiki, NuclearWinner, Looxix~enwiki, Ellywa, Ahoerstemeier, Docu, Glenn, Palfrey, Hectorthebat, Rl, Mxn, Laussy, Tpbradbury, Phys, Head, Bevo, Mignon~enwiki, Raul654, UninvitedCompany, Donarreiskoffer, Korath, Sanders muc, Calmypal, Rorro, Rholton, DHN, Gnomon Kelemen, LX, Fuelbottle, Alan Liefting, SimonMayer, Matt Gies, Dominick, Giftlite, Barbara Shack, Lupin, Orpheus, Dmmaus, Jason Quinn, Djegan, Matt Crypto, JRR Trollkien, Bodhitha, Andycjp, Mako098765, Mamizou, Karol Langner, APH, Lumidek, Deglr6328, Physics~enwiki, Urvabara, Discospinster, Rich Farmbrough, FT2, Ylai, Bender235, Bennylin, El C, Edward Z. Yang, Haxwell, Bobo192, Jung dalglish, Maurreen, I9Q79oL78KiL0QTFHgyc, Giraffedata, Rje, Fatphil, Gbrandt, Alansohn, Gary, Arthena, Atlant, Lightdarkness, Kocio, Hdeasy, Bucephalus, Velella, Tycho, Henry W. Schmitt, DV8 2XL, Redvers, Pcd72, Novacatz, Kurzon, Mpatel, Isnow, Palica, Awmarcz, Graham87, Qwertyus, Sjakkalle, Mayumashu, Mattmartin, R.e.b., RE, Klortho, Lor772, Who, Lmatt, Goudzovski, Srleffler, OpenToppedBus, Md7t, Chobot, Agerom, Bgwhite, YurikBot, Wavelength, Bambaiah, Ohwilleke, Techraj, Stephenb, Gaius Cornelius, CambridgeBayWeather, NawlinWiki, SCZenz, Ragesoss, Jpowell, Voidxor, Zwobot, Scottfisher, Bota47, Dna-webmaster, Emijrp, Le sacre, Ilmari Karonen, Archer7, Selkem, Physicsdavid, GrinBot~enwiki, Eog1916, SmackBot, Incnis Mrsi, Erwinrossen, Bggoldie~enwiki, Melchoir, Mcneile, Unyoyega, CRKingston, Jagged 85, AndreasJS, Dauto, Master Jay, MK8, MalafayaBot, Silly rabbit, Csgwon, DHN-bot~enwiki, Can't sleep, clown will eat me, QFT, Voyajer, Cybercobra, Jgwacker, Savidan, JonasRH, Valenciano, Penarestel, Drphilharmonic, Ohconfucius, Kuru, Robofish, Goodnightmush, Physis, Бью, Mets501, Ravi12346, MTSbot~enwiki, Ch2pgj, Iridescent, UncleDouggie, CapitalR, Battlemage~enwiki, George100, CRGreathouse, Van helsing, Comrade42, BeenAroundAWhile, Mato, Tfnewman, Chrislk02, Kozuch, Thijs!bot, Mojo Hand, Smarcus, Headbomb, Arcresu, MichaelMaggs, Jomoal99, Escarbot, Austin Maxwell, AntiVandalBot, Bm gub, Gnixon, Olexandr Kravchuk, Qwerty Binary, Res2216firestar, MER-C, The Transhumanist, Jameskeates, Magioladitis, Celithemis, Bongwarrior, VoABot II, Swpb, El Snubbe, Ling.Nut, Allstarecho, DerHexer, JaGa, Mermaid from the Baltic Sea, CommonsDelinker, J.delanoy, Maurice Carbonaro, MoogleEXE, EmanCunha, Vanished user 342562, Shawn in Montreal, LordAnubisBOT, Ryan Postlethwaite, Joshmt, Kenneth M Burke, Jamesontai, Inwind, Jxzj, Lseixas, VolkovBot, CWii, SarahLawrence Scott, TXiKiBoT, Docanton, Pandacomics, Someguy1221, Dev 176, JhsBot, ^demonBot2, BurtPeck, CloudNineAC, Wiae, Aroodman, Complex (de), SwordSmurf, Synthebot, Falcon8765, Trecool12, Monty845, Raphtee, Munkay, News0969, Kbrose, Ghalhud, SaltyBoatr, SieBot, Sonicology, Caltas, Bamkin, Oxymoron83, Lightmouse, Almostcrime, Stfg, Mike2vil, Denisarona, Poopfacer, Martarius, ClueBot, Snigbrook, Donzzz77, VsBot, TallMagic, Pet3r, Boing! said Zebedee, DragonBot, Howie Goodell, BobertWABC, Brews ohare, Jotterbot, PhySusie, JamieS93, Maine12329, Jimbill4321, Kaskofonous, Joe N, Party, Eik Corell, Oldnoah, Saeed.Veradi, WikHead, NellieBly, Pchapman47879, Truthnlove, Falconkhe, Addbot, Non-dropframe, Boomur, Fieldday-sunday, Vishnava, Download, PranksterTurtle, Favonian, LinkFA-Bot, AgadaUrbanit, SPat, Zorrobot, Legobot, Luckas-bot, Fraggle81, Planlips, Orion11M87, AnomieBOT, Khcf6971, Mouse7525, Materialscientist, USConsLib, Citation bot, Howdychicken, Richard Jay Morris, Xqbot, Plastadity, Witguiota, Brandonlovescrashincastles, JimVC3, Mark Schierbecker, RibotBOT, Metrictensor, A. di M., A.amitkumar, FrescoBot, Paine Ellsworth, Ironboy11, Joe iNsecure, Steve Quinn, Citation bot 1, Aknochel, Micraboy, CodeTheorist, FoxBot, Lotje, Jesse V., Tiki843, Zanzerjewel, Cjc38, EmausBot, Immunize, Racerx11, Bengt Nyman, Winner 42, Dcirovic, AsceticRose, Langsytank, JSquish, Mullactalk, D.Lazard, AManWithNoPlan, VictorFlaushenstein, Vanished user fijw983kjaslkekfhj45, Olhp, L Kensington, PhoenixFlentge, MonoAV, Rangoon11, ChuispastonBot, Ebehn, BR84, ClueBot NG, Elodzinski, Andyfreeberg, IOPhysics, Moritz37, Navasj, Helpful Pixie Bot, Electriccatfish2, Abid931, Bibcode Bot, BG19bot, MusikAnimal, AvocatoBot, Kirananils, Rclsa~enwiki, CimanyD, Wizardjr9o, Will Gladstone, Sunshine Warrior04, Hepforever, Klilidiplomus, BattyBot, StarryGrandma, Blondietroll, Th4n3r, Thepwninglol, ChrisGualtieri, Macko74, Sweet55033, Dexbot, Frognyanya, Wickedwondrous, AHusain314, Telfordbuck, Reatlas, Jeremymichaelmcvey, Euan Richard, Stephan Linn, SakeUPenn, Tangy Lemonz, Prokaryotes, NottNott, Cypherquest, Воображение, Triolysat, AddWittyNameHere, Susan.grayeff, Db9199 24, Stamptrader, Kdmeaney, Abitslow, Monstersmash10000, Philipphilip0001, Hexidominus, Englishcomptest, Comptest, Nathaniel 84, IiKkEe, Englishtest, TheMagikCow, Plaguetest, Englishtest3, Hachimods, Ryan Errico, Jpskycak, CAPTAIN RAJU, Matthewpatrickbarry, Wikiedittest, Samprv6, Valenciaproof and Anonymous: 367

- **Standard Model** *Source:* https://en.wikipedia.org/wiki/Standard_Model?oldid=697582361 *Contributors:* AxelBoldt, Derek Ross, CYD, Bryan Derksen, The Anome, Ed Poor, Andre Engels, Roadrunner, Anthere, David spector, Isis~enwiki, Youandme, Ram-Man, Stevertigo, Edward, Patrick, Boud, Michael Hardy, SebastianHelm, Looxix~enwiki, Julesd, Glenn, AugPi, Mxn, Raven in Orbit, Reddi, Phr, Tpbradbury, Populus, Haoherb428, Phys, Floydian, Bevo, Pierre Boreal, AnonMoos, BenRG, Jeffq, Dmytro, Drxenocide, Robbot, Nurg, Securiger, Texture, Roscoe x, Fuelbottle, Mattflaschen, Tobias Bergemann, Alan Liefting, Ancheta Wis, Giftlite, Dbenbenn, Harp, Herbee, Monedula, LeYaYa, Xerxes314, Dratman, Alison, JeffBobFrank, Dmmaus, Pharotic, Brockert, Bodhitha, Andycjp, Sonjaaa, HorsePunchKid,

IW.HG, AnomieBOT, Rubinbot, Materialscientist, Citation bot, ArthurBot, Xqbot, Nickkid5, Beeline23, Gap9551, GrouchoBot, Omnipaedista, Nathanielvirgo, ⁇⁇, Craig Pemberton, Cognitivelydissonant, Relke, RedBot, Tjlafave, Hickorybark, Halteres, Korepin, EmausBot, Ethereal-Blade, RA0808, H3llBot, Quondum, ChuispastonBot, Mikhail Ryazanov, ClueBot NG, Gareth Griffith-Jones, Asalrifai, Helpful Pixie Bot, Bibcode Bot, Krishnaprasaths, BG19bot, GKFX, Alexander1102, BattyBot, JYBot, Tony Mach, Frosty, Polentarion, Monochrome Monitor, Kharkiv07, Septate, Yakamashi, Sp20136761, Kfitzell29 and Anonymous: 200

- **List of particles** *Source:* https://en.wikipedia.org/wiki/List_of_particles?oldid=697214131 *Contributors:* AxelBoldt, Danny, Rmhermen, Stevertigo, Bdesham, Ahoerstemeier, Stan Shebs, Docu, Salsa Shark, Nikai, Evercat, Schneelocke, Charles Matthews, Jitse Niesen, CBDunkerson, Bevo, Raul654, Donarreiskoffer, Robbot, Sanders muc, Merovingian, Pengo, Giftlite, Herbee, Xerxes314, Dratman, Jeremy Henty, Alensha, Bodhitha, Physicist, Hayne, Quadell, RetiredUser2, Mysidia, Icairns, Asbestos, D6, Urvabara, Discospinster, Rich Farmbrough, FT2, Qutezuce, ArnoldReinhold, Neko-chan, El C, Laurascudder, Susvolans, EmilJ, Physicistjedi, Minghong, Gbrandt, Eddideigel, Axl, Mac Davis, David Ko, Radical Mallard, RJFJR, Count Iblis, Dirac1933, TenOfAllTrades, LFaraone, Oleg Alexandrov, Linas, JarlaxleArtemis, Duncan.france, GregorB, Cedrus-Libani, Karam.Anthony.K, Palica, Rjwilmsi, Zbxgscqf, JLM~enwiki, Strait, Ems57fcva, Krash, Dan Guan, DannyWilde, Lmatt, Goudzovski, Chobot, YurikBot, Bambaiah, Vuvar1, Madkayaker, Hydrargyrum, Presscorr, Chaos, Salsb, Tavilis, SCZenz, Lexicon, TUSHANT JHA, Dna-webmaster, Tomvds, Poulpy, Cstmoore, TLSuda, NeilN, MacsBug, Tom Lougheed, McGeddon, Bazza 7, WookieInHeat, Derdeib, Yamaguchi⁇⁇, Betacommand, Bluebot, Master of Puppets, DHN-bot~enwiki, Raistuumum, Juancnuno, Kittybrewster, Acepectif, Ligulembot, TriTertButoxy, ArglebargleIV, Khazar, John, FrozenMan, JorisvS, 041744, Dr Greg, Slakr, Mets501, Scorpion0422, Cbuckley, Iridescent, TwistOfCain, Happy-melon, JRSpriggs, Flickboy, Van helsing, Lithium6, Neelix, Rotiro, Cydebot, Quibik, Christian75, Omicronpersei8, Thijs!bot, Qwyrxian, TauLibrus, Headbomb, Inner Earth, 49, Guptasuneet, Scottmsg, WinBot, Elmoosecapitan, Tyco.skinner, AubreyEllenShomo, Arch dude, Johnman239, Mwarren us, TheEditrix2, CalamusFortis, MartinBot, Sadisticsuburbanite, Bissinger, Anaxial, CommonsDelinker, Maurice Carbonaro, Zojj, OliverHarris, Joshmt, Adanadhel, Lseixas, Graphite Elbow, VolkovBot, Jmrowland, Quilbert, Anonymous Dissident, Dstary, Escalona, JPMasseo, Figureskatingfan, Inx272, Meters, Antixt, Hamish a e fowler, GoddersUK, Bluetryst, SieBot, Ishvara7, WereSpielChequers, Audrius u, VovanA, Paolo.dL, RSStockdale, Anchor Link Bot, StewartMH, Explicit, ClueBot, Unbuttered Parsnip, Nolimitownass, DragonBot, Atomic7732, TimothyRias, SkyLined, Addbot, DOI bot, Jojhutton, Favonian, LinkFA-Bot, OlEnglish, Teles, Legobot, Luckas-bot, Yobot, Dov Henis, Azcolvin429, AnomieBOT, Götz, Icalanise, Flewis, Materialscientist, OllieFury, Vuerqex, ArthurBot, Vulcan Hephaestus, Blennow, Reality006, Coretheapple, Jcimorra, RibotBOT, Ernsts, A. di M., Axelfoley12, Zosterops, FrescoBot, Paine Ellsworth, Citation bot 1, JIK1975, Tom.Reding, Diffequa, WikitanvirBot, Racerx11, 112358sam, Aegnor.erar, Hops Splurt, HESUPERMAN, Hhhippo, AvicBot, JSquish, StringTheory11, Chharvey, Waperkins, Bamyers99, Suslindisambiguator, L Kensington, DennisIsMe, RockMagnetist, ClueBot NG, Snotbot, Primergrey, Vio45lin, Widr, MsFionnuala, Oklahoma3477, Bibcode Bot, CityOfSilver, Cap'n G, BML0309, Dan653, Twocount, Penguinstorm300, Dexbot, LightandDark2000, Ohiggy, TwoTwoHello, Andyhowlett, Printersmoke, Orion 2013, ARUNEEK, Seino van Breugel, AspaasBekkelund, TheMagikCow, Vyom27, ParkersComments, DERPALERT, Selva Ganapathy, Benoit schillings and Anonymous: 292

- **Elementary particle** *Source:* https://en.wikipedia.org/wiki/Elementary_particle?oldid=695842630 *Contributors:* CYD, Mav, Bryan Derksen, XJaM, Heron, Stevertigo, Patrick, Fbjon, Looxix~enwiki, Александър, Julesd, Glenn, AugPi, Mxn, Timwi, Reddi, Tpbradbury, Furrykef, Bevo, Donarreiskoffer, Robbot, Craig Stuntz, Nurg, Papadopc, Wikibot, Jimduck, Anthony, Ancheta Wis, Giftlite, DavidCary, Mikez, Haselhurst, Monedula, Xerxes314, Alison, Guanaco, Greydream, Anythingyouwant, Bodnotbod, Kate, Brianjd, Mormegil, Urvabara, Rich Farmbrough, Guanabot, Qutezuce, Hidaspal, Dmr2, Goplat, RJHall, RoyBoy, Robotje, Neonumbers, ליאור, Dirac1933, DV8 2XL, Azmaverick623, Blaxthos, Kay Dekker, Joriki, Simetrical, TomTheHand, Mpatel, Isnow, Ggonnell, Palica, Strait, Miserlou, Ligulem, Naraht, DannyWilde, Lmatt, Srleffler, Chobot, Cactus.man, Roboto de Ajvol, YurikBot, Hairy Dude, NTBot~enwiki, Ohwilleke, Albert Einsteins pipe, Stephenb, Chaos, Vibritannia, SCZenz, Edwardlalone, Larsobrien, Bota47, BraneJ, Dna-webmaster, Arthur Rubin, Oyvind, GrinBot~enwiki, SmackBot, Mrcoolbp, Bomac, GrGBL~enwiki, Chris the speller, MalafayaBot, George Rodney Maruri Game, Silly rabbit, Complexica, MovGP0, Fmalan, Scwlong, Amazins490, Cybercobra, EPM, Garry Denke, Drphilharmonic, Sadi Carnot, ArglebargleIV, Tktktk, NongBot~enwiki, WhiteHatLurker, Jonhall, Dekaels~enwiki, Jynus, Newone, Courcelles, Laplace's Demon, SchmittM, J Milburn, Fordmadoxfraud, Cydebot, Bvcrist, Kozuch, Thijs!bot, Lord Hawk, Headbomb, MichaelMaggs, Escarbot, Ssr, JAnDbot, Eurobas, Acroterion, VoABot II, Appraiser, BatteryIncluded, R'n'B, Sgreddin, MikeBaharmast, Lk69, Acalamari, DraakUSA, TomasBat, Joshua Issac, Kenneth M Burke, Ken g6, Idioma-bot, VolkovBot, SarahLawrence Scott, Nxavar, JhsBot, Abdullais4u, Lejarrag, Antixt, PGWG, SieBot, Timb66, Sonicology, PlanetStar, Bamkin, Dhatfield, Byrialbot, Svick, Perfectapproach, Thorncrag, Big55e, ClueBot, Jmorris84, Maxtitan, Alexbot, Dekisugi, Paradoxalterist, Saintlucifer2008, Cockshut12345, Rreagan007, RP459, Truthnlove, Addbot, Yakiv Gluck, Draco 2k, Mac Dreamstate, Funky Fantom, CarsracBot, HerculeBot, Legobot, Blah28948, Luckas-bot, Zhitelew, KamikazeBot, Kulmalukko, Orion11M87, AnomieBOT, Girl Scout cookie, Templatehater, Icalanise, Citation bot, Onesius, Vuerqex, Bci2, ArthurBot, Rightly, Xqbot, Phazvmk, Kirin13, FrescoBot, Pepper, Delphinus1997, Steve Quinn, Robo37, SuperJew, HRoestBot, Sthyne, Hellknowz, Yahia.barie, Skyerise, Tobi - Tobsen, FoxBot, Physics therapist, Think!97, Bj norge, RjwilmsiBot, Beyond My Ken, EmausBot, John of Reading, Mnkyman, GoingBatty, Mthorndill, ZéroBot, Bollyjeff, StringTheory11, Markinvancouver, Quantumor, Maschen, RolteVolte, Negovori, NTox, I hate whitespace, ClueBot NG, CocuBot, Widr, Micah.yannatos1, Helpful Pixie Bot, Guzman.c, Bibcode Bot, BG19bot, Spaceawesome, Rainbot, Leaverward, Let'sBuildTheFuture, Eduardofeld, Sha-256, Dr.RobertTweed, ZX95, Joeinwiki, Mark viking, Cephas Atheos, Yo butt, Snakeboy666, Psyruby42, Haminoon, Sardeth42, TaiSakuma, LadyCailin, Morph dtlr, Delbert7, Karam adel, Liance, Isambard Kingdom, Vegitō-λmericium, KasparBot, Are you freaking kidding me, Kurousagi and Anonymous: 189

- **Spin–statistics theorem** *Source:* https://en.wikipedia.org/wiki/Spin%E2%80%93statistics_theorem?oldid=696731382 *Contributors:* Michael Hardy, Agremon, Charles Matthews, Dysprosia, Phys, Tobias Bergemann, Giftlite, Dratman, Waltpohl, Fpahl, Pjacobi, Luqui, Pinzo, Mdf, Army1987, Guy Harris, Keenan Pepper, PAR, Rjwilmsi, Godzatswing, John Baez, Wavelength, Hairy Dude, Wigie, Dhollm, Teply, SmackBot, Incnis Mrsi, Kmarinas86, Sbharris, Tesseran, Dan Gluck, Jorbesch, JRSpriggs, CBM, Thijs!bot, Gamebm, Paakun, Igodard, Shikasta.net, Pohara, Liometopum, SieBot, Likebox, Jasondet, Melcombe, El bot de la dieta, Rostheskunk, Addbot, Mathieu Perrin, Chemuser, Jthrush, Bob K31416, Luckas-bot, Yobot, JBancroftBrown, AnomieBOT, Xqbot, Baz.77.243.99.32, LucienBOT, Craig Pemberton, Codwiki, MastiBot, Seattle Jörg, Pavithransiyer, Sm00th101, Super48paul, Iamfullofspam, ZéroBot, Quondum, L Kensington, QuantumSquirrel, Mikhail Ryazanov, Bibcode Bot, Brad7777, BattyBot, ChrisGualtieri, Tatata56 and Anonymous: 53

- **Quantum state** *Source:* https://en.wikipedia.org/wiki/Quantum_state?oldid=698191835 *Contributors:* RTC, Michael Hardy, Julesd, Andres, Laussy, Patrick0Moran, Bevo, BenRG, Bkalafut, Rorro, Papadopc, Tobias Bergemann, Giftlite, MathKnight, MichaelHaeckel, CSTAR, H Padleckas, Elroch, Mschlindwein, Chris Howard, Freakofnurture, Hidaspal, Slipstream, Geschichte, Alansohn, Cortonin, Dan East, Ott,

Woohookitty, Mpatel, Dzordzm, Colin Watson, Rjwilmsi, Mathbot, Margosbot~enwiki, Fresheneesz, Bgwhite, Wavelength, RobotE, Bambaiah, Agent Foxtrot, Hydrargyrum, PoorLeno, Larsobrien, Modify, Sbyrnes321, A13ean, Incnis Mrsi, Ptpare, Jutta234, Physis, Erwin, CapitalR, Petr Matas, BeenAroundAWhile, Mct mht, Phatom87, Dragon's Blood, Waxigloo, Thijs!bot, Colincmr, Headbomb, Second Quantization, Iviney, Eleuther, Bizzon, Magioladitis, Tercer, B. Wolterding, R'n'B, Hans Dunkelberg, Maurice Carbonaro, ARTE, Hulten, Sheliak, VolkovBot, LokiClock, Kinneytj, Thurth, TXiKiBoT, V81, Spinningspark, Kbrose, SieBot, Phe-bot, OKBot, StewartMH, ClueBot, Alksentrs, EoGuy, Rockfang, SchreiberBike, The-tenth-zdog, Dragonfi, SilvonenBot, RealityDysfunction, Stephen Poppitt, Addbot, Bob K31416, Luckasbot, Yobot, JTXSeldeen, AnomieBOT, Götz, Xqbot, Pvkeller, J04n, GrouchoBot, Omnipaedista, Nathanielvirgo, Waleswatcher, WaysToEscape, 🔢, Chjoaygame, FrescoBot, Freddy78, Steve Quinn, Machine Elf 1735, Oxonienses, RedBot, RobinK, BasvanPelt, Heurisko, Lotje, Eagleclaw6, RjwilmsiBot, Pierluigi.taddei, EmausBot, Gaurav biraris, Solomonfromfinland, Harddk, Zephyrus Tavvier, Maschen, Xronon, ClueBot NG, MelbourneStar, Theopolisme, Helpful Pixie Bot, Bibcode Bot, F=q(E+v^B), Ganitvidya, DrBugKiller, Chetan666, Jochen Burghardt, W. P. Uzer, Noix07, 7Sidz, Monkbot, Pratixit and Anonymous: 73

- **Quantum field theory** *Source:* https://en.wikipedia.org/wiki/Quantum_field_theory?oldid=697015959 *Contributors:* AxelBoldt, CYD, Mav, The Anome, XJaM, Roadrunner, Stevertigo, Michael Hardy, Tim Starling, IZAK, TakuyaMurata, SebastianHelm, Looxix~enwiki, Ahoerstemeier, Cyp, Glenn, Rotem Dan, Stupidmoron, Charles Matthews, Timwi, Jitse Niesen, Kbk, Rudminjd, Wik, Phys, Bevo, BenRG, Northgrove, Robbot, Bkalafut, Gandalf61, Rursus, Fuelbottle, Tobias Bergemann, Ancheta Wis, Giftlite, Lethe, Dratman, Alison, St3vo, Mboverload, DefLog~enwiki, ConradPino, Amarvc, Pcarbonn, Karol Langner, APH, AmarChandra, D6, CALR, Urvabara, Discospinster, Guanabot, Igorivanov~enwiki, Masudr, Pjacobi, Vsmith, Nvj, MuDavid, Bender235, Pt, El C, Shanes, Sietse Snel, Physicistjedi, KarlHallowell, PWilkinson, Helix84, Thialfi, Varuna, Gcbirzan, Docboat, Count Iblis, Egg, Mpatel, Marudubshinki, Graham87, Opie, Vanderdecken, Rjwilmsi, MarSch, Earin, R.e.b., RE, Strobilomyces, Arnero, Itinerant1, Alfred Centauri, Srleffler, Chobot, UkPaolo, Wavelength, Bambaiah, Hairy Dude, RussBot, TimNelson, Archelon, CambridgeBayWeather, SCZenz, Odddmonster, E2mb0t~enwiki, Semperf, Tetracube, Garion96, Erik J, Robert L, Banus, RG2, SmackBot, Stephan Schneider, Tom Lougheed, Melchoir, KocjoBot~enwiki, Mcld, Dauto, Chris the speller, Complexica, Threepounds, RuudVisser, QFT, Jmnbatista, Cybercobra, Rebooted, Victor Eremita, DJIndica, Lambiam, Mgiganteus1, Zarniwoot, Jim.belk, Stwalkerster, SirFozzie, Hu12, Dan Gluck, Iridescent, Joseph Solis in Australia, Albertod4, Van helsing, BeenAroundAWhile, Witten Is God, Cydebot, Jamie Lokier, Meno25, Michael C Price, The 80s chick, Mendicus~enwiki, AstroPig7, Msebast~enwiki, Mbell, Headbomb, Nick Number, Mentifisto, AntiVandalBot, Bt414, Bananan~enwiki, Martin Kostner, Moltrix, Kasimann, Kromatol, Puksik, Lerman, LLHolm, RogueNinja, Tlabshier, JEH, Nikolas Karalis, Storkk, JAnDbot, Igodard, Four Dog Night, N shaji, Bongwarrior, Andrea Allais, Soulbot, Etale, Maliz, Custos0, HEL, J.delanoy, Maurice Carbonaro, Acalamari, Jeepday, Policron, Blckavnger, Juliancolton, Skou, Telecomtom, GrahamHardy, Sheliak, Cuzkatzimhut, VolkovBot, Pleasantville, Bktennis2006, Marksr, HowardFrampton, Oshwah, The Original Wildbear, Dj thegreat, Markisgreen, TBond, Lejarrag, Moose-32, Raphtee, Sue Rangell, Neparis, Drschawrz, YohanN7, SieBot, TCO, Yintan, Likebox, Paolo.dL, Tugjob, Henry Delforn (old), Jecht (Final Fantasy X), OKBot, StewartMH, ClueBot, EoGuy, Wwheaton, The Wild West guy, Shvav~enwiki, Bob108, Brews ohare, Thingg, Count Truthstein, XLinkBot, PSimeon, SilvonenBot, Truthnlove, HexaChord, Addbot, ConCompS, Pinkgoanna, Leapold~enwiki, Dmhowarth26, Glane23, Hanish.polavarapu, Lightbot, Scientryst, R.ductor, Ettrig, Yndurain, Legobot, Luckas-bot, Yobot, Ht686rg90, Niout, Tamtamar, AnomieBOT, Ciphers, Palpher, IRP, Gjsreejith, Materialscientist, Citation bot, Bci2, ArthurBot, Northryde, LilHelpa, Caracolillo, Amareto2, MIRROR, Professor J Lawrence, Plasmon1248, Omnipaedista, RibotBOT, Spellage, JayJay, FrescoBot, Kenneth Dawson, D'ohBot, Knowandgive, N4tur4le, Hyqeom, Newt Winkler, Hickorybark, Lotje, Dinamik-bot, LilyKitty, Fortesque666, Reaper Eternal, Minimac, Marie Poise, Yaush, Dylan1946, EmausBot, Racerx11, GoingBatty, Carbosi, Thecheesykid, ZéroBot, Cogiati, Jjspinorfield1, Suslindisambiguator, Quondum, Maschen, Zueignung, Davidaedwards, RockMagnetist, Lom Konkreta, ClueBot NG, Gilderien, Iloveandrea, Vacation9, Heyheyheyhohoho, Fortune432, The ubik, Zak.estrada, Widr, Helpful Pixie Bot, Guy vandegrift, Evanescent7, Ykentluo, Martin.uecker, Walterpfeifer, Pfeiferwalter, Klilidiplomus, W.D., CarrieVS, Khazar2, Momo1381, Dexbot, Cerabot~enwiki, Garuda0001, AHusain314, Thepalerider2012, A.entropy, Mark viking, Faizan, Aj7s6, संजीव कुमार, Lemnaminor, BerFinelli, Axel.P.Hedstrom, Kclongstocking, Mutley1989, I art a troler, Liquidityinsta, Prokaryotes, DemonThuum, Dingdong2680, Asherkirschbaum, Monkbot, Gjbayes, Thedarkcheese, BradNorton1979, UareNumber6, Teelaskeletor, YeOldeGentleman, Mret81, KasparBot, CAPTAIN RAJU and Anonymous: 302

- **Spin (physics)** *Source:*https://en.wikipedia.org/wiki/Spin_(physics)?oldid=697570893*Contributors:*AxelBoldt, CYD, The Anome, Larry_Sanger, Andre Engels, XJaM, David spector, Stevertigo, Xavic69, Michael Hardy, Tim Starling, Dominus, Cyp, Stevenj, Glenn, AugPi, Rossami, Nikai, Andres, Med, Mxn, Charles Matthews, Timwi, Kbk, 4lex, Reina riemann, E23~enwiki, Phys, Wtrmute, Bevo, Elwoz, Robbot, Gandalf 61, Blainster, DHN, Hadal, Papadopc, Jheise, Anthony, Diberri, Xanzzibar, Giftlite, Smjg, Lethe, Lupin, MathKnight, Xerxes314, Average Earthman, AlistairMcMillan, Ato, Andycjp, Gzuckier, Beland, Karol Langner, Spiralhighway, Elroch, B.d.mills, Tsemii, Frau Holle, Mike Rosoft, Igorivanov~enwiki, FT2, MuDavid, Paul August, Pt, Susvolans, Army1987, Wood Thrush, SpeedyGonsales, Physicistjedi, Obradovic Goran, Neonumbers, Keenan Pepper, Count Iblis, Egg, Linas, Palica, Torquil~enwiki, Ashmoo, Grammarbot, Zoz, Rjwilmsi, Zbxgscqf, Drrngrvy, FlaBot, Mathbot, TheMidnighters, Itinerant1, Ewlyahoocom, Gurch, Fresheneesz, Srleffler, Kri, Chobot, DVdm, YurikBot, Bambaiah, Hairy Dude, JabberWok, Rsrikanth05, NawlinWiki, Buster79, Hwasungmars, Kkmurray, Werdna, Djdaedalus, Simen, Netrapt, Mpjohans, KSevcik, GrinBot~enwiki, Joshronsen, Bo Jacoby, Sbyrnes321, DVD R W, Shanesan, KasugaHuang, That Guy, From That Show!, SmackBot, Unyoyega, Bluebot, Complexica, DHN-bot~enwiki, Sergio.ballestrero, Vladislav, QFT, Voyajer, Terryeo, Ryanluck, Radagast83, Jgrahamc, Michael-Billington, Richard001, DMacks, Daniel.Cardenas, Bidabadi~enwiki, Sadi Carnot, Bdushaw, Andrei Stroe, Tesseran, SashatoBot, Leo C Stein, Vanished user 9i39j3, UberCryxic, Jonas Ferry, Vgy7ujm, Loodog, Jaganath, Terry Bollinger, Wierdw123, Inquisitus, Beefyt, Jc37, Dreftymac, Newone, RokasT~enwiki, Jaksmata, Aepryus, JRSpriggs, Joostvandeputte~enwiki, CRGreathouse, David s graff, Ahmes, JasonHise, Eric Le Bigot, Bmk, Myasuda, Mct mht, FilipeS, Cydebot, A876, Thijs!bot, Barticus88, Headbomb, Brichcja, Davidhorman, Oreo Priest, Widefox, Orionus, Tlabshier, Accordionman, Astavats, JAnDbot, Em3ryguy, MER-C, Igodard, PhilKnight, .anacondabot, Sangak, Magioladitis, Swpb, Dirac66, LorenzoB, Monurkar~enwiki, TechnoFaye, Brilliand, R'n'B, CommonsDelinker, Victor Blacus, J.delanoy, Numbo3, Sackm, Maurice Carbonaro, Klatkinson, Cmichael, Uberdude85, Craklyn, CardinalDan, VolkovBot, Error9312, JohnBlackburne, Bolzano~enwiki, TXiKiBoT, Hqb, Anonymous Dissident, Costela, BotKung, Kganjam, Petergans, Kbrose, SieBot, BotMultichill, The way, the truth, and the light, RadicalOne, Flyer22 Reborn, Jasondet, Paolo.dL, R J Sutherland, Lightmouse, JackSchmidt, Martarius, ClueBot, JonnybrotherJr, Warbler271, Mild Bill Hiccup, David Trochos, Outerrealm, Sbian, Peachypoh, SchreiberBike, Ant59, Crowsnest, XLinkBot, Addbot, Mathieu Perrin, Narayansg, Imeriki al-Shimoni, Sriharsha.karnati, Numbo3-bot, Tide rolls, Lightbot, Luckas-bot, Yobot, Nallimbot, AnomieBOT, Lendtuffz, Citation bot, Nepahwin, ArthurBot, Obersachsebot, Xqbot, Sionus, WandringMinstrel, Francine Rogers, Pradameinhoff, Tom1936, Ernsts, A. di M., NoldorinElf, Daleang, Baz.77.243.99.32, LucienBOT, Paine Ellsworth, Tobby72, Freddy78, Craig Pemberton, C.Bluck,

Jondn, Pokyrek, Citation bot 1, I dream of horses, Adlerbot, Casimir9999, Kallikanzarid, Jkforde, Trappist the monk, Michael9422, Miracle Pen, Sgravn, 8af4bf06611c, Garuh knight, EmausBot, Beatnik8983, GoingBatty, JustinTime55, Zhenyok 1, Atomicann, JSquish, ZéroBot, Harddk, Neh0000, Quondum, Jacksccsi, Maschen, Zueignung, Rasinj, RockMagnetist, Eg-T2g, ClueBot NG, Paolo328, Gilderien, Frietjes, Widr, PhiMAP, Helpful Pixie Bot, Bibcode Bot, BG19bot, PUECH P.-F., Mark Arsten, Yudem, F=q(E+v^B), Blaspie55, Halfb1t, Robertwilliams2011, Dexbot, Foreverascone, ScitDei, Mark viking, Pedantchemist, YiFeiBot, W. P. Uzer, Francois-Pier, Mathphysman, Aidan Clark, Brotter121, KasparBot and Anonymous: 234

- **Superfluidity** *Source:* https://en.wikipedia.org/wiki/Superfluidity?oldid=696002973 *Contributors:* DragonflySixtyseven, Discospinster, Isaac Rabinovitch, Kmarinas86, Sbharris, John, Chrumps, CuriousEric, Cydebot, Lamro, Coldcreation, Addbot, Linket, PianoDan, Materialscientist, Citation bot, Abce2, Tom.Reding, Trappist the monk, LilyKitty, Nickjf22, AManWithNoPlan, David C Bailey, Vippylaman, Bibcode Bot, Ugncreative Usergname, 220 of Borg, BattyBot, Adwaele, Dexbot, GyaroMaguus, Agonbroke, Reatlas, Jonpao523, Ybidzian, JamesMoose, HamiltonFromAbove, KasparBot, Loapsodiap, Chemistry1111 and Anonymous: 16

- **Superconductivity** *Source:* https://en.wikipedia.org/wiki/Superconductivity?oldid=698221745 *Contributors:* AxelBoldt, Lee Daniel Crocker, CYD, Bryan Derksen, AstroNomer~enwiki, Andre Engels, DavidLevinson, Quintanilla, Jqt, Azhyd, Waveguy, David spector, Heron, Olivier, Edward, Michael Hardy, Fred Bauder, DopefishJustin, Dominus, Karada, Tiles, Egil, Ahoerstemeier, Stevenj, Theresa knott, Snoyes, Julesd, Glenn, Cimon Avaro, GCarty, Cryoboy, Mxn, Tantalate, Reddi, Stone, Joerg Reiher~enwiki, Hao2lian, DJ Clayworth, E23~enwiki, Furrykef, Taxman, LMB, Fibonacci, Omegatron, Traroth, Topbanana, Pstudier, Pakaran, Phil Boswell, Donarreiskoffer, Robbot, Stephan Schulz, Rorro, Bkell, Hadal, UtherSRG, Robinh, Diberri, Cyberpunks~enwiki, Connelly, Giftlite, DocWatson42, MarkPNeyer, Harp, Tom harrison, Ferkelparade, Fastfission, Xerxes314, Leonard G., Foobar, Bobblewik, Wmahan, Irarum, Geni, Quadell, Spiralhighway, Icairns, Peter bertok, Gerrit, Deglr6328, Deeceevoice, Moxfyre, Reflex Reaction, Zowie, CALR, Discospinster, FT2, Rama, Vsmith, Pavel Vozenilek, Paul August, Andrejj, Kaisershatner, CanisRufus, Kwamikagami, PhilHibbs, Haxwell, Simonbp, Femto, Dalf, Bobo192, BrokenSegue, Enric Naval, Slicky, Kjkolb, Nk, Merope, PaulHanson, GiantSloth, Lightdarkness, Sligocki, Pion, Hu, Velella, Wtshymanski, Evil Monkey, RJFJR, Cmapm, Dfalkner, Gene Nygaard, Aeronautics, RHaworth, Dandv, StradivariusTV, Oliphaunt, Jeff3000, Jwanders, Alfakim, Firien, Triddle, Someone42, GregorB, Eras-mus, CharlesC, SeventyThree, Christopher Thomas, Graham87, Magister Mathematicae, Jan van Male, Josh Parris, Sjö, Sjakkalle, Rjwilmsi, Seidenstud, Fish and karate, FlaBot, PhilipSargent, Jeepo~enwiki, Gurch, Leslie Mateus, Fosnez, Goudzovski, Skierpage, Chobot, DVdm, Ahpook, Takaaki, Roboto de Ajvol, The Rambling Man, Wavelength, Mollsmolyneux, Bhny, JabberWok, Netscott, Hydrargyrum, CambridgeBayWeather, Salsb, GeeJo, Harksaw, Długosz, RyanLivingston, Ino5hiro, Mkouklis, Nineteenthly, Mccready, Dhollm, Scottfisher, Quarky2001, DeadEyeArrow, Oliverdl, Tonym88, Codell, Searchme, Light current, 2over0, DaveOinSF, Theda, Closedmouth, Bamse, Filou~enwiki, Petri Krohn, JoanneB, Alias Flood, Wylie440, Chaiken, SkerHawx, Kgf0, Children of the dragon, SmackBot, Melchoir, Gilliam, Oscarthecat, Chaojoker, Kmarinas86, Chris the speller, RevenDS, NCurse, Thumperward, Papa November, Complexica, AtmanDave, Kostmo, Dual Freq, Trekphiler, KaiserbBot, TheKMan, LouScheffer, Elendil's Heir, Toomontrangle, Pwjb, Smokefoot, Eynar, DMacks, Paulish, Simon Arnold, Lester, Nbishop, Breadbox, Kuru, John, JorisvS, Smartyllama, Manjish, IronGargoyle, Spiel496, Citicat, Majormcmuffin, Kvng, Astrobradley, JarahE, KJS77, Brienanni, Japhet, Hmtamza, Tawkerbot2, Chetvorno, CmdrObot, Van helsing, MorkaisChosen, CBM, WMSwiki, Tim1988, Lokal Profil, Phatom87, Britannic~enwiki, Cydebot, Kam42705, Neil Froschauer, Chasingsol, Myscrnnm, Lee, IComputerSaysNo, Arwen4014, Editor at Large, TrevorRC, Matwilko, Raschd, Epbr123, Kubanczyk, Dasacus, Headbomb, Dgies, Cyclonenim, Courtjester555, Mojohaza1, Casomerville, Yellowdesk, JAnDbot, Quentar~enwiki, Smartcat, Bongwarrior, VoABot II, Ginga2, SineWave, Jjasi, Web-Crawling Stickler, Dirac66, Coolkoon, Limtohhan, Joshua Davis, Schmloof, Xantolus, CommonsDelinker, Pharaoh of the Wizards, Jtw11, Dmrmatt19, Hans Dunkelberg, Uncle Dick, Maurice Carbonaro, Nigholith, MrBell, Eliz81, Bakkouz, Rod57, Bot-Schafter, TomyDuby, Anatoly larkin, Wimox, Equazcion, Tevonic, Useight, Qaz123qaz, Bertiethecat, Idioma-bot, JeffreyRMiles, VolkovBot, TXiKiBoT, Neha simon, Calwiki, Hqb, Liquidcentre, JosephJohnCox, OlavN, Sodapopinski, Robert1947, Burntsauce, Elecwikiman, Fischer.sebastian, AlleborgoBot, Shanmugammpl, Runewiki777, Steven Weston, SieBot, Yintan, Vanished User 8a9b4725f8376, FSUlawalumni, Keilana, Hzh, Henry Delforn (old), Onopearls, Anchor Link Bot, Hamiltondaniel, Geoff Plourde, Elliott-rhodes, TubularWorld, Tegrenath, LarRan, ClueBot, Trojancowboy, Fuzzylunkinz, Ctiefel, Techdawg667, VsBot, YBCO, Niceguyedc, Rotational, Cousins.inc, CohesionBot, Jeck1335, Doctorpsi, PixelBot, Bob man801, Lartoven, Brews ohare, Neucleon, Natty sci~enwiki, Doprendek, SchreiberBike, Aitias, Subash.chandran007, SoxBot III, HumphreyW, LSTech, Tarlneustaedter, Wertuose, BodhisattvaBot, Rror, Ngebbett, Ost316, WikHead, Noctibus, ElMeBot, Addbot, Forscite, AVand, DOI bot, Melab-1, Travisoto, Flning, Jncraton, CanadianLinuxUser, Leszek Jańczuk, CarsracBot, Dr. Universe, K Eliza Coyne, Gwcdt, Lightbot, SPat, Luckas-bot, Yobot, Fraggle81, THEN WHO WAS PHONE?, CinchBug, Csmallw, MassimoAr, AnomieBOT, Cryogenics, Guff2much, Materialscientist, Citation bot, Xqbot, Eep not for fat people, Waleswatcher, NinjaDreams, Janolaf30, Dave3457, GliderMaven, FrescoBot, WikiMcGowan, Tobby72, AlanDewey, Citation bot 1, ASchwarz, Pinethicket, HRoestBot, Schrodingers rabbit, 10metreh, Tom.Reding, Gruntler, Richardc03, Mikespedia, Heller2007, Felix0411, Anoop ranjan, Aleitner, Ahsbenton, Agnel P.B., Catcamus, Akoufos, Jiyojolly, Dick Chu, Noommos, Haj33, EmausBot, John of Reading, WikitanvirBot, DonyG, JasonSaulG, Mathew10111, Pascalf, Hhhippo, ManosHacker, Medeis, A930913, Tls60, Sailsbystars, Nothingbutdreamer, ChuispastonBot, AndyTheGrump, DASHBotAV, WikiBaller, ClueBot NG, Gilderien, Hightc, Widr, Names are hard to think of, Helpful Pixie Bot, Sina.zapf, Mightyname, Nightenbelle, Jubobroff, Bibcode Bot, BG19bot, Virtualerian, Island Monkey, Ymblanter, Andol, WikiHacker187, Mark Arsten, 52 6f 62, Pong711, BattyBot, Bv.vasiliev, Chim02, MahdiBot, Jimw338, Embrittled, Adwaele, Protectionwi, Dexbot, Anandaraja, Oliver brookes, Fittold27, TwoTwoHello, Andyhowlett, Reatlas, Ruby Murray, François Robere, Rabbitflyer, Asik Ram, Monkbot, Jrafner, Laurencejwolf, Scipsycho, OzRamos, KasparBot, Superspin, Shao xc and Anonymous: 525

- **Baryon** *Source:* https://en.wikipedia.org/wiki/Baryon?oldid=697569634 *Contributors:* AxelBoldt, Tobias Hoevekamp, Bryan Derksen, Ben-Zin~enwiki, Heron, Tim Starling, Alan Peakall, Paul A, Salsa Shark, Glenn, Mxn, Charles Matthews, The Anomebot, ElusiveByte, Phys, Bevo, Traroth, Donarreiskoffer, Robbot, Korath, Kristof vt, Merovingian, Ojigiri~enwiki, Sunray, Wikibot, Giftlite, DocWatson42, ShaunMacPherson, Herbee, Xerxes314, Dą,ugosz, Kaldari, OwenBlacker, Icairns, JohnArmagh, Rich Farmbrough, Guanabot, Mani1, E2m, Tompw, El C, Bobo192, I9Q79oL78KiL0QTFHgyc, Giraffedata, Physicistjedi, Jumbuck, Gary, ABCD, Oleg Alexandrov, Woohookitty, Tevatron~enwiki, BD2412, Kbdank71, Nightscream, Ae77, MZMcBride, Chekaz, R.e.b., Erkcan, Maxim Razin, Oo64eva, Chobot, Roboto de Ajvol, YurikBot, Bambaiah, Jimp, Salsb, Ergzay, DragonHawk, SCZenz, E2mb0t~enwiki, Bota47, Simen, Sbyrnes321, Lainagier, Timotheus Canens, Bluebot, Colonies Chris, Kingdon, Shadow1, Bigmantonyd, Drphilharmonic, Kseferovic, Wierdw123, Physicsdog, Torrazzo, Verdy p, Michael C Price, Thijs!bot, Headbomb, Hcobb, Orionus, QuiteUnusual, Spartaz, Plantsurfer, Amateria1121, Diamond2, Swpb, BatteryIncluded, David Eppstein, Hveziris, Saxophlute, Gwern, Ben MacDui, R'n'B, Ash, Tgeairn, Maurice Carbonaro, STBotD, VolkovBot, GimmeBot, NoiseEHC, Tearmeapart, BotKung, BrianADesmond, Antixt, AlleborgoBot, Lou427, SieBot, VVVBot, Gerakibot, LeadSong-

Dog, Keilana, Paolo.dL, Doctorfluffy, TrufflesTheLamb, OKBot, Hamiltondaniel, TubularWorld, ClueBot, Artichoker, ChandlerMapBot, CalumH93, Addbot, LaaknorBot, CarsracBot, Jonhstone12, Legobot, Luckas-bot, Bugbrain 04, AnomieBOT, JackieBot, Materialscientist, Citation bot, ArthurBot, Xqbot, Omnipaedista, SassoBot, Spellage, WaysToEscape, FrescoBot, Citation bot 1, FoxBot, Noommos, EmausBot, John of Reading, JSquish, ZéroBot, StringTheory11, Stibu, Ethaniel, Markinvancouver, ClueBot NG, Koornti, Kasirbot, Rezabot, Bibcode Bot, Atomician, Zedshort, Marioedesouza, ChrisGualtieri, WorldWideJuan, CoolHandLouis, Monkbot, KasparBot and Anonymous: 108

- **Lepton** *Source:* https://en.wikipedia.org/wiki/Lepton?oldid=696804823 *Contributors:* Bryan Derksen, Andre Engels, PierreAbbat, Ben-Zin~ enwiki, Heron, Xavic69, Fruge~enwiki, Fwappler, Ahoerstemeier, Julesd, Glenn, Mxn, A5, Wikiborg, Dysprosia, Radiojon, Imc, Morwen, Fibonacci, Bcorr, Phil Boswell, Donarreiskoffer, Robbot, Merovingian, Wikibot, Giftlite, Smjg, DocWatson42, Harp, Herbee, Xerxes314, Sysin, Knutux, LiDaobing, LucasVB, ClockworkLunch, RetiredUser2, Icairns, Mike Rosoft, Chris j wood, MartinI~enwiki, Smalljim, Giraffedata, Jumbuck, RobPlatt, Neonumbers, Ahruman, Computerjoe, Simon M, Woohookitty, Mindmatrix, Rjwilmsi, Strait, Erkcan, FlaBot, DannyWilde, Mas-torrent, Celebere, Peterl, YurikBot, Bambaiah, Jimp, Salsb, Spike Wilbury, Jaxl, SCZenz, DeadEyeArrow, Tetracube, Smoggyrob, Dmuth, Jaysbro, Sbyrnes321, That Guy, From That Show!, SmackBot, Bazza 7, KocjoBot~enwiki, Jrockley, Mom2jandk, Cool3, Hmains, Com-plexica, DHN-bot~enwiki, Mesons, Yevgeny Kats, TriTertButoxy, SashatoBot, Ouzo~enwiki, Happy-melon, Kurtan~ enwiki, Myasuda, Cy-debot, Meno25, Photocopier, Michael C Price, Casliber, Thijs!bot, Headbomb, Newton2, Mentifisto, Autotheist, Steveprutz, NeverWorker, NicoSan, MartinBot, Arjun01, HEL, J.delanoy, Numbo3, Gombang, Num1dgen, Ceoyoyo, VolkovBot, Macedonian, Mocirne, TXiKiBoT, Anonymous Dissident, Abdullais4u, Antixt, Jhb110, Thanatos666, AlleborgoBot, SieBot, ToePeu.bot, RadicalOne, Ngexpert7, Jacob.jose, Hamiltondaniel, TubularWorld, Muhends, ClueBot, ICAPTCHA, UniQue tree, Snigbrook, Fyyer, IceUnshattered, Cmj91uk, LieAfterLie, Manu-ve Pro Ski, TimothyRias, Addbot, Betterusername, AgadaUrbanit, Ehrenkater, OlEnglish, Zorrobot, Andy2308, Legobot, Luckas-bot, Ptbotgourou, Maxim Sabalyauskas, Planlips, JackieBot, Icalanise, Citation bot, .مدحم.د.ع.امدي 24, ArthurBot, Almabot, Omnipaedista, Alex-eymorgunov, ⁇⁇, Tormine, MathFacts, Citation bot 1, MastiBot, Trappist the monk, Earthandmoon, EmausBot, John of Reading, Az29, Galaktiker, StringTheory11, Quondum, Surajt88, I hate whitespace, ClueBot NG, Scimath Genius, Braincricket, Widr, Helpful Pixie Bot, Bib-code Bot, Tyler6360534, Katagun5, Melenc, Me, Myself, and I are Here, DerekWinters, Prasanna4s, Machosquirrel, Devinhorn, KasparBot, Chemistry1111 and Anonymous: 150

- **Quark** *Source:* https://en.wikipedia.org/wiki/Quark?oldid=697155177 *Contributors:* AxelBoldt, Derek Ross, Vicki Rosenzweig, Mav, Bryan Derksen, The Anome, Gareth Owen, Andre Engels, PierreAbbat, Peterlin~enwiki, Ben-Zin~enwiki, Zoe, Heron, Montrealais, Hfastedge, Ed- ward, Dante Alighieri, Ixfd64, CesarB, Card~enwiki, NuclearWinner, Looxix~enwiki, Ahoerstemeier, Elliot100, Docu, J-Wiki, Nanobug, Aarchiba, Julesd, Glenn, Schneelocke, Jengod, A5, Timwi, Dysprosia, DJ Clayworth, Phys, Ed g2s, Bevo, Olathe, MD87, Jni, Phil Boswell, Sjorford, Donarreiskoffer, Robbot, Sanders muc, Moncrief, Merovingian, PxT, Texture, Bkell, UtherSRG, Widsith, Ancheta Wis, Giftlite, ShaunMacPherson, Harp, Nunh-huh, Lupin, Herbee, Leflyman, Monedula, 0x6D667061, Xerxes314, Anville, Hoho~enwiki, Alison, Beardo, Moogle10000, Wronkiew, Jackol, Bobblewik, Bodhitha, Piotrus, Kaldari, Elroch, Icairns, Zfr, TonyW, Ukexpat, BrianWilloughby, Grunt, O'Dea, Jiy, Discospinster, Rich Farmbrough, Guanabot, T Long, Vsmith, Saintswithin, SocratesJedi, Mani1, Bender235, Lancer, RJHall, Mr. Billion, El C, Kwamikagami, Laurascudder, Susvolans, Triona, Axezz, Bobo192, Army1987, C S, Ziggurat, Rangelov, Matt McIrvin, Jojit fb, Nk, Pentalis, Obradovic Goran, Fwb22, Lysdexia, Benjonson, Alansohn, Gary, Gintautasm, Guy Harris, Keenan Pepper, MonkeyFoo, Lectonar, Mac Davis, Wdfarmer, Snowolf, Schapel, Knowledge Seeker, Evil Monkey, VivaEmilyDavies, CloudNine, Kusma, Kazvorpal, Kay Dekker, Crosbiesmith, Mogigoma, Linas, Mindmatrix, JarlaxleArtemis, ScottDavis, LOL, Wdyoung, Before My Ken, Tylerni7, Jwanders, Dataphiliac, AndriyK, Noetica, Wayward, Wisq, Palica, Marudubshinki, Calréfa Wéná, GSlicer, Graham87, Deltabeignet, Kbdank71, Yurik, Crzrussian, Rjwilmsi, Bremen, Marasama, SpNeo, Mike Peel, Bubba73, DoubleBlue, Matt Deres, Yamamoto Ichiro, Algebra, Dsnow75, RobertG, Nihiltres, Jeff02, RexNL, TeaDrinker, Chobot, DVdm, Jpacold, Gwernol, Elfguy, Roboto de Ajvol, YurikBot, Wavelength, Bamba- iah, Sceptre, Hairy Dude, Jimp, Phantomsteve, TheDoober, Dobromila, JabberWok, CambridgeBayWeather, Chaos, Salsb, Wimt, Ugur Basak, NawlinWiki, Spike Wilbury, Bossrat, SCZenz, Randolf Richardson, Danlaycock, Tony1, DRosenbach, Robertbyrne, Dna-webmaster, WAS 4.250, Closedmouth, Pietdesomere, Heathhunnicutt, Kevin, Banus, RG2, Kamickalo, That Guy, From That Show!, Veinor, MacsBug, Smack- Bot, Aigarius, BBandHB, Incnis Mrsi, InverseHypercube, C.Fred, Bazza 7, Ikip, Anastrophe, Jrockley, Eskimbot, AnOddName, Jonathan Karlsson, Edgar181, Gilliam, Dauto, NickGarvey, Vvarkey, Bluebot, KaragouniS, Keegan, Dahn, Bigfun, Miquonranger03, OrangeDog, Silly rabbit, Metacomet, Tripledot, Nbarth, DHN-bot~enwiki, Sbharris, Colonies Chris, Hallenrm, Scwlong, Gsp8181, Can't sleep, clown will eat me, Mallorn, Jeff DLB, Konczewski, TKD, Addshore, Mqjjb30e, Cybercobra, Khukri, B jonas, Jdlambert, Lpgeffen, Nrcprm2026, Akriasas, Zadignose, Jóna Þórunn, Bdushaw, Beyazid, TriTertButoxy, SashatoBot, SciBrad, Doug Bell, Soap, Richard L. Peterson, John, Mgiganteus1, SpyMagician, Edconrad, Loadmaster, 2T, Waggers, SandyGeorgia, Ravi12346, Dbzfrk15146, Peyre, Newone, GDallimore, Happy-melon, Ma- jora4, Chovain, Tawkerbot2, Cryptic C62, JForget, Vaughan Pratt, Hello789, ZICO, SUPRATIM DEY, Ruslik0, CuriousEric, Paulfriedman7, Logical2u, Myasuda, RoddyYoung, Typewritten, Cydebot, Abeg92, Mike Christie, Grahamec, Gogo Dodo, Jayen466, 879(CoDe), Michael C Price, Tawkerbot4, Ameliorate!, Akcarver, Gimmetrow, SallyScot, Casliber, Thijs!bot, Epbr123, NeoPhyteRep, LeBofSportif, Markus Pössel, Anupam, Sopranosmob781, Headbomb, Marek69, John254, KJBurns, MichaelMaggs, Escarbot, Eleuther, Ice Ardor, Aadal, AntiVandalBot, SmokeyTheCat, Tyco.skinner, Exteray, RobJ1981, Rsocol, Ke garne, Deflective, Husond, MER-C, CosineKitty, Andonic, East718, Pkop- penb, DanPMK, Magioladitis, WolfmanSF, Thasaidon, Bongwarrior, VoABot II, باسم, Inertiatic076, Kevinmon, Christoph Scholz~enwiki, Aka042, Giggy, Tanvirzaman, Johnbibby, Cyktsui, ArchStanton69, Ace42, Allstarecho, Shijualex, DerHexer, Elandra, Denis tarasov, Mar- tinBot, Poeloq, Dorvaq, CommonsDelinker, HEL, J.delanoy, Nev1, Ops101ex, DrKay, Hgpot, Ferdyshenko, Jigesh, DJ1AM, Tarotcards, Coppertwig, TomasBat, Nikbuz, SJP, FJPB, Vainamainien, Tiggydong, Robprain, Sheliak, Cuzkatzimhut, Lights, X!, VolkovBot, Off-shell, CWii, ABF, John Darrow, Holme053, Nousernamesleft, Ryan032, Philip Trueman, GimmeBot, Davehi1, A4bot, Captain Courageous, Guil- laume2303, Anonymous Dissident, Drestros power, Qxz, Anna Lincoln, Eldaran~enwiki, Leafyplant, Don4of4, PaulTanenbaum, Abdullais4u, Jbryancoop, Mbalelo, Gilisa, Eubulides, Chronitis, Seresin, Dustybunny, Insanity Incarnate, Upquark, Edge1212, Ollieho, AOEU Warrior, SieBot, Graham Beards, WereSpielChequers, Csmart287, Guguma5, Winchelsea, Jbmurray, Caltas, Vanished User 8a9b4725f8376, Keilana, Bentogoa, Aillema, RadicalOne, Arbor to SJ, Elcobbola, Physics one, Dhatfield, Hello71, RSStockdale, Son of the right hand, Ngexpert5, Ng- expert6, Ngexpert7, Psycherevolt, Sean.hoyland, Mygerardromance, Dabomb87, Nergaal, Muhends, Romit3, SallyForth123, Atif.t2, ClueBot, The Thing That Should Not Be, Wwheaton, Xeno malleus, Harland1, Piledhigheranddeeper, Maxtitan, DragonBot, Glopso, Choonkiat.lee, Himynameisdumb, Worth my salt, Arthur Quark, Estirabot, Brews ohare, Jotterbot, PhySusie, Brianboulton, Dekisugi, ANOMALY-117, Sallicio, Yomangan, Jtle515, Katanada, DumZiBoT, TimothyRias, XLinkBot, Vayalir, Oldnoah, Saintlucifer2008, Nathanwesley3, Dragon- firemage, Devilist666, Mancune2001, Jbeans, WikiDao, SkyLined, Truthnlove, Airplaneman, Eklipse, Addbot, Eric Drexler, AVand, Some jerk on the Internet, Captain-tucker, Giants2008, Iceblock, Ronhjones, Quarksci, Mseanbrown, Looie496, LaaknorBot, Peti610botH, AgadaUrbanit, Tide rolls, Vicki breazeale, Gail, Extruder~enwiki, Abduallah mohammed, Dealer77, Luckas-bot, Yobot, Fraggle81, Cflm001, Legobot II, Amble, Mmxx, Superpenguin1984, Worm That Turned, The Vector Kid, Planlips, Fangfyre, TestEditBot, Azcolvin429, Vroo, Synchronism, Bility, Orion11M87, AnomieBOT, Xi rho, Rubinbot, Jim1138, Bookaneer, Yotcmdr, Crystal whacker, Sonic h, Materialscientist, Citation bot, Pitke, Vuerqex, Bci2, ArthurBot, LilHelpa, Xqbot, Jeffrey Mall, AbigailAbernathy, Srich32977, Alex2510, Almabot, Uscbino, Pmlineditor, RibotBOT, Shmomuffin, Gunjan verma81, Chotarocket, Ernsts, Renverse, A. di M., Weekendpartier, ⁇, FrescoBot, Paine Ellsworth, DelphinidaeZeta, Steve Quinn, Citation bot 1, AstaBOTh15, Pinethicket, Jonesey95, Calmer Waters, Skyerise, Pmokeefe, Jschnur, Searsshoesales, Jrobbinz123, Lissajous, Turian, Lando Calrissian, Trappist the monk, Wotnow, Ansumang, Reaper Eternal, 564dude, Jackvancs, Bobotast, MINTOPOINT, TjBot, DexDor, Антон Гліністы, Daggersteel10, Chiechiecheist, EmausBot, John of Reading, Wikitanvir-Bot, Duskbrood, FergalG, Slightsmile, Barak90, Wikipelli, TheLemon1234, Manofgrass, Brazmyth, H3llBot, Stoneymufc29, GeorgeBarnick, Brandmeister, Ego White Tray, RockMagnetist, TYelliot, ClueBot NG, Rtucker913, Gilderien, A520, Cheeseequalsyum, Timothy jordan, 123Hedgehog456, Maplelanefarm, 336, Helpful Pixie Bot, Jeffreyts11, 123456789malm, Bibcode Bot, BG19bot, Hurricanefan25, MusikAnimal, Davidiad, MosquitoBird11, Mydogpwnsall, MrBill3, Njavallil, Glacialfox, Walterpfeifer, Thebannana, CE9958, Marioedesouza, Mediran, Dexbot, Rishab021, TwoTwoHello, Cjean42, Sriharsh1234, Sam boron100, Wankybanky, Wikitroll12345, RojoEsLardo, Jwratner1, NottNott, Saebre, JNrgbKLM, KheltonHeadley, AspaasBekkelund, HectorCabreraJr, Hazinho93, Quadrupedi, QuantumMatt101, Philipphilip0001, Monkbot, RiderDB, Dhm44444, Egfraley, Tetra quark, Isambard Kingdom, Weed305, KasparBot, Sheep killer 123, Unitdraws and Anonymous: 709

- **Up quark** *Source:* https://en.wikipedia.org/wiki/Up_quark?oldid=696868229 *Contributors:* Bryan Derksen, Alfio, Jni, Giftlite, Xerxes314, Kjoonlee, Bookandcoffee, CharlesC, Rjwilmsi, Mike Peel, Chobot, Hairy Dude, Rt66lt, Spike Wilbury, SCZenz, Poulpy, Eog1916, Bluebot, Tamfang, T-borg, Eric Saltsman, Hetar, Lottamiata, Laplace's Demon, Merryjman, CmdrObot, Myasuda, Raoul NK, Headbomb, JAnDbot, Abyssoft, I310342~enwiki, Idioma-bot, Sheliak, Wilmot1, VolkovBot, TXiKiBoT, Anonymous Dissident, Gekritzl, AlleborgoBot, SieBot, Muhends, Bobathon71, DragonBot, DumZiBoT, TimothyRias, Addbot, Eivindbot, LaaknorBot, ChenzwBot, Naidevinci, Ehrenkater, Lightbot, Luckas-bot, Citation bot, ArthurBot, Xqbot, DSisyphBot, Ditimchanly, Almabot, A. di M., Paine Ellsworth, Citation bot 1, Trappist the monk, TjBot, Ripchip Bot, EmausBot, WikitanvirBot, StringTheory11, Quondum, Helpful Pixie Bot, Bibcode Bot, BG19bot, P76837, Oznitecki, Alexzhang2, The Great Leon, Monkbot and Anonymous: 38

- **Down quark** *Source:* https://en.wikipedia.org/wiki/Down_quark?oldid=696752389 *Contributors:* Bryan Derksen, Alfio, Timwi, Jni, Herbee, Xerxes314, Rich Farmbrough, Kjoonlee, Rjwilmsi, Mike Peel, Chobot, YurikBot, Rt66lt, Acidsaturation, Spike Wilbury, SCZenz, Poulpy, Otto ter Haar, Skizzik, Bluebot, Tamfang, Llwang, Eric Saltsman, MTSbot~enwiki, Hetar, Laplace's Demon, Myasuda, Raoul NK, Thijs!bot, Headbomb, Davidhorman, JAnDbot, Abyssoft, MartinBot, Jvineberg, I310342~enwiki, Sheliak, VolkovBot, TXiKiBoT, Anonymous Dissident, SieBot, Ngexpert6, Muhends, Bobathon71, Lawrence Cohen, Daigaku2051, Auntof6, NuclearWarfare, TimothyRias, Addbot, Lightbot, Luckas-bot, Yobot, Citation bot, ArthurBot, Xqbot, DSisyphBot, Paine Ellsworth, Citation bot 1, Tim1357, Trappist the monk, EmausBot, ZéroBot, Quondum, Rezabot, Helpful Pixie Bot, Bibcode Bot, TheMan4000, 786b6364, Monkbot and Anonymous: 22

- **Strange quark** *Source:* https://en.wikipedia.org/wiki/Strange_quark?oldid=696863744 *Contributors:* Bryan Derksen, Alfio, Jni, Owain, Xerxes 314, Soman, Kjoonlee, Kwamikagami, Rsholmes, Esb82, Neonumbers, Rjwilmsi, Mike Peel, Gurch, Erik4, Chobot, YurikBot, Jimp, Salsb, SCZenz, Poulpy, SmackBot, Bluebot, NCurse, Vina-iwbot~enwiki, Yevgeny Kats, Zzzzzzzzzz, Laplace's Demon, MightyWarrior, Myasuda, Thijs!bot, Headbomb, Chillysnow, JAnDbot, Abyssoft, Bongwarrior, Albmont, McSly, I310342~enwiki, Pdcook, Sheliak, VolkovBot, SieBot, Muhends, Auntof6, Iohannes Animosus, TimothyRias, IngerAlHaosului, Addbot, ProbablyAmbiguous, Luckas-bot, Yobot, AnomieBOT, Citation bot, Sarah12sarah, Erik9bot, Thehelpfulbot, Paine Ellsworth, Rkr1991, Citation bot 1, Skyerise, Johann137, Trappist the monk, Puzl bustr, Agrasa, Wikiborg4711, EmausBot, Hhhippo, ZéroBot, Quondum, CocuBot, Helpful Pixie Bot, Bibcode Bot, Vkpd11, P76837, Matthew gib, Glaisher, RhinoMind and Anonymous: 38

- **Charm quark** *Source:* https://en.wikipedia.org/wiki/Charm_quark?oldid=696885690 *Contributors:* Bryan Derksen, Alfio, Bogdangiusca, Xerxes314, Bodhitha, Perey, Kjoonlee, Rjwilmsi, Mike Peel, Chobot, YurikBot, Bambaiah, Conscious, Salsb, SCZenz, Scottfisher, Poulpy, SmackBot, Delldot, Warhol13, Rezecib, Vina-iwbot~enwiki, Happy-melon, Laplace's Demon, CRGreathouse, Michael C Price, Thijs!bot, Headbomb, Nisselua, JAnDbot, Abyssoft, Uncle.wink, Bryanhiggs, HEL, I310342~enwiki, Qoou.Anonimu, Idioma-bot, Sheliak, Anonymous Dissident, Kumorifox, BeIsKr, AlleborgoBot, SieBot, Muhends, TimothyRias, Addbot, Mjamja, Lightbot, Luckas-bot, Yobot, Nallimbot, Citation bot, ArthurBot, Quebec99, Xqbot, DSisyphBot, GrouchoBot, RibotBOT, SassoBot, A. di M., Paine Ellsworth, Dogposter, D'ohBot, Citation bot 1, Citation bot 4, RedBot, MastiBot, Trappist the monk, EarthCom1000, Alph Bot, EmausBot, ZéroBot, Quondum, Anita5192, CocuBot, Rezabot, Helpful Pixie Bot, Bibcode Bot, Penguinstorm300, Hoppeduppeanut, Leowestland and Anonymous: 35

- **Bottom quark** *Source:* https://en.wikipedia.org/wiki/Bottom_quark?oldid=696005304 *Contributors:* Bryan Derksen, Xerxes314, Bodhitha, Icairns, Kjoonlee, Bobo192, Pinar, WadeSimMiser, Rjwilmsi, Mike Peel, Erkcan, FlaBot, Itinerant1, Chobot, YurikBot, Bambaiah, Jimp, Conscious, Ozabluda, SpuriousQ, Salsb, SCZenz, Lexicon, Poulpy, Physicsdavid, SmackBot, Hmains, Luís Felipe Braga, Laplace's Demon, CmdrObot, Outriggr (2006-2009), Niubrad, הסרפד, Thijs!bot, Headbomb, JAnDbot, Abyssoft, Pkoppenb, Dr. Morbius, I310342~enwiki, Joshmt, Idioma-bot, Sheliak, VolkovBot, Antixt, AlleborgoBot, BartekChom, Muhends, Auntof6, TimothyRias, Lockalbot, Addbot, Mr Sme, Luckas-bot, THEN WHO WAS PHONE?, Citation bot, ArthurBot, Xqbot, GrouchoBot, StevenVerstoep, Thehelpfulbot, Paine Ellsworth, Citation bot 1, Jonesey95, Double sharp, TjBot, EmausBot, Barak90, TuHan-Bot, ZéroBot, StringTheory11, Quondum, Chris857, ChuispastonBot, Widr, Helpful Pixie Bot, Bibcode Bot, P76837, ChrisGualtieri, Ajd268, Mfb, Monkbot, Axel Azzopardi, Kenijr, Chemistry1111 and Anonymous: 33

- **Top quark** *Source:* https://en.wikipedia.org/wiki/Top_quark?oldid=695598329 *Contributors:* Damian Yerrick, Bryan Derksen, HPA, Haryo, Bkell, Giftlite, Xerxes314, Edcolins, Bodhitha, David Schaich, Kjoonlee, Axl, Woohookitty, Rjwilmsi, Strait, Mike Peel, Vegaswikian, Wikiliki, Goudzovski, Chobot, YurikBot, Bambaiah, JabberWok, Gaius Cornelius, Salsb, Howcheng, SCZenz, Emijrp, Physicsdavid, SmackBot, Incnis Mrsi, ZerodEgo, Mr.Z-man, Jgwacker, Pulu, Stikonas, Mets501, Peyre, RekishiEJ, Banedon, הסרפד, Headbomb, Davidhorman, Oreo Priest, AntiVandalBot, JAnDbot, Abyssoft, Maliz, HEL, Fatka, I310342~enwiki, Idioma-bot, Sheliak, Biggus Dictus, TXiKiBoT, Reibot, Kachuak, Ptrslv72, SieBot, Hatster301, Muhends, ClueBot, Niceguyedc, Noca2plus, Choonkiat.lee, Brews ohare, Kakofonous, Jtle515, TimothyRias, Prostarplayer321, SkyLined, Cockatoot, Addbot, Mr0t1633, Mjamja, ChenzwBot, Ginosbot, Zorrobot, Luckas-bot, Naudefjbot~enwiki, Dreamer08, AnomieBOT, Icalanise, Citation bot, ArthurBot, LilHelpa, DSisyphBot, Unready, GrouchoBot, RibotBOT, Soandos, Paine Ellsworth, Citation bot 1, Jonesey95, Thinking of England, Nomis2k, Higgshunter, RjwilmsiBot, Mophoplz, EmausBot, John of Reading, WikitanvirBot, Barak90, StringTheory11, Peter.poier, Quondum, Samlever, Whoop whoop pull up, Reify-tech, Helpful Pixie Bot, Bibcode Bot, Glevum, Kephir, Mmitchell10, Quadrupedi, Monkbot, BrunoUbaldo, Chemistry1111 and Anonymous: 65

- **Electron** *Source:* https://en.wikipedia.org/wiki/Electron?oldid=697811995 *Contributors:* AxelBoldt, CYD, Mav, Bryan Derksen, AstroNomer Ap, Ed Poor, Andre Engels, Ryrivard, William Avery, SimonP, Peterlin~enwiki, Heron, Camembert, Stevertigo, Bdesham, Patrick, D, JohnOwens, Michael Hardy, Tim Starling, Ixfd64, Fruge~enwiki, Arpingstone, PingPongBoy, Egil, NuclearWinner, Ahoerstemeier, Suisui, Jebba, JWSchmidt, Kingturtle, Aarchiba, Glenn, Scott, Kwekubo, Andres, Jordi Burguet Castell, Mxn, Agtx, Timwi, Wikiborg, Reddi, Rednblu, Markhurd, Maximus Rex, E23~enwiki, Omegatron, Secretlondon, Jusjih, BenRG, Jeffq, Donarreiskoffer, Gentgeen, Robbot, Sanders muc, Vespristiano, Merovingian, Pingveno, Blainster, Hadal, Wikibot, Wereon, Widsith, HaeB, Diberri, Dmn, Dina, Giftlite, Christopher Parham, Ferkelparade, Fastfission, Zigger, Herbee, Dissident, Xerxes314, Curps, Michael Devore, Bensaccount, Ssd, Gilgamesh~enwiki, Vadmium, Gdr, Knutux, Slowking Man, Yath, Gzuckier, Pcarbonn, Joizashmo, Karol Langner, Anythingyouwant, RetiredUser2, Bbbl67, Elroch, Icairns, JohnArmagh, JimQ, Mike Rosoft, Mindspillage, Patrick L. Goes, Discospinster, Brianhe, Rich Farmbrough, Guanabot, Hidaspal, Vsmith, Deh, Ardonik, Roybb95~enwiki, Xezbeth, Zazou, Mani1, SpookyMulder, Dmr2, Bender235, ZeroOne, Kjoonlee, Goplat, Calair, Nabla, Brian0918, RJHall, Pt, Jaques O. Carvalho, El C, Huntster, Edward Z. Yang, Susvolans, Art LaPella, RoyBoy, ~K, Bobo192, Army1987, Asierra~enwiki, Flxmghvgvk, AtomicDragon, Evgeny, AllyUnion, Bert Hickman, Deryck Chan, PeterisP, Beetle B., Obradovic Goran, (aeropagitica), Pearle, Mpulier, HasharBot~enwiki, Confusedmiked, Mote, Jumbuck, Gary, ChristopherWillis, Ricky81682, Benjah-bmm27, Riana, AzaToth, DonJStevens, BernardH, Malo, David Hochron, Bart133, EagleFalconn, Schapel, Omphaloscope, RainbowOfLight, RichBlinne, H2g2bob, DV8 2XL, Gene Nygaard, Redvers, StuTheSheep, Linas, Mindmatrix, GrouchyDan, StradivariusTV, Uncle G, BillC, Kurzon, Jeff3000, HcorEric X, Eleassar777, Ozielke, Wayward, Palica, Omega21, FreplySpang, Enzo Aquarius, Rjwilmsi, Shaadow, Strait, Mike Peel, Chekaz, Bubba73, Dar-Ape, Yamamoto Ichiro, FlaBot, RobertG, Latka, DannyWilde, Nihiltres, RexNL, Kolbasz, Thecurran, Srleffler, Physchim62, Chobot, DVdm, Unclevortex, Eric B, YurikBot, Wavelength, RobotE, Bambaiah, AcidHelmNun, Jimp, Peter G Werner, Wolfmankurd, Wigie, Ventolin, JabberWok, SpuriousQ, Lucinos~enwiki, Akamad, Ori Livneh, Gaius Cornelius, Shaddack, Eleassar, Rsrikanth05, Salsb, Hawkeye7, Spike Wilbury, Jaxl, Welsh, DarthVader, Długosz, BirgitteSB, SCZenz, Retired username, Ravedave, PhilipO, Adam Rock, Mlouns, Chichui, BOT-Superzerocool, Gadget850, Bota47, Kkmurray, James Trotter~enwiki, Dna-webmaster, Ms2ger, Light current, Lycaon, Imaninjapirate, Josh3580, Kriscotta, JoanneB, Peyna, Lpm, JLaTondre, Heavy bolter, RG2, GrinBot~enwiki, Sbyrnes321, ChemGardener, Itub, SmackBot, Zazaban, Incnis Mrsi, KnowledgeOfSelf, Royalguard11, Melchoir, J.Sarfatti, KocjoBot~enwiki, Stepa, Pandion auk, Jrockley, JoeMarfice, ZerodEgo, Edgar181, Yamaguchi??, Skizzik, Dauto, JSpudeman, Kurykh, Rajeevmass~enwiki, Persian Poet Gal, Pieter Kuiper, Jprg1966, Acrinym, Miquonranger03, MalafayaBot, Droll, Complexica, DHN-bot~enwiki, Sbharris, RAlafriz, Vladislav, Vanished User 0001, Darthgriz98, Voyajer, Addshore, Percommode, Krich, DavidStern, Theonlyedge, Nakon, Nrcprm2026, DMacks, Daniel.Cardenas, Zeamays, Jonnyapple, Sadi Carnot, Bdushaw, Wilt, TriTertButoxy, Chymicus, UberCryxic, Bagel7, Mattfont, Heimstern, Jaganath, Ocatecir, Mr. Lefty, Ckatz, 16@r, Omnedon, Owlbuster, Waggers, SandyGeorgia, Spiel496, Funnybunny, HappyVR, Iridescent, Newone, NativeForeigner, J Di, Amakuru, Tawkerbot2, Chetvorno, Thermochap, CmdrObot, Ale jrb, Megaboz, RedRollerskate, Ruslik0, MrZap, McVities, WMSwiki, Bakanov, RobertLovesPi, Equendil, Cydebot, Acelor, Reywas92, Cantras, Bvcrist, LouisBB, Travelbird, Llort, David edwards, Tawkerbot4, Christian75, Narayanese, Ssilvers, Thijs!bot, Epbr123, Mbell, Dougsim, Nonagonal Spider, Headbomb, Yzmo, Marek69, West Brom 4ever, Tellyaddict, Cool Blue, Greg L, Sean William, VictorP, KrakatoaKatie, AntiVandalBot, WinBot, Skymt, Voyaging, Opelio, Tyco.skinner, Gef756, Chill doubt, Naturalnumber, Gdo01, Spencer, Leuko, CosineKitty, J-stan, Smith Jones, Acroterion, Magioladitis, WolfmanSF, Bennybp, Bongwarrior, VoABot II, A4, Nyq, JNW, JamesBWatson, بلاسم, Drondent, Slartibartfast1992, Jackal irl, Animum, Dirac66, 28421u2232nfenfcenc, David Eppstein, Hveziris, User A1, Maliz, PoliticalJunkie, DerHexer, GregU, PEBill, MartinBot, BetBot~enwiki, Mermaid from the Baltic Sea, WizendraW, Xantolus, Thereen, CommonsDelinker, AlexiusHoratius, J.delanoy, DrKay, Rgoodermote, Numbo3, Acalamari, TheChrisD, Dispenser, LordAnubisBOT, JayMars, Lathrop, AntiSpamBot, TomasBat, NewEnglandYankee, Nwbeeson, SmoothK, Sunderland06, MetsFan76, Joshmt, Cometstyles, STBotD, RB972, Treisijs, D-Kuru, Dineshextreeme, Martial75, CardinalDan, Idioma-bot, Sheliak, Bondslave777, FeralDruid, X!, VolkovBot, ABF, Thisisborin9, Jacroe, Ryan032, Philip Trueman, DoorsAjar, TXiKiBoT, Oshwah, GimmeBot, Kriak, Hqb, GDonato, Anonymous Dissident, Crohnie, Monkey Bounce, Voorlandt, Mr. Hallman, Michael H 34, TBond, Wiae, Suriel1981, Rbdebole, Graymornings, Synthebot, Enviroboy, Rurik3, Generalguy11, !dea4u, Insanity Incarnate, Ceranthor, Yoos~enwiki, AlleborgoBot, Kalivd, EmxBot, Neparis, Swimallday, Ponyo, EJF, SieBot, Graham Beards, Scarian, CircafuciX, BotMultichill, Jauerback, Dawn Bard, Joncam, Caltas, Sergeanthuggy, Bentogoa, RadicalOne, Arbor to SJ, Prestonmag, Thadaddy3233, Oxymoron83, Antonio Lopez, KPH2293, Lightmouse, WingkeeLEE, Ealdgyth, BenoniBot~enwiki, Stustjohn, Dabomb87, PlantTrees, Dolphin51, Nergaal, Tomdobb, Muhends, WikipedianMarlith, ClueBot, Trojancowboy, GorillaWarfare, Artichoker, PipepBot, UniQue tree, The Thing That Should Not Be, Hongthay, Unbuttered Parsnip, GreenSpigot, Liekmudkipz, Mild Bill Hiccup, Correcting nonsense, NovaDog, Blanchardb, Richerman, RandomTREES, Rotational, Piledhigheranddeeper, Inala, DragonBot, Almcaeobtac, Jusdafax, MEJG, Gtstricky, Rhododendrites, Brews ohare, NuclearWarfare, Lunchscale, Jotterbot, PhySusie, Tonyfey, Lkruijsw, Kaiba, SchreiberBike, Stepheng3, Thingg, Jamyricks, Aitias, Melibarr05, Kurtcobain321, Scalhotrod, Versus22, Johnuniq, MasterOfHisOwnDomain, DumZiBoT, TimothyRias, Sjodenenator, XLinkBot, Maky, Rror, Avoided, Mitch Ames, Ilikepie2221, WikHead, Mgaarafan, SkyLined, Addbot, Chizkiyahuavraham, AVand, Some jerk on the Internet, Hurleymann1, Uruk2008, DOI bot, Tcncv, Booba5, AkhtaBot, Jessepfrancis, Ronhjones, Jncraton, Moosehadley, CanadianLinuxUser, WFPM, LaaknorBot, Chamal N, CarsracBot, FiriBot, Omnipedian, LinkFA-Bot, Ehrenkater, Pnacitum, Tide rolls, Lightbot, Potekhin, UPS Truck Driver, VP-bot, Luckas-bot, Yobot, Nergality, Kan8eDie, THEN WHO WAS PHONE?, Eric-Wester, Tonyrex, AnomieBOT, Shootbamboo, DemocraticLuntz, Rubinbot, Götz, Jim1138, IRP, Piano non troppo, Icalanise, Kingpin13, Mydickishuge24, Materialscientist, The High Fin Sperm Whale, Citation bot, Neurolysis, ArthurBot, LovesMacs, Mrhellcool, Rightly, Xqbot, IrishChemistPride, IrishChemistPride2, GeometryGirl, Restu20, Gap9551, Srich32977, S0aasdf2sf, John5955, Alan8, ProtectionTaggingBot, Omnipaedista, RibotBOT, TonyHagale, Phillycheesesteaks, LyleHoward, A. di M., RyanOrdemann, Peter470, Thehelpfulbot, Al Wiseman, FrescoBot, Surv1v4l1st, Eadoncom, Paine Ellsworth, Tobby72, Gauravdce07, Steve Quinn, C.Bluck, Citation bot 1, MarB4, Galmicmi, Gil987, Pinethicket, HRoestBot, Voltron Hax, Raen79, Hoo man, Yos233, Allthingstoallpeople, MastiBot, Kuririmo, Noel Streatfield, Ezhuttukari, Swifterthenyou, Noisalt, Jujutacular, Euchanels, Lissajous, Dude1818, December21st2012Freak, IJBall, Jauhienij, Utility Monster, FoxBot, Sheogorath, Jdlawlis, Diannaa, Odatus, Bestcallumuk, Sampathsris, DARTH SIDIOUS 2, Mean as custard, RjwilmsiBot, TjBot, MinicheddarsandelephantsFTW, Benjadow, Mcmonsterbrothers, Priceracks, Csilcock, Sohaib360, Androstachys, Techhead7890, EmausBot, Optiguy54, GoingBatty, Jjasharpe, Pcorty, TuHan-Bot, Hhhippo, HiW-Bot, John Cline, Harddk, Fæ, Josve05a, StringTheory11, Wackywace, Quondum, GianniG46, Fizicist, Wayne Slam, Raynor42, Arnaugir, Jacksccsi, Brandmeister, Donner60, Negovori, RockMagnetist, Mni9791, ClueBot NG, HLachman, Hermajesty21, Jacobkh, Letoya123, Samsau ninjaguy, Ggonzalm, Moritz37, Braincricket, Helpful Pixie Bot, Geo7777, SzMithrandir, Bibcode Bot, Dfbowsmountainer, Ymblanter, Vagobot, Paolo Lipparini, Wzrd1, Lk00la1dl, JacobTrue, Socal212, Begman5, Mark Arsten, Cadiomals, Jikepaddy, Caterpillar111, Macymae, 06seagsa, BEEPTHENOOB, ItzzRevolution, Shawn Worthington Laser Plasma, Duxwing, Klilidiplomus, Uopchem251, Joe0x7F, BattyBot, Justincheng12345-bot, Cyberbot II, ChrisGualtieri, GoShow, Ankap~enwiki, Glenzo999, Barant2,

BrightStarSky, Mohanp06, Dexbot, Astromango2215, Webclient101, Mogism, 331dot, Spray787, Vanquisher.UA, Lugia2453, Kondormari, Reatlas, JellyBean4.1, Prof.Professer, The User 111, Bluemanyoung, Rohitgunturi, Ugog Nizdast, The Herald, Jwratner1, Zahid2233, MorshusApprentice, 2005-Fan, Phub Dorji, Epic Failure, Gindor, Ian98989898, Monkbot, SkateTier, Waldmannevan, Wiki1098, Wulfiedude14, Mario Castelán Castro, Fleivium, Crystallizedcarbon, Mcwikigeek, Jesus is the Light of my life, Acesoli, Soumilm, Lemmegetyou, Flying g shot, SirLagsalott, Tetra quark, Skipfortyfour, Stim 2.0, KasparBot, Fazbear7891, Nog642, Anantharamanraja, Rewdon1, Pocko19, Brandonleepick and Anonymous: 941

- **Electron neutrino** *Source:*https://en.wikipedia.org/wiki/Electron_neutrino?oldid=674917553*Contributors:*Bryan Derksen,Bobrayner,Rjwilmsi ,Strait,Bgwhite,Jeffhoy,Lockesdonkey,Dna-webmaster,GDallimore,Thijs!bot,Headbomb,Oreo Priest,Magioladitis,Maurice Carbonaro, Nwbeeson,Ggenellina,FourteenDays,SieBot,SkyLined,Addbot,LaaknorBot,PieterJanR,Luckas-bot,Rubinbot,Citation bot,ArthurBot,Xqbot, Carlog3,Paine Ellsworth,Citation bot1,TjBot,EmausBot,John of Reading,Optiguy54,TuHan-Bot,JSquish,Quondum,Kasirbot,Helpful Pixie Bot, Bibcode Bot,Love's Labour Lost,Tpaine krk,Svebert,Makecat-bot,DD4235,Scipsycho and Anonymous:10• **Muon***Source:*https://en.wikipedia. org/wiki/Muon?oldid=695122712*Contributors:*AxelBoldt, The Epopt, CYD, Mav, Bryan Derksen, Zun-dark, Roadrunner, Bkellihan, Youandme, Tim Starling, EddEdmondson, Looxix~enwiki, Ahoerstemeier, Angela, Rob Hooft, Kbk, Donar-reiskoffer, AlexPlank, Robbot, Merovingian, Rholton, Ojigiri~enwiki, Auric, Roscoe x, Bkell, Millosh, Wikibot, Ruakh, Diberri, Giftlite,Wizzy, Herbee, Xerxes314, Bodhitha, LiDaobing, Pcarbonn, DragonflySixtyseven, Deglr6328, Eb.hoop, Rich Farmbrough, Pjacobi, Vsmith,Mani1, STGM, Kjoonlee, RJHall, Army1987, Danski14, Anthony Appleyard, RobPlatt, Keenan Pepper, RJFJR, Ceyockey, Falcorian, Woohookitty,Xinghuei, Bennetto, Graham87, Vanderdecken, Rjwilmsi, Strait, Mike Peel, Bubba73, DoubleBlue, Dougluce, FlaBot, Fivemack, Danny-Wilde, Sp00n, Lmatt, Goudzovski, Srleffler, Chobot, YurikBot, Wavelength, Bambaiah, Hairy Dude, Limulus, JabberWok, Hellbus, Salsb,SCZenz, Ravedave, Scott fisher, Tetracube, Lt-wiki-bot, E Wing, Roberto DR, CrniBombarder!!!, Sbyrnes321, Eog1916, SmackBot, IncnisMrsi, Melchoir, Stifle, Gilliam, Dauto, Bluebot, Tigerhawkvok, Sbharris, Colonies Chris, Can't sleep, clown will eat me, Yevgeny Kats, DaneSorensen, JorisvS, Mets 501, JoeBot, CapitalR, SchmittM, Ruslik0, Ken Gallager, Rotiro, A876, Corpx, Thijs!bot, Headbomb, D.H, Bm gub,Andrew Carlssin, Spencer, Kariteh, Deflective, Belg4mit, Swpb, Mother.earth, Nono64, HEL, Hans Dunkelberg, 5Q5, Tarotcards, Copper-twig, Bermy88, Jarry1250, Thecinimod, Sheliak, Cuzkatzimhut, VolkovBot, Larryisgood, VasilievVV, TXiKiBoT, Anonymous Dissident,Mihaip, Graymornings, SalomonCeb, SieBot, Csmart287, Gerakibot, Statue2, Mhouston, StewartMH, ClueBot, Polyamorph, DnetSvg, Esb-boston, Saritepe, Stefan Ritt, BarretB, Kajabla, Addbot, Roentgenium111, Toyokuni3, Download, Ehrenkater, Lightbot, Luckas-bot, Yobot,Evaders99, Kulmalukko, AnomieBOT, Icalanise, Kingpin13, Materialscientist, Citation bot, Kotika98, ArthurBot, Xqbot, Cjxc92, Gilo1969,Srich32977, Misterigloo, Kyng, A. di M., Paine Ellsworth, Citation bot 1, Citation bot 4, Rameshngbot, Isofox, Jetstoknowhere, TobeBot, Trappist the monk, Puzl bustr, RjwilmsiBot, EmausBot, John of Reading, WikitanvirBot, Dewritech, GoingBatty, Milledit, Naviguessor, StringTheory11, Medeis, Suslindisambiguator, Quondum, Timetraveler3.14, Layona1, Aerthis, Mikhail Ryazanov, Frietjes, CaroleHenson, Kebil, Bibcode Bot, Jesusmonkey, NotWith, BattyBot, Kisokj, Layzeeboi, Liam135, MuonRay, TwoTwoHello, Tony Mach, Telfordbuck, SJ Defender, Krotera, Ajdigregorio, Seoman2snowlock, Monkbot, Kjerish, Jromerofontalvo, KasparBot, Corrupt Titan, QzPhysics and Anonymous: 193

- **Muon neutrino** *Source:* https://en.wikipedia.org/wiki/Muon_neutrino?oldid=680138955 *Contributors:* Bryan Derksen, The Anome, Twang, Goudzovski, Eleassar, Dna-webmaster, SmackBot, Ruslik0, Thijs!bot, Headbomb, Magioladitis, Ggenellina, FourteenDays, SieBot, Thesavagenorwegian, Muhends, Ajoykt, SkyLined, Addbot, PieterJanR, Luckas-bot, AnomieBOT, Rubinbot, Icalanise, ArthurBot, RibotBOT, Paine Ellsworth, Citation bot 1, TjBot, Mithril, EmausBot, ZéroBot, StringTheory11, Quondum, Kasirbot, Bibcode Bot, Sunitharay, Artdk, DaveW51, QzPhysics and Anonymous: 7

- **Tau (particle)** *Source:* https://en.wikipedia.org/wiki/Tau_(particle)?oldid=685074264 *Contributors:* Bryan Derksen, Iluvcapra, Ahoerstemeier, Bueller 007, Schneelocke, Dysprosia, Donarreiskoffer, Merovingian, Rorro, Davidl9999, Millosh, Harp, Herbee, Codepoet, Xerxes314, Bodhitha, CryptoDerk, Icairns, Rich Farmbrough, Pjacobi, Martpol, Sunborn, Kjoonlee, El C, Reuben, JellyWorld, RobPlatt, RJFJR, Falcorian, Dmitry Brant, Christopher Thomas, Palica, Rjwilmsi, Strait, Mike Peel, FlaBot, DannyWilde, Goudzovski, Chobot, RobotE, Bambaiah, AcidHelmNun, JabberWok, Eleassar, Salsb, SCZenz, Zwobot, Ospalh, PS2pcGAMER, Bota47, Someones life, Poulpy, Physicsdavid, Incnis Mrsi, Dauto, Pieter Kuiper, Loodog, JorisvS, MTSbot~enwiki, WISo, Q43, Thijs!bot, Headbomb, Davidhorman, Hcobb, Escarbot, RogueNinja, Yill577, Soulbot, Kostisl, STBotD, Sheliak, Joyko~enwiki, VolkovBot, Fences and windows, TXiKiBoT, Awl, Jba138, SieBot, OKBot, ImageRemovalBot, Plastikspork, Djr32, Alexbot, TimothyRias, Assosiation, BodhisattvaBot, SkyLined, J Hazard, Addbot, Eric Drexler, Ronhjones, ChenzwBot, Jklukas, Theozzfancometh, Skippy le Grand Gourou, Luckas-bot, Yobot, Grebaldar, AnomieBOT, Icalanise, Citation bot, Xqbot, Blennow, Franco3450, 四, Paine Ellsworth, Jonesey95, Three887, Plasticspork, 3ph, Miracle Pen, RjwilmsiBot, TjBot, Ripchip Bot, EmausBot, Dcirovic, Suslindisambiguator, Quondum, Rezabot, Helpful Pixie Bot, Bibcode Bot, BG19bot, Sudsguest, YFdyh-bot, Redcliffe maven, TwoTwoHello, Akro7, Nøkkenbuer, KasparBot, JPPepper, QzPhysics and Anonymous: 48

- **Tau neutrino** *Source:* https://en.wikipedia.org/wiki/Tau_neutrino?oldid=644666618 *Contributors:* B.d.mills, Kfitzner, Rjwilmsi, Salsb, TriTert-Butoxy, Newone, Thijs!bot, Headbomb, Fences and windows, Ggenellina, FourteenDays, SieBot, SkyLined, Addbot, Ronhjones, Luckas-bot, Rubinbot, JackieBot, Icalanise, Citation bot, ArthurBot, Carlog3, LucienBOT, Paine Ellsworth, Citation bot 1, TjBot, EmausBot, K6ka, Joe Gazz84, ZéroBot, Kasirbot, Bibcode Bot and Anonymous: 5

- **Antiparticle** *Source:* https://en.wikipedia.org/wiki/Antiparticle?oldid=693880154 *Contributors:* AxelBoldt, CYD, Mav, Bryan Derksen, Andre Engels, Josh Grosse, Stevertigo, Mrwojo, Patrick, RTC, Paddu, CesarB, Nikai, Nikola Smolenski, Charles Matthews, The Anomebot, Wik, Omegatron, Bevo, Altenmann, Merovingian, Intangir, Wikibot, Martinwguy, Giftlite, Bogdanb, Harp, BenFrantzDale, Herbee, Spencer195, Fleminra, Jason Quinn, Zeimusu, Mako098765, Karol Langner, Mike Rosoft, Helohe, Rich Farmbrough, Guanabot, Pjacobi, Guanabot2, Mr. Billion, Joanjoc~enwiki, Kghose, Cmdrjameson, Giraffedata, Matt McIrvin, HasharBot~enwiki, Pediddle, Deror avi, Woohookitty, Mindmatrix, Wdyoung, GregorB, SeventyThree, Justin Ormont, Palica, Marudubshinki, Tevatron~enwiki, Rjwilmsi, Ae77, MZMcBride, KaiMartin, FlaBot, Krackpipe, Commander Nemet, Roboto de Ajvol, YurikBot, Borgx, Bambaiah, Zhaladshar, Spike Wilbury, Bota47, Terbospeed, Mkossick, Tim314, 四四四 robot, SmackBot, FocalPoint, Alsandro, Srnec, Dauto, Octahedron80, Drphilharmonic, Marcus Brute, Vinaiwbot~enwiki, Jake-helliwell, Grumpyyoungman01, Newone, Mellery, Van helsing, Tim1988, Myasuda, Gogo Dodo, Goldencako, Thijs!bot, Headbomb, Tyco.skinner, JAnDbot, Steveprutz, Ferritecore, Jpod2, Singularity, Dbiel, TomasBat, Eternalmatt, Joshmt, DorganBot, Cuckooman4, VolkovBot, TXiKiBoT, Red Act, Anonymous Dissident, AlleborgoBot, SieBot, Likebox, RadicalOne, Flyer22 Reborn, KoenDelaere, Thomega, RW Marloe, BrightRoundCircle, Davidmosen, Jacob.jose, Anyeverybody, ClueBot, Diagramma Della Verita, Alexbot, Eeekster, Rishi.bedi, SilvonenBot, NellieBly, Lilaspastia, SkyLined, AkhtaBot, CarsracBot, Lightbot, Legobot, Luckas-bot, Yobot, Planlips, Csmallw,

AnomieBOT, Citation bot, Vuerqex, ArthurBot, Xqbot, Omnipaedista, RibotBOT, Muhwang, EmausBot, John of Reading, L Kensington, Benazhack, ClueBot NG, Geekingreen, Mesoderm, Bibcode Bot, BG19bot, B wik, Mark Arsten, Rm1271, Penguinstorm300, Robotsheepboy, YFdyh-bot, 77Mike77, संजीव कुमार, Dert567, Monkbot, Dhm4444, Iapec, Nazo!nin, KasparBot and Anonymous: 102

- **Weyl semimetal** *Source:* https://en.wikipedia.org/wiki/Weyl_semimetal?oldid=697644661 *Contributors:* Rjwilmsi, Lfstevens, KylieTastic, Niceguyedc, AnomieBOT, I dream of horses, Trappist the monk, Primefac, Mz7, BG19bot, RaoOfPhysics, Bianguang and Anonymous: 9

- **Dirac fermion** *Source:* https://en.wikipedia.org/wiki/Dirac_fermion?oldid=693375177 *Contributors:* Phys, Flauto Dolce, Herbee, Foobaz, Roboto de Ajvol, SCZenz, Modify, SmackBot, Incnis Mrsi, Whatever1111, Lfstevens, Pamputt, BartekChom, Addbot, OlEnglish, Luckas-bot, AnomieBOT, Erik9bot, ⁇⁇, MastiBot, Amonet, ZéroBot, ChuispastonBot and Anonymous: 5

- **Majorana fermion***Source:*https://en.wikipedia.org/wiki/Majorana_fermion?oldid=698226532*Contributors:*Pablo Mayrgundter, Jerzy, BenRG,Finlay McWalter, Lumos3, Phil Boswell, Giftlite, Dmmaus, Chris Howard, ArnoldReinhold, Bender235, Sburke, Rjwilmsi, Bubba73, HairyDude, Chris Capoccia, Buster79, SCZenz, Larsobrien, Thnidu, Teply, Incnis Mrsi, Modest Genius, JorisvS, Brienanni, Tmangray, Vttoth,Cydebot, Ntsimp, Difluoroethene, Quibik, Whatever1111, Headbomb, Davidhorman, Tjmayerinsf, Bpmullins, R sirahata, HEL, MistyMorn,Ljgua124, Satani, Pamputt, Afernand74, Coinmanj, Jwfvalle, Another Believer, Dthomsen8, Addbot, Roentgenium111, GDK, Luckas-bot,Yobot, Ptbotgourou, Amirobot, AnomieBOT, Citation bot, Obersachsebot, MIRROR, Omnipaedista, The Interior, Astiburg, Nicolas PerraultIII, Paine Ellsworth, Abductive, Tom.Reding, Amonet, Trappist the monk, RRBiswas, RobinPolt, JSquish, Quondum, AlbertusmagnusOP,Brandmeister, KarlsenBot, Claradea, ClaudeDes, Editør, Tyzoid, Bibcode Bot, BG19bot, Ymblanter, Moguns, ⁇⁇⁇⁇, YFdyh-bot, M Krikke,Neutrinomajorana, MrCondense, Anton.akhmerov, Kowtje, Kolen Cheung, Anrnusna, Alien Putsch resistant, Mgaak, IRW0, ScrapIronIV,DrKitts and Anonymous: 56

- **Skyrmion** *Source:* https://en.wikipedia.org/wiki/Skyrmion?oldid=693450794 *Contributors:* Michael Hardy, Charles Matthews, Phys, Icairns, Lumidek, Brianhe, Pjacobi, Jag123, Fwb22, Rjwilmsi, Conscious, Wikid77, Headbomb, Widefox, Lfstevens, Lincoln F. Stern, Tarotcards, KylieTastic, LeadSongDog, PixelBot, Doprendek, XLinkBot, Addbot, Luckas-bot, Yobot, Citation bot, Obersachsebot, Omnipaedista, FrescoBot, Citation bot 1, Merongb10, Meier99, Korepin, EmausBot, JSquish, Grondilu, ZéroBot, StringTheory11, AManWithNoPlan, Isocliff, Parcly Taxel, Bibcode Bot, BG19bot, BattyBot, ChrisGualtieri, Andyhowlett, 1andreasse, Zimboras, Nicohoho, NorskMaelstrom, Noah Van Horne, Jingxia0817, Farank olamaeian and Anonymous: 9

- **Soliton** *Source:* https://en.wikipedia.org/wiki/Soliton?oldid=693284477 *Contributors:* AxelBoldt, William Avery, DrBob, Olivier, Patrick, Michael Hardy, Kku, StephanWehner, Suisui, Notheruser, Lancevortex, Charles Matthews, Timwi, Jitse Niesen, Nickg, Colin Marquardt, Phys, Twang, Naddy, Decumanus, Giftlite, Sietse, AndrewTheLott, Mozzerati, Mamdu, Chris Howard, Jkl, Bender235, Billlion, Liberatus, Shanes, Mairi, Foobaz, I9Q79oL78KiL0QTFHgyc, Physicistjedi, Guy Harris, Atlant, Bart133, Raygirvan, Linas, Shreevatsa, MFH, Ashraful, SDC, Gerbrant, Eteq, Enzo Aquarius, Rjwilmsi, Phillip Jordan, MarSch, R.e.b., Williamborg, Alejo2083, FlaBot, Zaheer~enwiki, Sodin, Srleffler, Chobot, YurikBot, Borgx, Hillman, Pacaro, Gaius Cornelius, Shaddack, Yahya Abdal-Aziz, CecilWard, E2mb0t~enwiki, Epipelagic, Salmanazar, 2over0, Nikkimaria, A bit iffy, SmackBot, Mitchan, InverseHypercube, Mjspe1, Unyoyega, Akanksh, MalafayaBot, Tsca.bot, Can't sleep, clown will eat me, Ackbeet, Whpq, Nakon, Akriasas, Sina2, Dr.saptarshi, E afshari, NongBot~enwiki, Yms, Kvng, WAREL, Mfrosz, Zaphody3k, CRGreathouse, CmdrObot, Fnfal, Xxanthippe, Headbomb, Bobblehead, D.H, LachlanA, Father Goose, Christophe.Finot, Allstarecho, David Eppstein, B9 hummingbird hovering, JTiago, Adavidb, Boghog, Dhaluza, Idioma-bot, Apssen, Speaker to wolves, 99DB-SIMLR, Piperh, Arwtee, Mbz1, StewartMH, Ei2g, ClueBot, JP.Martin-Flatin, Ajayravivarma, BakuMatt, SchreiberBike, Rswarbrick, Crowsnest, Ch7ic2ke5n, XLinkBot, Kaycopperfield, Nicoguaro, Trabelsiismail, Twilighttome, Addbot, DOI bot, Auspex1729, Aboctok, Low-frequency internal, Plitter, CosmiCarl, Lightbot, سعد, Yobot, Vectorsoliton, Amirobot, NotHugo, Citation bot, ArthurBot, MauritsBot, Bdmy, Tripodian, ManasShaikh, GrouchoBot, Gballan, Constructive editor, FrescoBot, Citation bot 1, Wiuser, Abductive, MastiBot, Empln2, Li.logogogo, John of Reading, Hhhippo, ZéroBot, Basheersubei, H3llBot, Suslindisambiguator, Wikfr, SBaker43, Anagogist, CocuBot, Helpful Pixie Bot, Bibcode Bot, $oliton, Trevayne08, Queen4thewin, Wiki potal, Anupama Srinivas, Isafiz, Monkbot, Noah Van Horne, KasparBot, Clinamental and Anonymous: 78

- **Fermionic condensate** *Source:* https://en.wikipedia.org/wiki/Fermionic_condensate?oldid=684258629 *Contributors:* CYD, SebastianHelm, Glenn, Cimon Avaro, Schneelocke, Charles Matthews, Fuzheado, Phys, Jeffq, Fredrik, Cedars, Xerxes314, Steuard, Gdr, Srbauer, Laurascudder, Prsephone1674, Lysdexia, Camw, Bluemoose, Nightscream, BradBeattie, YurikBot, Hellbus, Długosz, Alain r, That Guy, From That Show!, KnightRider~enwiki, SmackBot, Hmains, Kmarinas86, Kcordina, Ohconfucius, WhiteHatLurker, Usgnus, WeggeBot, Difluoroethene, Thijs!bot, Mbell, Headbomb, AntiVandalBot, Majorly, Maliz, MartinBot, Philip Trueman, ClueBot, The Help Fishy, Vanished user uih38riiw4hjlsd, MystBot, Addbot, Mjamja, Luckas-bot, AnomieBOT, Citation bot, Cowgoesmoo2, ProtectionTaggingBot, Tom.Reding, Full-date unlinking bot, EmausBot, KHamsun, ZéroBot, Alpha Quadrant (alt), Aeonx, Quondum, ClueBot NG, Yen-Tzu and Anonymous: 45

- **Fermionic field** *Source:* https://en.wikipedia.org/wiki/Fermionic_field?oldid=692935053 *Contributors:* SimonP, BlackGriffen, SebastianHelm, Charles Matthews, Giftlite, Fropuff, Bender235, Shenme, RJFJR, Mpatel, Bachrach44, Sandlarino, Larsobrien, SmackBot, Melchoir, Colonies Chris, QFT, Yevgeny Kats, David edwards, AlphaNumeric, DiracAttack, DrKay, Dextercioby, Mild Bill Hiccup, Nilradical, SchreiberBike, Addbot, Yobot, Omnipaedista, FrescoBot, Puzl bustr, Quondum, Maschen, Davidaedwards, Anagogist, Ghartshaw, Asi013, Helpful Pixie Bot, Bibcode Bot, YFdyh-bot, Paritto, Dimension10, MarkovianStumble and Anonymous: 11

- **Kogut–Susskind fermion** *Source:* https://en.wikipedia.org/wiki/Kogut%E2%80%93Susskind_fermion?oldid=655421414 *Contributors:* Xezbeth, RHaworth, SmackBot, Gilliam, Hebrides, Headbomb, Choihei, Omnipaedista, Erik9bot, Dorcasloss, Bibcode Bot and Anonymous: 2

37.3.2 Images

- **File:AirShower.svg** *Source:* https://upload.wikimedia.org/wikipedia/commons/2/2c/AirShower.svg *License:* CC BY 3.0 *Contributors:* originally from nl.wikipedia; description page is/was here. *Original artist:* Mpfiz

- **File:Ambox_important.svg** *Source:* https://upload.wikimedia.org/wikipedia/commons/b/b4/Ambox_important.svg *License:* Public domain *Contributors:* Own work, based off of Image:Ambox scales.svg *Original artist:* Dsmurat (talk · contribs)

- **File:Stickstoff_gekühlter_Supraleiter_schwebt_über_Dauermagneten_2009-06-21.jpg** *Source:* https://upload.wikimedia.org/wikipedia/commons/3/3c/Stickstoff_gek%C3%BChlter_Supraleiter_schwebt_%C3%BCber_Dauermagneten_2009-06-21.jpg *License:* CC BY-SA 3.0 *Contributors:* Own work *Original artist:* Henry Mühlpfordt

- **File:Strong_force_charges.svg** *Source:* https://upload.wikimedia.org/wikipedia/commons/b/b6/Strong_force_charges.svg *License:* CC BY-SA 3.0 *Contributors:* Own work, Created from Garret Lisi's Elementary Particle Explorer *Original artist:* Cjean42

- **File:Stylised_Lithium_Atom.svg** *Source:* https://upload.wikimedia.org/wikipedia/commons/e/e1/Stylised_Lithium_Atom.svg *License:* CC-BY-SA-3.0 *Contributors:* based off of Image:Stylised Lithium Atom.png by Halfdan. *Original artist:* SVG by Indolences. Recoloring and ironing out some glitches done by Rainer Klute.

- **File:Symbol_list_class.svg** *Source:* https://upload.wikimedia.org/wikipedia/en/d/db/Symbol_list_class.svg *License:* Public domain *Contributors:* ? *Original artist:* ?

- **File:The_incomplete_circle_of_everything.svg** *Source:* https://upload.wikimedia.org/wikipedia/commons/0/0d/The_incomplete_circle_of_everything.svg *License:* CC BY 3.0 *Contributors:* Own work *Original artist:* Zhitelew

- **File:Top_antitop_quark_event.svg** *Source:* https://upload.wikimedia.org/wikipedia/commons/3/35/Top_antitop_quark_event.svg *License:* Public domain *Contributors:* Own work *Original artist:* Raeky

- **File:Upwelling.svg** *Source:* https://upload.wikimedia.org/wikipedia/commons/e/ee/Upwelling.svg *License:* Public domain *Contributors:*

- File:Upwelling.jpg *Original artist:* Lichtspiel

- **File:Virtual_pairs_near_electron.png** *Source:* https://upload.wikimedia.org/wikipedia/commons/6/6f/Virtual_pairs_near_electron.png *License:* CC BY-SA 3.0 *Contributors:* Own work by uploader; rendered using Bryce 6. *Original artist:* RJHall

- **File:Weyl_Balents.png** *Source:* https://upload.wikimedia.org/wikipedia/commons/1/16/Weyl_Balents.png *License:* CC BY-SA 4.0 *Contributors:* Own work *Original artist:* Bianguang

- **File:Wiki_letter_w_cropped.svg***Source:*https://upload.wikimedia.org/wikipedia/commons/1/1c/Wiki_letter_w_cropped.svg*License:*CC-BY-SA-3.0*Contributors:* This file was derived fromWiki letter w.svg: <a href='//commons.wikimedia.org/wiki/File:Wiki_letter_w.svg'class='image'><img alt='Wiki letter w.svg' src='https://upload.wikimedia.org/wikipedia/commons/thumb/6/6c/Wiki_letter_w.svg/50px-Wiki_letter_w.svg.png' width='50' height='50' srcset='https://upload.wikimedia.org/wikipedia/commons/thumb/6/6c/Wiki_letter_w.svg/75px-Wiki_letter_w.svg.png 1.5x, https://upload.wikimedia.org/wikipedia/commons/thumb/6/6c/Wiki_letter_w.svg/100px-Wiki_letter_w.svg.png 2x' data-file-width='44' data-file-height='44' /></a>
*Original artist:*Derivative work byThumperwardfromfhjmnhtkuztikhzrhjhgrftirddfjzteghfn,jhklktrtujbgdfghgfdfvhnbcxashjhwghntrzzzhjkljjlj

- **File:Wikisource-logo.svg** *Source:* https://upload.wikimedia.org/wikipedia/commons/4/4c/Wikisource-logo.svg *License:* CC BY-SA 3.0 *Contributors:* Rei-artur *Original artist:* Nicholas Moreau

- **File:Wikiversity-logo.svg** *Source:* https://upload.wikimedia.org/wikipedia/commons/9/91/Wikiversity-logo.svg *License:* CC BY-SA 3.0 *Contributors:* Snorky (optimized and cleaned up by verdy_p) *Original artist:* Snorky (optimized and cleaned up by verdy_p)

- **File:Wiktionary-logo-en.svg** *Source:* https://upload.wikimedia.org/wikipedia/commons/f/f8/Wiktionary-logo-en.svg *License:* Public domain *Contributors:* Vector version of Image:Wiktionary-logo-en.png. *Original artist:* Vectorized by Fvasconcellos (talk · contribs), based on original logo tossed together by Brion Vibber

- **File:Wolfgang_Pauli_young.jpg** *Source:* https://upload.wikimedia.org/wikipedia/commons/4/43/Wolfgang_Pauli_young.jpg *License:* Public domain *Contributors:* ? *Original artist:* ?

37.3.3 Content license

Made in the USA
Lexington, KY
14 January 2016